Foundations of Engineering Mechanics

Series editors

More information about this series at http://www.springer.com/series/3582

Igor T. Selezov · Yuriy G. Kryvonos
Ivan S. Gandzha

Wave Propagation and Diffraction

Mathematical Methods and Applications

Springer

Igor T. Selezov
Institute of Hydromechanics
National Academy of Sciences of Ukraine
Kyiv
Ukraine

Ivan S. Gandzha
Institute of Physics
National Academy of Sciences of Ukraine
Kyiv
Ukraine

Yuriy G. Kryvonos
Institute of Cybernetics
National Academy of Sciences of Ukraine
Kyiv
Ukraine

ISSN 1612-1384 ISSN 1860-6237 (electronic)
Foundations of Engineering Mechanics
ISBN 978-981-10-4922-4 ISBN 978-981-10-4923-1 (eBook)
DOI 10.1007/978-981-10-4923-1

Library of Congress Control Number: 2017941462

Printed on acid-free paper

This Springer imprint is published by Springer Nature
The registered company is Springer Nature Singapore Pte Ltd.
The registered company address is: 152 Beach Road, #21-01/04 Gateway East, Singapore 189721, Singapore

Foreword

This monograph is devoted to topical problems of the contemporary theory of wave propagation and diffraction. It starts from the brief account of mathematical methods related to studying the problems of wave diffraction theory. Then spectral methods in the theory of wave propagation are considered in more detail. The refraction of surface gravity waves is studied with the use of the ray method that originates from geometrical optics. Various problems pertaining to the diffraction and scattering of hydrodynamic, acoustic, electromagnetic, and elastic waves by local inhomogeneities in infinite and semi-infinite domains are considered and their solutions are provided. Some aspects of the problem of generation and propagation of tsunami waves are analysed. Finally, the evolution of wave trains in a two-layer fluid is studied by the method of multiple scales.

We hope that this work would be of interest for the researchers working in the field of applied and engineering mechanics, mathematical and computational physics, hydrodynamics, etc. It would also be useful for lecturers and postgraduate students of relevant specialties.

Preface

Wave phenomena represent a fairly intriguing area of the contemporary applied mathematics and physics. Wave processes have numerous applications in hydrodynamics, electromagnetism, magnetohydrodynamics, biophysics and biomechanics, acoustics, etc. This monograph is devoted to two distinct aspects of wave dynamics: wave propagation and diffraction, with the main focus put on the wave diffraction.

Wave interaction with rough bottom surfaces (topography), offshore drilling platforms, and wave energy collectors is accompanied by the diffraction of waves. The diffraction theory lies at the interface between physics and applied mathematics. In the broad sense, wave diffraction means any deviation of the wave motion from the laws of geometrical optics. From a mathematical point of view, the purpose of the diffraction theory is to develop analytical and numerical methods for solving diffraction problems, to classify the corresponding solutions, and to investigate their properties.

It is possible to distinguish three stages in the development of the diffraction theory: (i) Fresnel (1818) formulated the Huygens–Fresnel principle that combines the geometric approach (Huygens 1690) and interference approach (Young 1800); (ii) Helmholtz (1859) gave a strict formulation of the Huygens principle and demonstrated that it results in an integral formula (as in the potential theory) that makes it possible to calculate the value of the field at some point in terms of the field values (including the field's normal derivative) on some auxiliary closed surface enclosing that point; (iii) Poincaré (1892) and Sommerfeld (1896) showed the diffraction problems to be ordinary boundary-value problems of mathematical physics. Sommerfeld (1912) also formulated the radiation conditions. Then Meixner (1948) established the boundary conditions on the edge.

The rigorous diffraction theory distinguishes three approaches: the method of surface currents, where the diffracted field is represented as a superposition of secondary spherical waves emitted by each element (the Huygens–Fresnel principle); Fourier method; method of separation of variables and Wiener–Hopf transformation. In this monograph, we apply various mathematical methods to the solution of typical problems in the theory of wave propagation and diffraction and analyse the results obtained.

Chapter 1 presents some of the methods that are useful for solving the problems of wave diffraction theory: method of separation of variables, method of power series, method of spline functions, and method of an auxiliary boundary. We also consider some algorithms for the numerical inversion of the Laplace transform, which is often used to solve the wave diffraction problems. Finally, we give a brief account of the method of multiple scales that is often used to study the propagation of transient waves.

Chapter 2 deals with spectral methods in the theory of wave propagation. The main focus is given to the Fourier methods in application to studying the Stokes (gravity) waves on the surface of an inviscid fluid. A spectral method for calculating the limiting Stokes wave with a corner at the crest is considered as well. We also briefly consider the evolution of narrow-band wave trains on the surface of an ideal finite-depth fluid. Finally, a two-parameter method for describing the nonlinear evolution of narrow-band wave trains is described by the example of the Klein–Gordon equation with a cubic nonlinearity. The problem is reduced to a high-order nonlinear Schrödinger equation for the complex amplitude of wave envelope. This equation is integrated numerically using a split-step Fourier technique to describe the evolution of quasi-solitons.

Chapter 3 presents some results of modelling the refraction of surface gravity waves in terms of the ray method that originates from geometrical optics. Wave refraction essentially depends on the bed topography. As a result, the convergence of wave energy (or its divergence) can be observed in some water areas. The ray method makes possible the discrete or analytical assignment of the bed topography. This approach can be realised in the form of a computational programme and allows the distribution of wave fronts, rays, and heights to be constructed and analysed for the case of the transition of regular waves from deep to shallow waters. The model is verified by comparing the results with exact analytical solutions and field observations. The asymptotic analysis of the nonlinear refraction theory extends the limits of applicability of the traditional theory and provides the prediction of ray bending at approaching the wave breaking conditions. Moreover, we study the anomalous refraction in caustics.

Chapter 4 is devoted to the diffraction of surface gravity waves. Main aspects of the wave diffraction theory are described. Specific aspects and methods used to solve the problems of the wave diffraction theory are described in brief. Wave diffraction by a partially submerged elliptical cylinder with elliptical front surface and by a circular submerged cylinder is considered. Ellipticity is demonstrated to have strong effect on the wave load and its extremums, depending on the wave number. Solutions to the problem of wave diffraction by a system of vertical cylinders are presented and analysed. In this case, the wave force does not attain its maximum on the front vertical cylinder because of the significant reconstruction of the diffracted wave fields in multiply connected regions. An exact solution to the problem of wave diffraction by an asymmetrically nonuniform cylindrical scatterer is derived in the case when the scatterer parameters depend on the two coordinates—

radial and angular. The scatterer inhomogeneity is demonstrated to affect the cross scattering. The method of auxiliary boundary is used to study the diffraction of waves by a vertical column of arbitrary revolution shape. The extremums of the wave force and overturning moment applied to a cone column are found as functions of the wave number. The numerical method of spline collocation is used to study the problem of diffraction of acoustic waves by an arbitrary body of revolution. The accuracy of the numerical solution is analysed. The problem of wave scattering by a truncated cone with smooth spherical ends is considered. The effect of wave incidence angle is studied.

Chapter 5 deals with the approach that is based on the repeated use of the method of images to solve the problems of stationary acoustic, electromagnetic, and elastic wave scattering and diffraction by cylindrical and spherical obstacles in a semi-infinite domain. The solution is written in terms of an infinite series of multiply diffracted fields. Explicit approximate asymptotic solutions are found and investigated for the case of distant scattered fields in the longwave approximation. The known solutions for point obstacles are obtained as special cases described by the first terms of the series.

Chapter 6 deals with some aspects of the initial-boundary-value problems of the initiation, generation, and propagation of tsunami waves. The generation of tsunami waves by bottom movements is considered. We formulate an appropriate initial-boundary-value problem and analyse the effect of the sharpness of vertical axisymmetric bottom disturbance and the disturbance duration on the generation of tsunami waves. The propagation of nonlinear waves on water and their evolution over a nonrigid elastic bottom are investigated. Some aspects and indeterminacy of the formulation of the initial-boundary-value problems dealing with the initiation and generation of tsunami waves are considered. We consider some typical types of tsunami waves that demonstrate the indeterminacy of their initiation in time because of the indeterminacy in the physical trigger mechanism of underwater earthquakes. Based on the three-dimensional formulation, evolution equations describing the propagation of nonlinear dispersive surface waves on water over a spatially inhomogeneous bottom are obtained with allowance for the bottom disturbances in time. We use the Laplace transform with respect to the time coordinate and the power series method with respect to the spatial coordinate to find a solution to the nonstationary problem of the diffraction of surface gravity waves by a radial bottom inhomogeneity that deviates from its initial position. The propagation and stability of nonlinear waves in a two-layer fluid with allowance for surface tension are analysed by the asymptotic method of multiscale expansions.

Some insights on the directions of further development of the wave diffraction theory are outlined in the conclusion.

Kyiv, Ukraine

Igor T. Selezov
Yuriy G. Kryvonos
Ivan S. Gandzha

Acknowledgements

The authors would like to express their sincere appreciation to Dr. Vladimir Babitsky, Dr. Christoph Baumann, and Mr. Ramamoorthy Rajangam for their support in the publication of this monograph. The authors thank Mrs. S.O. Savchenko for her help in the technical work on the preparation of the manuscript for publication. We are also grateful to Dr. O.D. Kocherga and Dr. V.V. Gozhenko for their careful work in the translation of some excerpts of the manuscript into English.

We regret to note recently departed Prof. Yuriy G. Kryvonos, who constantly and actively encouraged researchers in the field of wave propagation and diffraction.

Contents

Chapter 1
Some Analytical and Numerical Methods in the Theory of Wave Propagation and Diffraction

1.1 Method of Separation of Variables

The wave propagation, diffraction, and scattering in continuous media are described by a system of partial differential equations with relevant initial and matching conditions [37]. In the case of inhomogeneous media, these equations have variable coefficients that depend on spatial coordinates. In many cases, such equations can be reduced to one higher order equation [12]. Following Ref. [18], consider a linear equation

$$Lu(\boldsymbol{x},\, t) = 0. \tag{1.1}$$

Here, $\boldsymbol{x} = (x_1,\, x_2,\, x_3)$ is a vector of spatial coordinates, $t \equiv x_4$ is the temporal coordinate, L is a differential operator of order m defined in some domain D of the three-dimensional Euclidean space $\mathbb{R}_3$ and in the range $t \in [t_0,\, T]$:

$$L = \sum_{j=0}^{m} \sum_{i_1,\, i_2 \ldots i_j = 1} a_{i_1,\, \ldots,\, i_j}(\boldsymbol{x}) \frac{\partial^j}{\partial x_{i_1} \ldots \partial x_{i_j}}. \tag{1.2}$$

The coefficients $a_{i_1,\, \ldots,\, i_j}(\boldsymbol{x})$ are assumed to be defined for $(x_1,\, x_2,\, x_3) \in D$, symmetric with respect to indices $i_1,\, i_2,\, \ldots,\, i_j$, and differentiable in $\mathbb{R}_3$ as many times as needed. For $j = 0$ the operator L in (1.2) is defined as the operator of multiplication by u.

In what follows, we limit our consideration to a class of such equations of type (1.1) that admit the separation of variables. This means that the solution can be looked for as follows

$$u(x_1,\, x_2,\, x_3,\, t) = X_1(x_1) X_2(x_2) X_3(x_3) T(t). \tag{1.3}$$

I.T. Selezov et al., *Wave Propagation and Diffraction*,
Foundations of Engineering Mechanics, DOI 10.1007/978-981-10-4923-1_1

Substituting (1.3) into (1.1), we obtain independent ordinary differential equations for each $X_k(x_k)$, $k = 1, 2, 3, 4$:

$$\sum_{j=0}^{m} b_{kj}(x_k) \frac{d^j}{dx_k^j} X_k(x_k) = 0. \tag{1.4}$$

Going over from Eq. (1.1) to (1.3) and (1.4) is possible only under significant restrictions imposed on the differential operator L and the coefficients $a_{i_1,\ldots,i_j}(\boldsymbol{x})$. Suppose that the variables can also be separated in the initial and matching conditions.

The authors of Ref. [65] showed how a linear partial differential equation with variable coefficients can be reduced to an equation with constant coefficients. Consider the following second-order partial differential equation:

$$\alpha(x)\frac{\partial^2 g(x,t)}{\partial x^2} + \beta(x)\frac{\partial g(x,t)}{\partial x} + \gamma(x)g(x,t) = a\frac{\partial g(x,t)}{\partial t} + b\frac{\partial^2 g(x,t)}{\partial t^2}. \tag{1.5}$$

It can be reduced to the canonical form

$$C(X)\frac{\partial}{\partial X}\left(\frac{1}{C(X)}\frac{\partial f}{\partial X}\right) = a\frac{\partial f}{\partial t} + b\frac{\partial^2 f}{\partial t^2} \tag{1.6}$$

by simple transformations:

$$X = \int^x |\alpha(\tau)|^{-1/2} d\tau \text{ and } f(X,t) = \frac{g(x,t)}{g_0(x)}, \tag{1.7}$$

provided that $g = g_0(x)$ is a nonzero equilibrium solution to Eq. (1.5) (when $a = b = 0$). The function $C(X)$ in (1.6) is completely determined by the functions $\alpha(x)$, $\beta(x)$, and $\gamma(x)$ in (1.5).

Equation (1.5) is rather general and includes the wave equation, heat conduction equation, Laplace equation, Schrödinger equation, and Fokker–Planck equation as particular cases. It was shown that for certain functions $C(X)$ any solution to Eq. (1.6) can be expressed in terms of solutions to the equation with constant coefficients,

$$\frac{\partial^2 F(X,t)}{\partial X^2} = a\frac{\partial F(X,t)}{\partial t} + b\frac{\partial^2 F(X,t)}{\partial t^2}, \tag{1.8}$$

namely,

$$f(X,t) = \sum_{n=0}^{N} f_n(X)\frac{\partial^n F(X,t)}{\partial X^n}, \tag{1.9}$$

where the functions $f_n(X)\,(n = 0, 1, 2, \ldots, N)$ and $C(X)$ satisfy a system of coupled nonlinear ordinary differential equations with constant $f_0 = f_0(X)$. The choice of N in (1.9) is rather broad, so that any given function $C(X)$ in (1.6) could be approximated by a function that satisfies the above system.

Note that solutions to ordinary differential equations of type (1.8) with variable coefficients cannot be expressed in terms of elementary or special functions in the general case, in particular in reference to distributed inhomogeneities that are of great practical interest. Below we consider some methods that allow solutions to such equations to be constructed.

1.2 Method of Power Series

Consider a localised inhomogeneity that occupies a closed domain $\Omega \subset \mathbb{R}^3$ in the $\mathbb{R}^3$ space. In this case, we deal with a typical problem of wave diffraction by a scatterer with variable properties. The difficulties of solving such a problem are related to the consideration of partial differential equations with variable coefficients even in the case of linear problems. The methods used to solve such problems include the method of exact analytical solutions that are possible in some cases [55]; the asymptotic method of Born approximations [7]; the Bremmer series method, in which the first term represents the WKB approximation [9]; and the method of power series [56].

The Born approximations (or approximations of weak scattering) can be used in the case when $(kR\,\delta p/p) \ll 1$, where the ratio $\delta p/p$ characterises the magnitude of the inhomogeneity and $kR = 2\pi R/\lambda$ is the ratio of the inhomogeneity size to the typical wavelength λ. This approach was particularly developed in Refs. [49, 73]. Two main approximations are usually considered in this regard. The Rayleigh approximation refers to the case when the inhomogeneity magnitude is not small, i.e. $\delta p/p = O(1)$, but the inhomogeneity size is small, $kR \ll 1$. The optical Rayleigh–Hans approximation holds true in the case of weak inhomogeneity whose size cannot be regarded small, $\delta p/p \ll 1, kR = O(1)$.

The Bremmer series method is discussed in Refs. [24, 68]. In this approach, the continuous functions characterising the inhomogeneity are approximated by piecewise constant functions, and the solutions for the reflected and transmitted waves are written for each particular layer. Then the passage to the limit for an infinite number of layers is performed with the separation of the first WKB term.

The method of generalised power series is most useful in analysing the structure of the scattered wave field. Under certain mild constraints on the inhomogeneity magnitude and size, solutions to the wave scattering problem can be obtained in the form of convergent power series. Thus, the method of power series can be used in the case of arbitrary inhomogeneities, in contrast to the Born approximations.

Consider a linear second-order differential equation with variable coefficients for some function $y(x)$:

$$y'' + f(x)y' + g(x)y = 0. \qquad (1.10)$$

Suppose that $f(x)$ and $g(x)$ can be represented as power series in terms of integer positive powers of x, so that Eq. (1.10) could be written as

$$y'' + \left(\sum_{n=0}^{\infty} a_n x^n\right) y' + \left(\sum_{n=0}^{\infty} b_n x^n\right) y = 0. \qquad (1.11)$$

We look for a solution to Eq. (1.11) in the form of power series with unknown coefficients:

$$y = \sum_{n=0}^{\infty} \alpha_n x^n. \qquad (1.12)$$

Substituting (1.12) into Eq. (1.11), we get

$$\sum_{n=2}^{\infty} n(n-1)\alpha_n x^{n-2} + \sum_{n=0}^{\infty} a_n x^n \sum_{n=1}^{\infty} n\alpha_n x^{n-1} + \sum_{n=0}^{\infty} b_n x^n \sum_{n=0}^{\infty} \alpha_n x^n = 0.$$

By equating the equations at the like powers of x in the lefthand side of the above equations to zero, we come to an infinite system of recurrence equations [57]:

$$\begin{array}{ll} x^0 & 2\cdot 1\alpha_2 + a_0\alpha_1 + b_0\alpha_0 = 0, \\ x^1 & 3\cdot 2\alpha_3 + 2a_0\alpha_2 + a_1\alpha_1 + b_0\alpha_1 + b_1\alpha_0 = 0, \\ x^2 & 4\cdot 3\alpha_4 + 3a_0\alpha_3 + 2a_1\alpha_2 + a_2\alpha_1 + b_0\alpha_2 + b_1\alpha_1 + b_2\alpha_0 = 0, \\ \vdots & \qquad\qquad \vdots \\ x^n & (n+2)(n+1)\alpha_{n+2} + Q_n(\alpha_0, \alpha_1, \ldots, \alpha_{n+1}) = 0, \\ \vdots & \qquad\qquad \vdots \end{array} \qquad (1.13)$$

Here, $Q_n(\alpha_0, \alpha_1, \ldots, \alpha_{n+1})$ is a homogeneous first-degree polynomial in variables $\alpha_0, \alpha_1, \ldots, \alpha_{n+1}$ (α_0 and α_1 being arbitrary constants).

The recurrence equations of system (1.13) can be used to determine consecutively all the coefficients $\alpha_2, \alpha_3, \ldots, \alpha_n, \ldots,$ but the questions dealing with the convergence of the obtained series and the existence of solution remain to be open. The following theorem holds true [57]:

Theorem *If the series*

$$f(x) = \sum_{n=0}^{\infty} a_n x^n,\ g(x) = \sum_{n=0}^{\infty} b_n x^n$$

are convergent for $|x| < R$, *then the power series constructed in the way described above is convergent for the same* x *and is a solution to Eq. (1.11).*

In particular, if $f(x)$ and $g(x)$ are the polynomials in x, then the obtained power series is convergent at any x.

Consider a linear second-order differential equation,

$$P_0(x)y'' + P_1(x)y' + P_2(x)y = 0, \tag{1.14}$$

that can be reduced to Eq. (1.10) after the division by $P_0(x)$ (when $P_0(x) \neq 0$ in some interval $[0, h]$) with

$$f(x) = \frac{P_1(x)}{P_0(x)},\ g(x) = \frac{P_2(x)}{P_0(x)}.$$

Then a solution to Eq. (1.14) can also be sought in the power series form. In this case, it is easier to substitute expression (1.12) directly into Eq. (1.14) and equate the unknown coefficients at the same powers of x to zero without reducing Eq. (1.14) to form (1.10).

When the coefficients in the equations of form (1.14) have poles, the use of power series approach is regulated by the Fuchs theorem [63]:

Theorem *If differential equation (1.10) is such that $f(x)$ and $g(x)$ have poles at $x = x_0$, then its solution can be found in the form of convergent generalised power series*

$$y(x) = (x - x_0)^{\nu} \sum_{n=0}^{\infty} \alpha_n (x - x_0)^n,$$

provided that the products $(x - x_0)f(x)$ and $(x - x_0)g(x)$ remain to be finite at $x = x_0$.

In this case, the pole x_0 can be translated to the point $x = 0$ by the change of variable and the following equation satisfying the Fuchs theorem can be obtained:

$$x^2y'' + f(x)xy' + g(x)y = 0,\ y = y(x). \tag{1.15}$$

This equation can be solved by the method of generalised Frobenius power series [63] when $f(x)$ and $g(x)$ are holomorphic functions at $x < R$ (R being the convergence radius) and $x = 0$ is a regular singular point. In the latter case, the functions $f(x)$ and $g(x)$ can be expressed in the power series form in terms of integer positive powers of x or be given by polynomials. Then Eq. (1.15) can be written as

$$x^2y'' + \left(\sum_{n=0}^{\infty} a_n x^n\right)xy' + \left(\sum_{n=0}^{\infty} b_n x^n\right)y = 0, \tag{1.16}$$

and the solution to Eq. (1.16) is written in the form of convergent power series:

$$y = x^{\nu} \sum_{n=0}^{\infty} \alpha_n x^n,\ \alpha_0 = 1. \tag{1.17}$$

Hence, the problem reduces to the determination of parameter ν and coefficients α_ν from the original equation and the recurrence relations for α_ν. The corresponding fundamental system of solutions is

$$w_1 = x^{\nu_1}\varphi_1, \quad w_2 = x^{\nu_2}\varphi_2,$$

where the first coefficient α_0 can be regarded to be nonzero in view of the indeterminacy of the exponent ν of the multiplier x^ν.

Substituting series (1.17) into Eq. (1.16) and equating the coefficients at each power of x to zero, we get an infinite system of coupled algebraic equations for the exponent ν and coefficients α_n:

$$\begin{array}{ll} x^{\nu-2} & \alpha_0 D(\nu) = \alpha_0 + [\nu(\nu-1) + a_0\nu + b_0] = 0, \\ x^{\nu-1} & \alpha_1 D(\nu+1) + \alpha_0(\nu a_1 + b_1) = 0, \\ x^{\nu} & \alpha_2 D(\nu+2) + \alpha_1[(\nu+1)a_1 + b_1] + \alpha_0(\nu a_2 + b_2) = 0, \\ \vdots & \vdots \\ x^{\nu+n-2} & \alpha_n D(\nu+n) + \alpha_{n-1}[(\nu+n-1)a_1 + b_1] + \cdots + \alpha_0(\nu a_n + b_n) = 0, \\ \vdots & \vdots \end{array} \tag{1.18}$$

The first equation of this system is called a governing equation. Let ν_1 and ν_2 be its roots. Suppose that they are different and their difference is a noninteger number. In this case, two sets of the coefficients α_n corresponding to each root of the governing equation can be found consecutively. Hence, we get two generalised power series of form (1.17) that represent linearly independent solutions to Eq. (1.16). These solutions converge inside of a circle reaching the nearest special point of the equation apart from the point $x = 0$. The coefficient α_0, which is present as a multiplier in all the terms of the series, remains to be arbitrary, i.e. each solution is determined to within a constant multiplier. The general solution to Eq. (1.16) is written as the linear combination of these two solutions.

When the governing equation has a double root, there exists only one solution of type (1.17). The second linearly independent solution can be determined as follows. Let $y_1(x)$ be a solution to Eq. (1.10). Then the Liouville formula [63]

$$y_2(x) = y_1(x)\left(B + A\int \frac{\mathrm{d}x}{y_1^2(x)e^{\int f(x)\mathrm{d}x}}\right)$$

implies that there exists a solution that is linearly independent of the solution of type (1.17)

$$y_1 \int \frac{e^{-\int f(x)\mathrm{d}x}}{y_1^2}\mathrm{d}x. \tag{1.19}$$

Finally, consider the case when the roots of the governing equation differ from one another by an integer, i.e. $\nu_2 = \nu_1 + n$ (n being a positive integer). The coefficients of the series corresponding to the root ν_2 can be all calculated, since the polynomials $D(\nu_2 + 1), D(\nu_2 + 2), \ldots$ take nonzero values. The coefficients of the series corresponding to the root ν_1 can be calculated only up to order $(n - 1)$ because the coefficient at α_n in the equation

$$\alpha_n D(\nu_1 + n) + \alpha_{n-1}[a_1(n + \nu_1 - 1) + b_1] + \cdots + \alpha_0(a_n\nu_1 + b_n) = 0$$

for α_n is equal to zero. This equation can be reduced to the equality

$$\alpha_{n-1}[a_1(n + \nu_1 - 1) + b_1] + \cdots + \alpha_0(a_n\nu_1 + b_n) = 0. \tag{1.20}$$

If this equality holds, the coefficients $\alpha_{n+1}, \alpha_{n+2}, \ldots$ can be expressed in terms of the coefficient α_n, which remains to be undetermined. This implies that the solution corresponding to ν_1 contains two arbitrary parameters: the coefficient α_0 that is a common factor in the solution and the coefficient α_n that determines the higher coefficients $(\alpha_{n+1}, \alpha_{n+2}, \ldots)$ of the series. If equality (1.20) does not hold, the second solution can be found using formula (1.19), namely,

$$y_2 = y_1 \ln x + x^{\nu_2} \sum_{n=0}^{\infty} \beta_n x^n. \tag{1.21}$$

The coefficients β_n can be found by substituting expression (1.21) into Eq. (1.16) and equating the coefficients at the like powers of x.

In particular, the method of power series was used to solve the problems of wave diffraction by radially inhomogeneous obstacles [56].

1.3 Method of Spline Functions

Solving the problems of wave diffraction by obstacles of complex shape is related to the approximation of functions. One of the most efficient approaches in this regard is the surface interpolation with piecewise polynomials, which is used instead of constructing the high-order interpolation polynomials. Third-order (cubic) polynomial interpolation splines $S_3(x)$ are most widely used in practical applications.

Spline approximation methods are closely related to solving partial differential equations numerically by the finite-difference method with the use of the Ritz method with specially selected basis functions. An account of using this method to the investigation of wave diffraction is given in Sect. 4.7 and Ref. [36]. Here, we briefly outline the basic concepts of spline function theory [75].

Consider some partition $\Delta\colon a = x_0 < x_1 < \cdots < x_k = b$ of the interval $[a, b]$. Let $C^k = C^k[a, b]$ be a set of k times continuously differentiable functions on $[a, b]$

(k being an integer), and $C^{-1}(a, b)$ be a set of piecewise-continuous functions with points of discontinuity of the first kind.

The function $S_{n,\nu}(x)$ is called an n-order spline of defect ν (ν is an integer such that $0 \leq \nu \leq n + 1$) with nodes on the mesh Δ if

(i) the function $S_{n,\nu}(x)$ is a polynomial of order n on each interval $[x_i, x_{i+1}]$, i.e.

$$S_{n,\nu}(x) = \sum_{\alpha=0}^{n} a_\alpha^i (x - x_i)^\alpha \text{ for } x \in [x_i, x_{i+1}], \quad i = 0, \ldots, N - 1; \tag{1.22}$$

(ii) $S_{n,\nu}(x) \in C^{n-\nu}[a, b]$.

The spline definition is valid on the whole real axis if we set $a = -\infty$, $b = +\infty$. Besides formula (1.22), the following representation is possible for the spline on each interval $[x_i, x_{i+1}]$:

$$S_{n,\nu}(x) = \sum_{\alpha=0}^{n} b_\alpha^i (x - x_{i+1})^\alpha, \; i = 0, \ldots, N - 1.$$

For cubic splines of class C^2, which have been used most frequently, we can introduce an interpolation spline function that is continuous together with its derivative on each of the intervals $[x_i, x_{i+1}]$.

Consider the spline collocation method by the example of an ordinary differential equation that can be obtained after the separation of variables in the wave diffraction problems described by partial differential equations [36]. We look for a solution to the equation [75]

$$L[y(x)] \equiv y''(x) + p(x)y'(x) + q(x)y(x) = r(x), \; x \in [a, b], \tag{1.23}$$

that satisfies the following boundary conditions:

$$\alpha_1 y(a) + \beta_1 y'(a) = \gamma_1, \; \alpha_2 y(b) + \beta_2 y'(b) = \gamma_2. \tag{1.24}$$

We assume that the two-point boundary problem (1.23), (1.24) has a unique solution $y(x)$. The requirements in regard to the smoothness of $y(x)$ and constraints imposed on the given coefficients $p(x)$, $q(x)$, $r(x)$; α_i, β_i, γ_i, $i = 1, 2$ should be stipulated in each particular case.

We introduce the partition $\Delta\colon a = x_0 < x_1 < \cdots < x_N = b$ on $[a, b]$ and look for an approximate solution to problem (1.22), (1.23) in the form of a cubic spline $S(x)$ of class C^2 with nodes on the mesh Δ. The spline $S(x)$ is required to satisfy Eq. (1.23) in the points $\xi_k \in [a, b]$, $k = 0, \ldots, N$ (collocation conditions) and boundary conditions (1.24):

$$L[S(\xi_k)] \equiv S''(\xi_k) + p(\xi_k)S'(\xi_k) + q(\xi_k)S(\xi_k) = r(\xi_k),\ k = 0, \ldots, N, \quad (1.25)$$

$$\alpha_1 S(a) + \beta_1 S'(a) = \gamma_1, \quad \alpha_2 S(b) + \beta_2 S'(b) = \gamma_2. \quad (1.26)$$

Relations (1.25), (1.26) represent the set of algebraic equations for the spline parameters. The points ξ_k are called the collocations nodes, and their number is determined by the dimension of the space formed by splines of class C^2. This dimension is equal to $N + 3$. Since $S(x)$ satisfies two boundary conditions (1.26), the number of collocation nodes should be equal to $N + 1$. Note that the positions of collocation nodes on the interval $[a, b]$ cannot be arbitrary.

1.4 Method of an Auxiliary Boundary

The idea behind this method is that a scatterer of an arbitrary convex shape is enclosed by an auxiliary canonical surface—cylindrical in the two-dimensional case or spherical in the three-dimensional case [47]. Then the external problem is solved exactly, and the problem in the interior domain between the scatterer and the canonical surface is solved numerically. The corresponding matching conditions need to be satisfied on the scatterer's boundary.

Suppose that we need to determine the field $\boldsymbol{u}(\boldsymbol{x}, t)$ scattered by a local inhomogeneity of an arbitrary (noncanonical) shape in the infinite exterior domain (Fig. 1.1). Let us enclose the inhomogeneity by an auxiliary surface B and employ a numerical method in the domain Ω between B and the surface S of the inhomogeneity. For the problem in Ω to be correctly defined, $\boldsymbol{u}$ has to satisfy the boundary conditions on the surface B that needs to be nonreflective. Such formulations, which involve the nonreflective boundary conditions on the auxiliary boundary, were considered, in particular, for elastic and electromagnetic waves in Refs. [26, 29–32] and for acoustic waves in Ref. [20]. The method of an auxiliary boundary was also used in the problem of scattering of electromagnetic waves by an obstacle of arbitrary shape [33] and in the problem of diffraction of surface gravity waves [47]. The diffraction of surface gravity waves by inhomogeneities was studied in Ref.

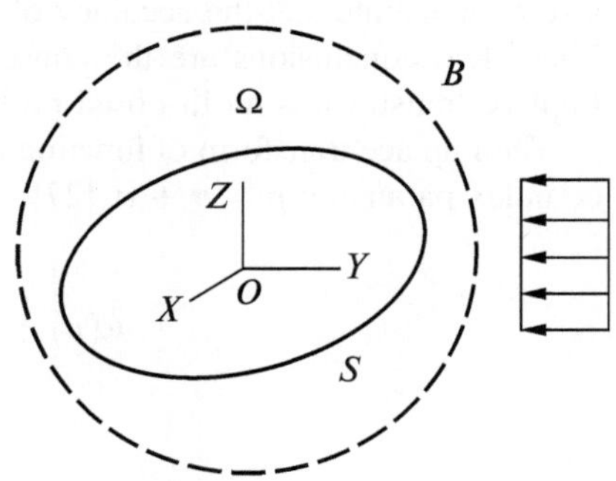

Fig. 1.1 An obstacle enclosed by an auxiliary boundary B (cylindrical or spherical)

[64] in the framework of an approximate model involving a canonical auxiliary surface. The auxiliary boundary approach was also used in studying the scattering of flexural waves by cavities in Ref. [44].

Noteworthy also are two other numerical methods that were developed in the theory of wave diffraction by obstacles of complex shape. The first one is based on the integral equation approach and was mainly used in the problems with a single obstacle [6]. The second method involves the so-called *T*-matrices (scattering matrices) [8]. The *T*-matrix method was proposed in Ref. [70]. In this approach, the field scattered by an inhomogeneity is sought in the form of expansion in terms of cylindrical (in 2D) or spherical (in 3D) functions. The matching conditions for the incident and scattered fields on the boundary result in infinite sets of equations that are represented in the form of *T*-matrices [60]. The *T*-matrix allows both the scattered and internal fields to be defined at an arbitrary point for a scatterer of arbitrary convex shape, size, and properties [71]. However, the more the scatterer shape deviates from the canonical one (cylindrical or spherical), the more terms in the expansions must be retained in order to obtain a solution with desired accuracy.

1.5 Some Algorithms for the Numerical Inversion of the Laplace Transform

The numerical inversion of the Laplace transform has been thoroughly studied, in particular, in Refs. [10, 13, 17, 19, 25, 45, 66, 74]. Some asymptotical methods were also discussed in Refs. [4, 5, 58]. The numerical inversion with the use of Laguerre polynomials was considered in Refs. [1, 22, 23, 72], and Ref. [39] deals with Jacobi polynomials. Refs. [3, 11, 14, 16, 21, 27, 28, 40, 41, 61, 67, 69] deal with the implementation of the numerical inversion of the Laplace transform in the problems of mechanics and physics.

Here, we analyse several algorithms for the numerical inversion of the Laplace transform and compare the results with reference originals and some exact solutions. It is shown that there is an optimal number of the expansion terms, i.e. the number of terms in the convergent series that can be taken into account is limited by the number of significant digits available in a particular numerical realisation. We also demonstrate that the accuracy of all the algorithms decreases for larger times. These two conclusions are the consequence of the fact that the inversion of the Laplace transform is an ill-posed problem [35, 38, 50].

The Laplace transform of function $f(t)$ is defined by the following operator for a complex parameter $p = \sigma + \mathrm{i}\tau$ [27]:

$$F(p) = \int_{0}^{\infty} f(t) e^{-pt} \mathrm{d}t, \tag{1.27}$$

provided that $F(p)$ is a uniformly convergent analytical function in the domain $\mathrm{Re}\,p > \sigma_c$ and $F(p) \to 0$ with respect to $\mathrm{Arg}\,p$ at $p \to \infty$. If (1.27) is absolutely convergent for all $\mathrm{Re}\,p > \sigma_c$, then there exists the operator of the inverse transform,

$$f(t) = \frac{1}{2\pi \mathrm{i}} \int_{\sigma_c - \mathrm{i}\infty}^{\sigma_c + \mathrm{i}\infty} F(p) e^{pt} \mathrm{d}p. \tag{1.28}$$

The inverse problem consists in finding a solution $f(t)$ to the integral equation of the first kind (1.27), with $F(p)$ being a given function of complex variable p. Since the kernel $\exp(-pt)$ is a smooth function of t and p, averaging of f with weight function $\exp(-pt)$ can considerably smooth out the singularities of $f(t)$. Therefore, to recover all the local irregularities of $f(t)$, one needs to use methods that are sensitive to minor peculiarities of $F(p)$.

Since $f(t)$ is unstable with respect to small variations of $F(p)$, the inversion problem is ill-posed: its solutions do not exist for some values of numerical or functional parameters, and small variations of these parameters can lead to large variations of the solution [59, 62]. This is the main reason why all the known inversion algorithms are limited. In order to keep the inversion accuracy high enough, one has to retain as many terms in the expansions as possible. However, this requires that the expansion coefficients be calculated with higher accuracy, which is usually limited by the available computational resources. For nonanalytic and discontinuous functions, the largest errors are expected to accumulate in the vicinity of points of discontinuity. Respectively, the numerical inversion results in a nonphysical peak at small t. On the other hand, asymptotic expansions should be involved at large t.

1.5.1 Shifted Legendre Polynomials

The change of variable $e^{-t} = \zeta$ transforms the interval $(0, \infty)$ of variable t into the interval $(0, 1)$ of variable ζ, and transform (1.27) takes the form

$$F(p) = \int_0^1 f(\zeta) \zeta^{p-1} \mathrm{d}\zeta, \tag{1.29}$$

where $f(\zeta)$ is expressed as a convergent series in terms of polynomials that are orthogonal on the segment $[0, 1]$. The orthogonal polynomials can be chosen in the form of the so-called shifted Legendre polynomials $P_n^*(\zeta)$ [38]:

$$f(\zeta) = \sum_{n=0}^{\infty} (2n+1) a_n P_n^*(\zeta), \quad a_n = \sum_{k=0}^{n} \alpha_k^{(n)} F(k+1), \tag{1.30}$$

where,

$$P_n^*(\zeta) = (-1)^n \sum_{k=0}^{n} \alpha_k^{(n)} \zeta^k, \quad \alpha_k^{(n)} = (-1)^k \binom{n}{k} \frac{(n+k)!}{n!k!}.$$

The coefficients a_n are calculated at a finite number k of points evenly spaced along the real axis of the complex parameter p.

Figures 1.2 and 1.3 demonstrate some results of our calculations performed with 5, 6, and 10 terms taken into account in series (1.30) (the authors are grateful to Dr. V.A. Tkachenko for his assistance in producing these plots). Ten terms proved to be optimal for the approximation with nine significant digits. In all the cases, the accuracy of calculations increases with n.

This algorithm was used to solve a number of problems in the theory of wave propagation and diffraction, including the initial-boundary-value problems for a hydraulic shock [53], pulse impact on an open spherical elastic shell [54], tsunami wave generation [48], etc.

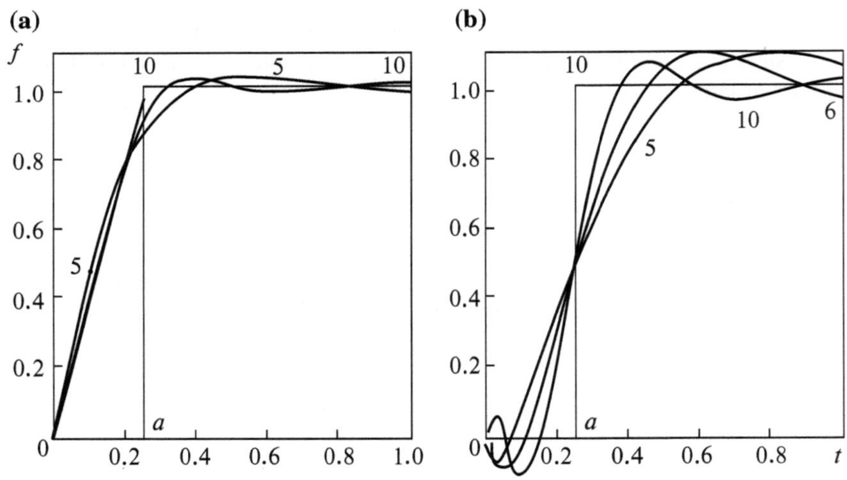

Fig. 1.2 Results of the numerical inversion in terms of shifted Legendre polynomials for **a** linearly increasing function $f(t) = \begin{cases} t/a, & 0 < t \le a, \\ 1, & t > a, \end{cases}$ $F(p) = \frac{1-\exp(-ap)}{ap^2}$ that reduces to a constant, **b** Heaviside step function $f(t) = H(t-a)$, $F(p) = \frac{\exp(-ap)}{p}$

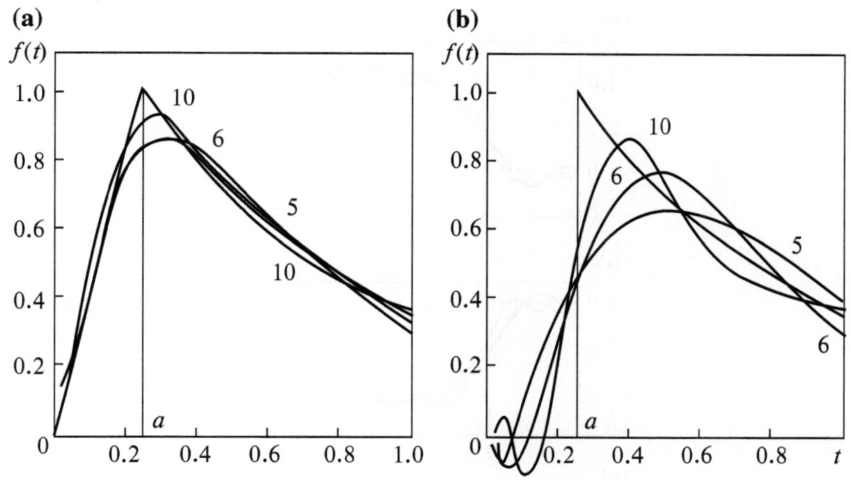

Fig. 1.3 Results of the numerical inversion in terms of shifted Legendre polynomials for **a** linearly increasing and then exponentially decreasing function $f(t) = \begin{cases} t/a, & 0 < t \le a, \\ \exp(-b(t-a)), & t > a, \end{cases}$ $F(p) = \frac{1 - \exp(-ap)}{ap^2} + \frac{\exp(-ap)}{p+b} - \frac{\exp(-ap)}{p}$, **b** $f(t) = \exp(-b(t-a))H(t-a)$, $F(p) = \frac{\exp(-ap)}{p-b}$

1.5.2 Fourier Sine Series

We introduce the variable θ instead of time t as

$$\cos\theta = e^{-\sigma t},\ p = (2n+1)\sigma,\ \sigma > 0,\ n = 0, 1, \ldots$$

and write the function $f(t)$ in (1.27) in the form of the sine series [15, 43]:

$$f(t) = f\left(-\frac{1}{\sigma}\ln(\cos\theta)\right) = \Phi(\theta) = \sum_{\nu=0}^{\infty} C_\nu \sin(2\nu+1)\theta. \tag{1.31}$$

Then we obtain a set of linear equations for the coefficients C_ν:

$$C_0 = \frac{4}{\pi}\sigma F(\sigma),\ C_0 + C_1 = \frac{4^2}{\pi}\sigma F(3\sigma),\ 2C_0 + 3C_1 + C_2 = \frac{4^3}{\pi}\sigma F(5\sigma),\ \ldots.$$

Note that the realisation of this algorithm essentially depends on the value of parameter σ and selecting it in the optimal way involves supplementary numerical calculations. Some results of our test calculations are shown in Fig. 1.4.

This algorithm was used to analyse the propagation of acoustic waves in a compressible fluid under the surface wave excitation by bottom movements [2]. The corresponding initial-boundary-value problem is formulated as follows

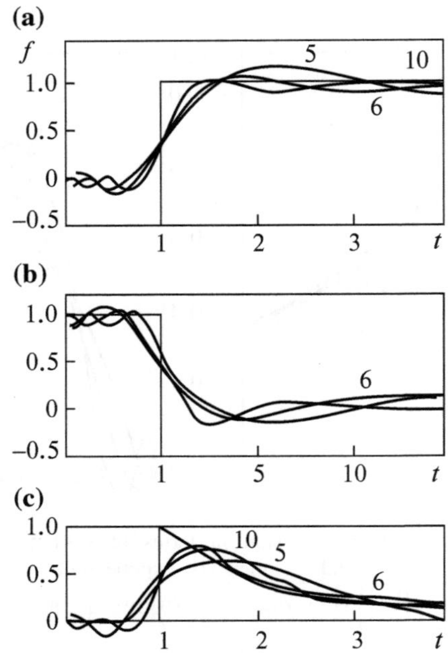

Fig. 1.4 Results of the numerical inversion with the use of the sine series for several reference functions: **a** $f(t) = H(t - a)$, $F(p) = p\exp(-ap)$; **b** $f(t) = \begin{cases} 1, & t < 1, \\ 0, & t > 1, \end{cases}$ $F(p) = \frac{1 - \exp(-p)}{p}$; **c** $f(t) = \begin{cases} 0, & 0 < t < 1, \\ \exp(-(t-1)), & t > 1, \end{cases}$ $F(p) = \frac{\exp(-p)}{p+1}$

$$\varphi_{zz} - c_0^{-2}\varphi_{tt} = 0, \quad -d \le z \le 0, \quad t > 0, \tag{1.32}$$

$$\left(\varphi_{tt} + g\varphi_z\right)_{z=0} = 0, \tag{1.33}$$

$$\varphi_z\big|_{z=-d} = \kappa_t, \tag{1.34}$$

$$\varphi(z,t)\big|_{t=0} = \varphi_z\big|_{t=0} = 0, \tag{1.35}$$

where φ is the velocity potential; $\kappa(t)$ is a given bottom excitation function,

$$\kappa(t) = (\alpha t)^4 e^{-2\alpha t}, \quad \alpha = \text{const},$$

g is the gravitation constant, and the sound velocity c_0 changes from a constant value (corresponding to a compressible fluid) to infinity (incompressible fluid).

An exact analytical solution to problem (1.32)–(1.35) was obtained in the form of an infinite sum of residues for the free surface displacement $\eta = -(1/g)\varphi_t|_{z=0}$ in Ref. [2]. At the same time, a numerical solution was found by the inversion using formula (1.31). The both solutions were compared at different α and c_0. Figure 1.5 shows the free surface displacement η normalised to the fluid depth d as a function of time t normalised to the factor $\sqrt{d/g}$. Figures 1.5a, b, c were plotted at fixed parameter $\alpha\,(\alpha^* = \alpha\sqrt{d/g} = 4)$ and variable sound velocities $c_0\,(c_0^* = c_0/\sqrt{gd} = 4, 6, \text{and } 8)$. Figures 1.5b, d were plotted at fixed $c_0\,(c_0^* = 6)$ and variable $\alpha\,(\alpha^* = 4, 6)$. The obtained results demonstrate that the accuracy of the numerical inversion degrades significantly with increasing t. This fact manifests itself in the deviation of the numerical solutions from the dashed curves that represent the exact solution. It can also be seen that the free surface displacement in the case of compressible fluid is larger than in the case of incompressible fluid.

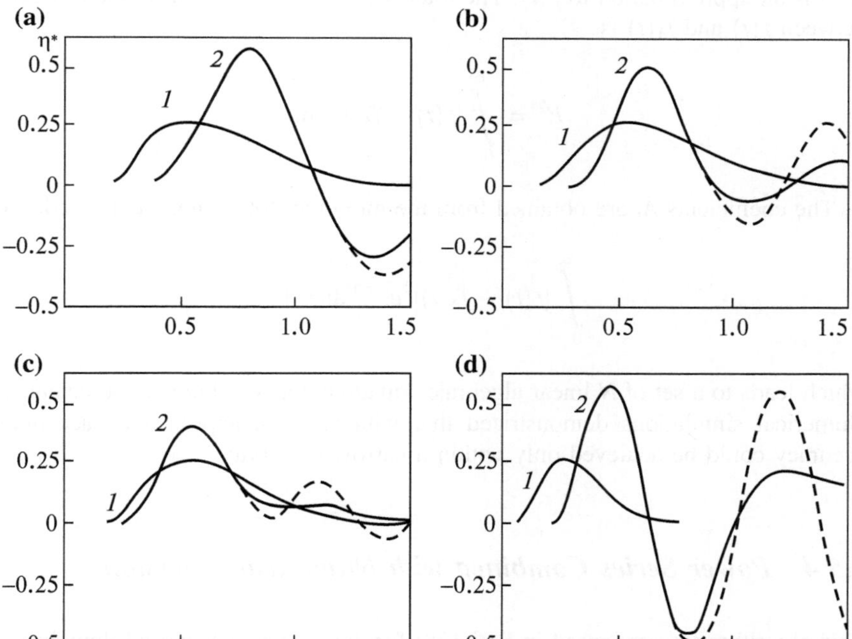

Fig. 1.5 Displacement of the free fluid surface $\eta^* = \eta/d$ disturbed by the bottom movement as a function of normalised time $t^* = t\sqrt{g/d}$: **a** $\alpha^* = 4$, $c_0^* = 4$; **b** $\alpha^* = 4$, $c_0^* = 6$; **c** $\alpha^* = 4$, $c_0^* = 8$; **d** $\alpha^* = 6$, $c_0^* = 6$. *Solid* curves depict the results of numerical inversion and *dashed* curves correspond to the exact solutions: **1** incompressible fluid, **2** compressible fluid (different sound velocities c_0 correspond to different compressibility)

Another example employing the numerical inversion of the Laplace transform with the use of the Fourier sine series refers to the propagation of pulse pressure waves in blood vessels, where the function approximating a cardiac impulse can be defined as $f(t) = t\ \exp(-\alpha t)$ [51].

1.5.3 Exponential Functions

The function $f(t)$ is represented as the following series [46]:

$$f_A(t) = \sum_{i=1}^{N} A_i e^{-t/t_i}, \tag{1.36}$$

where A_i are undetermined coefficients and t_i are given positive constants, so that $f_A(t)$ is an approximation to $f(t)$. The total squared error defined by the difference between $f(t)$ and $f_A(t)$ is

$$E^2 = \int_0^\infty [f(t) - f_A(t)]^2 \mathrm{d}t.$$

The coefficients A_i are obtained from the minimum total squared error criterion

$$\int_0^\infty [f(t) - f_A(t)]^2 e^{-t/t_i} \mathrm{d}t = 0,$$

which leads to a set of N linear algebraic equations for N unknown coefficients A_j. Numerical simulations demonstrated that with such an approach an acceptable accuracy could be achieved only within a narrow t interval.

1.5.4 Power Series Combined with Numerical Inversion

This algorithm was proposed in Ref. [28] for the case of a localised domain with variable properties (e.g. an underwater island). The corresponding initial-boundary-value problem for a partial differential equation with variable coefficients in the localised domain is solved by the power series method resulting in an infinite set of algebraic equations. The corresponding reduced system contains the coefficients present in the power series and the coefficients arising from the problem of the Laplace transform inversion. This approach was used to solve an initial-boundary-value problem of wave diffraction by a local bottom inhomogeneity.

1.5.5 Fourier–Bessel Series

The majority of algorithms for the numerical inversion of the Laplace transform are based on the expansion of the sought original function in terms of orthogonal polynomials. The numerical realisation of such algorithms involves significant complications because the expansion coefficients are determined from infinite sets of equations with triangular matrices that are unstable to error accumulation. Respectively, the accuracy of calculating the original function is rather limited. The inversion problem is similar to the finite-interval moment problem and is thereby related to the theory of approximations in Hilbert spaces [35].

Here, we consider a regular method for the numerical inversion of the Laplace transform proposed in Ref. [52]. It is based on the expansions of the original functions in terms of the Fourier–Bessel functions and the regularisation of the corresponding ill-posed problem. The coefficients of such expansions are expressed by well-converged series, so that smooth original functions could be calculated with high accuracy.

Let $D = \{\mathrm{Re}\, p > \gamma\}$ be an open domain in the complex plane p, $A(D)$ be a set of analytical functions in D. Let $Z(D)$ denote a class of Laplace image functions $F(p)$ and $L(0, T)$ denote a set of original functions $f(t)$ that are Lebesgue integrable on $[0, T)$. The necessary condition for $F(p)$ to belong to $Z(D)$ is its analyticity: $F(p) \in A(D)$; the sufficient conditions can be formulated in different ways [35]. Then the Laplace transform inversion reduces to solving a first-kind Fredholm integral equation

$$\int_0^{\infty} e^{-pt} f(t)\mathrm{d}t = F(p), \tag{1.37}$$

where $f(t) \in L(0, \infty)$, $F(p) \in Z(D)$. The following theorem holds true.

Theorem *Let a function $f(t)$ be continuous on the interval $[0, \infty)$ except for a set of measure zero. If $f(0) = 0, f(t)\exp(-3t/2) \in L(0, \infty)$, and $F(p) \in Z(\mathrm{Re}\, p > 3/2)$, then the following expansion is valid for all $t \in [0, \infty)$:*

$$f(t) = \sum_{n=1}^{\infty} c_{n\nu} J_\nu(k_{n\nu} e^{-\sigma t}), \quad J_\nu(k_{n\nu}) = 0, \tag{1.38}$$

$$c_{n\nu} = \frac{2\sigma}{J_{\nu+1}^2(k_{n\nu})} \sum_{m=0}^{\infty} \frac{(-1)^m}{m!\Gamma(m+\nu+1)} \left(\frac{k_{n\nu}}{2}\right)^{2m+\nu} F(\sigma(2m+2+\nu)), \tag{1.39}$$

with $\nu > -1/2$ and σ being an arbitrary parameter.

This theorem and formulas (1.38) and (1.39) can be proven by making the substitution $\zeta = \exp(-\sigma t)$ in (1.27) and expanding the function $f^*(\zeta) = f(-\ln\zeta/\sigma)$ in Fourier–Bessel series on the interval [0, 1]. Note that the condition $f(0) = 0$ does not pose an essential constraint because it can always be satisfied by making a transform

$$f_1(t) = f(t) - f(0)e^{-t},\ F_1(p) = F(p) - \frac{f(0)}{p+1}.$$

Expansions (1.38) and (1.39) provide an exact solution to the above-formulated inversion problem in the framework of the finite-interval moment problem [35]. Nevertheless, the inversion problem of operational calculus is known to be ill-posed (e.g., in the Lebesgue space L_2). When the original function is expanded in series in terms of the given orthonormal set of functions $\varphi_n(t)$ with given coefficients a_n,

$$f(t) = \sum_{n=1}^{\infty} a_n \varphi_n(t), \tag{1.40}$$

the ill-posedness of the inverse problem results in the fact that the summation of this series is also an ill-posed problem. In this case, the use of the well-developed regularisation techniques [62] proves to be really helpful.

Let $b = \{b_n\}$ be a sequence of approximate values of the expansion coefficients that is known instead of their exact values $a = \{a_n\}$ and $\rho_{L_2}(a, b) \leq \varepsilon$ (ρ being a metric on L_2). Suppose that we need to find a function$\tilde{f}(t)$ defined in the $L(0, \infty)$ space that approximates $f(t)$ in terms of the sequence b so that $\rho_{L(0,\infty)}(\tilde{f}, f) \to 0$ at$\varepsilon \to 0$. Such a problem can be solved using Tikhonov's classical regularisation scheme. As a result, the function $\tilde{f}(t)$ that satisfies all the above-mentioned conditions can be found in the form of the following series:

$$\tilde{f}(t) = \sum_{n=1}^{\infty} b_n r(n, \delta) \varphi_n(t), \tag{1.41}$$

where $r(n, \delta) = (1 + \delta\theta_n)^{-1}$ is the stabilising factor and δ is the regularisation parameter. The sequence θ_n is defined in terms of the set of eigenvalues of the Sturm–Liouville problem for $\varphi_n(t)$ with $\theta_n = O(n^{2+\alpha})$, $n \to \infty$, $\alpha \geq 0$.

In reference to the Fourier–Bessel series (1.38) used in the numerical inversion of the Laplace transform, the regularised expansion is given by [52]

$$\tilde{f}(t) = \sum_{n=1}^{\infty} \frac{\tilde{c}_{n\nu}}{1 + \delta k_{n\nu}^2} J_\nu(k_{n\nu} e^{-\alpha t}), \tag{1.42}$$

where $\tilde{c}_{n\nu}$ is the set of approximate coefficients. The regularisation parameter δ is determined by the condition [62]

$$\sum_{n=1}^{\infty} \frac{\delta^2 k_{n\nu}^4}{\left(1 + \delta k_{n\nu}^2\right)^2} = \varepsilon^2, \tag{1.43}$$

where ε is the given error of calculating the coefficients $\tilde{c}_{n\nu}$. A special note should be made regarding the stable computation of the coefficients $c_{n\nu}$ in expansion (1.38). Without loss of generality, we put $\nu = 0$ and rewrite formula (1.39) as

$$c_{n0} = \frac{2\sigma}{J_1^2(k_n)} \sum_{m=0}^{\infty} (-1)^m b_{mn} F((2m+2)\sigma),\ b_{mn} = \left(\frac{k_n}{2}\right)^{2m} \frac{1}{m!^2},\ k_n \equiv k_{n,0}.$$

The sequence elements $\{b_{mn}\}$ satisfy the following recurrence relation at fixed $n \geq 1$:

$$b_{m+1,n} = q_{mn} b_{mn},\ q_{mn} = \left(\frac{k_n}{2}\right)^2 \frac{1}{(m+1)^2},\ b_{0n} = 1. \tag{1.44}$$

For relations (1.44) to be computationally stable, the calculations should start from $b_{\max} = b_{m_0 n}$, $m_0 \approx [(k_n - 1)/2]$, and then go backwards for $1 \leq m \leq m_0 - 1$ and upwards for $m \geq m_0 + 2$ [21]. In this case, the error does not build up and the computation of $b_{mn}$is stable.

The algorithm based on relations (1.38), (1.39), and (1.42) was realised in the form of working computer program for the case $\nu = 0$. The numerical values of k_n with an accuracy of up to 19 significant digits were taken from Ref. [25]. The test calculations were performed with double precision for a wide range of reference functions whose Laplace transforms are well known. The number of terms in (1.38) that can be summed up without the loss of accuracy is limited because of the coefficients in the expansion (1.39) for c_{n0} rapidly increase. Their growth rate is $O((\pi e)^n)$, therefore we could retain only 10–12 terms in (1.38). The regularisation parameter determined from condition (1.43) was $\delta = 10^{-3}$. The proposed algorithm proved to be stable and allowed the smooth original functions to be calculated with maximum error of no more than 5%.

Note that the use of the numerical method based on formulas (1.38) and (1.39) does not exclude the preliminary processing of the image function in order to eliminate its singularities and to obtain asymptotic expressions for the original function at small and large t.

1.6 Method of Multiple Scales

Method of multiple scales is used in the solution of many problems described by ordinary and partial differential equations [34, 42]. A typical feature of the method of multiple scales is the use of rescaling of the temporal variable t to construct solutions with respect to time in the form of expansions in small parameter ε. Introduce a sequence of time variables $T_k = \varepsilon^k t$ so that

$$\frac{\partial}{\partial t} = \frac{\partial}{\partial T_0} + \varepsilon\frac{\partial}{\partial T_1} + \varepsilon^2\frac{\partial}{\partial T_2} + \varepsilon^m\frac{\partial}{\partial T_m}, \tag{1.45}$$

where $T_0, T_1, T_2, \ldots$ are the new temporal variables (T_0 being the fast time and $T_1, T_2, \ldots$ being the slow times), m determines the time scale on which the corresponding expansion holds.

In the case of partial differential equations, the unknown function $\phi(\boldsymbol{x}, t, \varepsilon)$ depends on spatial variables and the solution can be represented in the form of power series in terms of parameter ε:

$$\phi = \phi_0 + \varepsilon\,\phi_1 + \varepsilon^2\phi_2 + \ldots. \tag{1.46}$$

The same expansions are also written for the spatial variables. In particular, in one-dimensional case we have

$$\frac{\partial}{\partial X} = \frac{\partial}{\partial X_0} + \varepsilon\frac{\partial}{\partial X_1} + \varepsilon^2\frac{\partial}{\partial X_2} + \cdots + \varepsilon^m\frac{\partial}{\partial X_m}, \tag{1.47}$$

where $X_0, X_1, X_2, \ldots, X_m$ are the spatial scales: X_0 is the short scale and $X_1, X_2, \ldots$ are the long scales.

Substituting the above expansions into the equation of motion and equating the coefficients at the same powers of ε leads to linear problems (chains of linear equations) in each approximation. The first-order problem is usually homogeneous, and higher order problems are usually inhomogeneous. In the general case, one should satisfy the solvability condition (elimination of secular terms) in each linear equation. For example, in the case of a set of linear algebraic equations

$$A\boldsymbol{x} = \boldsymbol{b}, \tag{1.48}$$

where A is some square matrix of order N and $\boldsymbol{x}$ and $\boldsymbol{b}$ are column vectors with N components, the solvability condition requires the right-hand side of system (1.48) to be orthogonal to any solution of the homogeneous adjoint system. If the homogeneous system has a nontrivial solution, then the right-hand side of system (1.48) satisfies the solvability condition.

The applications of the method of multiple scales to the problems of wave propagation are considered in more detail in Chaps. 2 and 6.

References

1. Amerbaev, V.M., Utembaev, N.A.: Chislennyi analiz lagerrovskogo spektra (Numerical Analysis of the Laguerre Spectrum). Nauka, Alma-Ata (1982)
2. Avramenko, O.V., Selezov, I.T.: Analiticheskii i chislennyi analiz szhimaemosti zhidkosti pri vozbuzhdenii voln pod'emom dna (Analytical and numerical analysis of the fluid compressibility under the wave excitation by bottom rising). Gidromekhanika **68**, 3–8 (1994)

3. Bellman, R., Kalaba, R.E., Lockett, J.A.: Numerical inversion of the laplace transform. In: Applications to Biology, Economics and Physics. Elsevier, Amsterdam (1966)
4. Belov, M.A.: Primenenie metoda asimptoticheskogo rasshireniya intervala dlya resheniya uravneniya teploprovodnosti (Using the method of asymptotic extension of the interval in solving the heat equation). In: Elektrodinamika i mekhanika sploshnykh sred (Electrodynamics and Mechanics of Continuous Media), pp. 101–108. Pēteris Stučka Latvian State University, Riga (1982)
5. Belov, M.A., Tsirulis, T.T.: Asimptoticheskie metody obrashcheniya integralnykh preobrazovanii (Asymptotic Methods of the Inversion of Integral Transforms). Zinatne, Riga (1985)
6. Berkhoff, J.C.W.: Linear wave propagation problems and finite element method. In: Finite Elements in Fluids: Viscous Flow and Hydrodynamics, vol. 1, pp. 251–264. Wiley, London (1975)
7. Born, M.: Optik. Springer, Berlin (1933)
8. Bostrom, A., Kristensson, C.: Scattering of a pulsed Rayleigh wave by a spherical cavity in an elastic half space. Wave Motion **2**, 137–143 (1983)
9. Bremmer, H.: The W.K.B. approximation as the first term of a geometric-optical series. Comm. Pure Appl. Math. **4**, 105–115 (1951)
10. Brigham, E.O.: The Fast Fourier Transform. Prentice-Hall, Englewood Cliffs (1974)
11. Chow, E.P., Jeffreson, C.P.: Application of numerical Laplace transform inversion techniques to the analysis of the dynamics of short parallel-plate packed beds. Chem. Eng. Austral. **2**, 11–15 (1977)
12. Courant, R.: Methods of mathematical physics. In: Partial Differential Equations, vol. 2. Interscience, New York (1962)
13. Crump, K.S.: Numerical inversion of Laplace transforms using a Fourier series approximation. J. Assoc. Comput. Math. **23**, 89–96 (1976)
14. Debnath, L., Thomas, J.G.: On finite Laplace transformation with applications. Zeitschrift für Angewandte Mathematik und Mechanik **56**(12), 559–563 (1976)
15. Doetsch, G.: Anleitung zum praktischen Gebrauch der Laplace-Transformation. R. Oldenbourg, München (1956)
16. Doyle, J.F.: Spectral analysis of coupled thermoelastic waves. J. Therm. Stresses **11**(3), 175–185 (1988)
17. Dubner, R., Abate, J.: Numerical inversion of Laplace transforms by relating them to the finite Fourier cosine transforms. J. Assoc. Comput. Math. **15**, 115–123 (1968)
18. Dunford, N., Schwartz, J.T.: Linear operators, part 2. In: Spectral Theory, Self Adjoint Operators in Hilbert space. Interscience, New York (1963)
19. Durbin, F.: Numerical inversion of Laplace transforms: an efficient improvement to Dubner and Abate's method. Comput. J. **17**, 371–376 (1974)
20. Engquist, B., Majda, A.: Radiation boundary conditions for acoustic and elastic wave calculations. Comm. Pure Appl. Math. **32**, 313–357 (1979)
21. Forsythe, G.E., Malcolm, M.A., Moler, C.B.: Mechanical Methods for Mathematical Computations. Prentice-Hall, Englewood Cliffs (1976)
22. Galazyuk, V.A.: Metod polinomov Chebysheva-Lagerra v smeshannoi zadache dlya linei-nogo differentsialnogo uravneniya vtorogo poryadka s postoyannymi koeffitsientami (Method of Chebyshev-Laguerre polynomials in a mixed problem for a second-order linear differential equation with constant coefficients). Dopovidi AN URSR. Ser. A **1**, 3–7 (1981)
23. Galazyuk, V.A., Gorechko, A.N.: Issledovanie nestatsionarnykh volnovykh protsessov v vyazkoi szhimaemoi zhidkosti metodom polinomov Chebysheva-Lagerra (Investigating nonstationary wave processes in a viscous compressible fluid by the method of Chebyshev-Laguerre polynomials). Dopovidi AN URSR. Ser. A **5**, 31–35 (1981)
24. Gautesen, A.K., de Hoop, M.V.: Uniform asymptotic expansion of the generalized Bremmer series. SIAM J. Appl. Math. **60**(4), 1302–1329 (2000)
25. Gerber, H.: First one hundred zeros of $J_0(x)$ accurate to 19 significant figures. Math. Comput. **18**, 319–322 (1964)

26. Givoli, D., Keller, J.B.: Non-reflecting boundary conditions for elastic waves. Wave Motion **12**(3), 261–279 (1990)
27. Gnoenskii, L.S., Kamenskii, G.A., Elsgolts, L.E.: Matematicheskie osnovy teorii upravlyaemykh sistem (Mathematical Foundations of the Theory of Controlled Systems). Nauka, Moscow (1969)
28. Gordon, P.F.: Accurate transform inversion in the elastodynamics of heterogeneous media. Trans. ASME. E **44**(2), 352–354 (1977)
29. Grote, M.J.: Nonreflecting boundary conditions for time dependent wave propagation. Research Report No. 2000–04, Eidgenössische Technische. Hochschule, Zürich (2000)
30. Grote, M.J., Keller, J.B.: Exact nonreflecting boundary conditions for the time dependent wave equation. SIAM J. Appl. Math. **55**, 280–297 (1995)
31. Grote, M.J., Keller, J.B.: Nonreflecting boundary conditions for time dependent scattering. J. Comput. Phys. **127**, 52–65 (1996)
32. Grote, M.J., Keller, J.B.: Nonreflecting boundary conditions for Maxwell's equations. J. Comput. Phys. **139**, 327–342 (1998)
33. Jones, D.S.: The scattering of long electromagnetic waves. Quart. J. Mech. Appl. Math. **33**(1), 105–122 (1980)
34. Kevorkian, J., Cole, J.D.: Multiple Scale and Singular Perturbation Methods. Springer, New York (1996)
35. Krylov, V.I., Skoblya, N.S.: A Handbook of Methods of Approximate Fourier Transformation and Inversion of the Laplace Transformation. Mir Publishers, Moscow (1977)
36. Kryvonos, Yu.G., Kornilov, I.E., Selezov, I.T.: Ob odnom metode issledovaniya dinamicheskoi ustoichivosti sistemy uprugaya poverkhnost–vyazkaya zhidkost (On one method of investigating the dynamic stability of the elastic surface–viscous fluid system). Preprint, Institute of Cybernetics AN USSR, Kiev (1982)
37. Ladyzhenskaya, O.A.: The Boundary Value Problems of Mathematical Physics. Springer, New York (1985)
38. Lanczos, C.: Applied Analysis. Prentice-Hall, Englewood Cliffs (1956)
39. Miller, M.K., Guy, W.T.: Numerical inversion of the Laplace transform by use of Jacobi polynomials. SIAM J. Numer. Anal. **3**, 624–635 (1966)
40. Moodie, E.B., Barday, D.W., Tait, R.T.: A boundary value problem for fluid-filled viscoelastic tubes. Math. Model. **4**, 195–207 (1983)
41. Narayanan, G.V., Beskos, D.E.: Use of dynamic influence coefficients in forced vibration problems with the aid of fast Fourier transform. Comput. Struct. **9**(2), 145–150 (1978)
42. Nayfeh, A.H.: Perturbation Methods. Wiley–VCH Verlag, Weinheim (2004)
43. Papoulis, A.: A new method of inversion of the Laplace transform. Quart. Appl. Math. **14**, 405–414 (1957)
44. Paskaramoorthy, R., Shah, A.H., Datta, S.K.: Scattering of flexural waves by cavities in a plate. Int. J. Solids Struct. **25**(10), 1177–1191 (1989)
45. Piessens, R.: A bibliography on numerical inversion of the Laplace transforms and its applications. J. Comp. Appl. Math. **1**(2), 115–126 (1975)
46. Schapery, R.A.: Approximate methods of transform inversion for viscoelastic stress analysis. In: Proceedings of the 4th U.S. National Congress of Applied Mechanics, vol. 2, pp. 1075–1085 (1962)
47. Selezov, I.T.: Numerical solution of the boundary value problems of the wave diffraction by arbitrary revolution surfaces. Numerical Methods. Colloquia Mathematica Societatis János Bolyai, vol. 50, pp. 473–480. North-Holland, Amsterdam (1987)
48. Selezov, I.T.: Modelirovanie volnovykh i difraktsionnykh protsessov v sploshnykh sredakh (Modeling the Wave and Diffraction Processes in Continuous Media). Naukova Dumka, Kiev (1989)
49. Selezov, I.T.: Diffraction of waves by radially inhomogeneous inclusions. Phys. Express **1**(2), 101–115 (1993)
50. Selezov, I.T.: Issledovanie neustanovivshykhsya volnovykh dvizhenii v gidrouprugikh sistemakh obolochka-zhidkost (Investigating the transient wave motions in hydroelastic

shell-fluid systems). In: Prikladnye problemy mekhaniki tonkostennykh konstruktsii (Applied Problems in the Mechanics of Thin-Wall Constructions), pp. 286–305. Izdatelstvo Moskovskogo Universiteta, Moscow (2000)
51. Selezov, I., Fratamico, G.: Pulse waves in arteries with a vessel joint. In: Mela, M., Pallotti, G. (eds.) International Summer School in Biophysics, pp. 89–94. Bellaria Igea Marina, Italy 9–16 Sept. (1996)
52. Selezov, I.T., Korsunskii, S.V.: Chislennoe obrashchenie preobrazovaniya Laplasa na osnove razlozhenii Fur'e–Besselya (Numerical inversion of the Laplace transform with the use of Fourier–Bessel expansions). Dopovidi AN URSR. Ser. A. **11**, 25–28 (1988)
53. Selezov, I.T., Nikulinskaya, S.N.: Rasprostranenie malykh vozmushchenii v zhidkosti, protekayushchei v uprugoi tsilindricheskoi obolochke (Propagation of small disturbances in the fluid flowing in an elastic cylindrical shell). Izvestiya AN SSSR. Mekhanika **1**, 173–175 (1965)
54. Selezov, I.T., Sorokina, V.V., Tsyganov, N.K., Iakovlev, V.V.: Dynamics of open spherical shell under impulsive excitation. Mechanics of Solids **13**(2), 131–135 (1978)
55. Selezov, I.T., Yakovlev, V.V.: Nekotorye zadachi difraktsii ploskikh voln na tsilindre s peremennoi plotnostyu (Some problems of plane wave diffraction by a cylinder with variable density). Akusticheskii zhurnal **23**(5), 805–809 (1977)
56. Selezov, I.T., Yakovlev, V.V.: Difraktsiya voln na simmetrichnykh neodnorodnostyakh (Wave Diffraction by Symmetric Inhomogeneities). Naukova Dumka, Kiev (1978)
57. Smirnov, V.I.: A Course of Higher Mathematics, vol. 2. Pergamon Press, Oxford (1964)
58. Smotrovs, J.A.: Obrashchenie preobrazovaniya Fur'e metodom asimptoticheskogo ras-shireniya intervala (Fourier transform inversion by the method of asymptotic extension of the interval). In: Voprosy elektrodinamiki i mekhaniki sploshnykh sred (Problems of Electrodynamics and Mechanics of Continuous Media), vol. 4, pp. 128–143. Pēteris Stučka Latvian State University, Riga (1978)
59. Starkov, V.N.: Konstruktivnye metody vychislitelnoi fiziki v zadachakh interpretatsii (Constructive Methods of Computational Physics in Interpretation Problems). Naukova Dumka, Kiev (2002)
60. Su, J.-H., Varadan, V.V., Varadan, V.K.: Acoustic wave scattering by a finite elastic cylinder in water. J. Acoust. Soc. Amer. **68**(2), 686–691 (1980)
61. Talbot, A.: The accurate numerical inversion of Laplace transforms. J. Inst. Math. Appl. **23**, 97–120 (1979)
62. Tikhonov, A.N., Arsenin, V.Ya.: Solutions of Ill-Posed Problems. VH Winston & Sons, New York (1977)
63. Tricomi, F.G.: Differential Equations. Blackie and Son Ltd., London (1961)
64. Tsay, T.-K., Zhu, W., Liu, P.L.-F.: A finite element model for wave refraction, diffraction, reflection and dissipation. Appl. Ocean Res. **11**(1), 33–38 (1989)
65. Varley, E., Seymour, B.R.: A method for obtaining exact solutions to partial differential equations with variable coefficients. Stud. Appl. Math. **78**, 183–225 (1988)
66. Vershinin, V.D.: K odnomu sposobu opredeleniya obratnogo preobrazovaniya Laplasa (On one way of the determination of the inverse Laplace transform). Izvestiya vuzov. Matematika **10**, 24–28 (1978)
67. Vilge, Ya.I., Zakharov, E.V.: K raschetu nestatsionarnogo akusticheskogo polya v skvazhine metodom chislennogo obrashcheniya preobrazovaniya Laplasa (On calculating the nonstationary acoustic field in a well by the method of the numerical inversion of the Laplace transform). Chislennye metody v geofizike **1**, 102–107 (1978)
68. Vissers, F.P.L.M., Kamp, L.P.J., Sluijter, F.W.: Some generalizations of radial Bremmers series. Wave Motion **13**(1), 43–51 (1991)
69. Voronenok, E.Ya.: Chislennyi metod obratnogo preobrazovaniya Laplasa i ego realizatsiya v odnoi zadache gidrouprugosti (Numerical method of the inverse Laplace transform and its realization in one problem of hydroelasticity). In: Problemy stroitelnoi mekhaniki korablya (Problems of Ship Construction Mechanics), pp. 43–51. Sudostroenie, Leningrad (1973)

70. Waterman, P.C.: New formulation of acoustic scattering. J. Acoust. Soc. Amer. **45**(6), 1417–1429 (1969)
71. Waterman, P.C.: Symmetry, unitarity, and geometry in electromagnetic scattering. Phys. Rev. D **3**, 825–839 (1970)
72. Weeks, W.T.: Numerical inversion of Laplace transforms using Laguerre functions. J. Assoc. Comput. Math. **13**, 419–426 (1966)
73. Wu, R.-S.: The perturbation method in elastic wave scattering. Pure. Appl. Geophys. **131**(4), 605–637 (1989)
74. Yakovlev, Yu.S.: Obshchii metod obrashcheniya integralnykh preobrazovanii Fur'e, Laplasa, Khankelya i Stiltesa funktsii klassa L_2 (A general method for the inversion of the integral Fourier, Laplace, Hankel, and Stieltjes transformations of L_2 class functions). Mekhanika Tverdogo Tela **5**, 162–165 (1977)
75. Zavyalov, Yu.S., Kvasov, B.I., Miroshnichenko, V.D.: Metody splain-funktsii (Methods of Spline Functions). Nauka, Moscow (1980)

Chapter 2
Spectral Methods in the Theory of Wave Propagation

2.1 Equations of Motion of an Ideal Fluid. Small-Amplitude Waves

Analysis of wave phenomena at the interface between two media is one of the topical problems in many areas of physics and mechanics. This chapter deals with studying the wave motion of the fluid surface when the density of the medium above the surface is much smaller than the density of the fluid under consideration. An example of such a system is the surface of sea or ocean. We limit our consideration to the wave dynamics caused by only one external factor—*force of gravity*. In hydrodynamics such waves are called *gravity waves*. These are the waves with a length of one metre and longer. On smaller scales, surface tension and viscosity start to play an essential role. Amongst the variety of surface waves, we put our focus on two-dimensional steady periodic waves that propagate along a given direction.

2.1.1 Euler and Laplace Equations with Boundary Conditions

In the general model of continuous medium, the fluid motion is described by the Navier–Stokes equation [40]. In the case of *ideal fluid*, when the effects due to viscosity and heat conduction can be ignored, the Navier–Stokes equation reduces to the Euler equation describing the conservation of the momentum of a fluid particle. For the fluid motion in the gravity field, the Euler equation is written as

$$\frac{\partial \boldsymbol{v}}{\partial t} + (\boldsymbol{v}\nabla)\boldsymbol{v} = -\frac{\nabla p}{\rho} + \boldsymbol{g}. \tag{2.1}$$

Here, $\boldsymbol{v}$ is the flow velocity vector field, p is the pressure inside the fluid, ρ is the fluid density and $\boldsymbol{g}$ is the acceleration due to gravity. This equation is essentially nonlinear owing to the term $(\boldsymbol{v}\nabla)\boldsymbol{v}$ that is a part of the material derivative $\frac{\mathrm{d}}{\mathrm{d}t} = \frac{\partial}{\partial t} + (\boldsymbol{v}\nabla)$. Another fundamental law of motion of an ideal fluid is the continuity equation

I.T. Selezov et al., *Wave Propagation and Diffraction*,
Foundations of Engineering Mechanics, DOI 10.1007/978-981-10-4923-1_2

$$\frac{\partial \rho}{\partial t} + \operatorname{div}(\rho \boldsymbol{v}) = 0, \tag{2.2}$$

which is a result of the mass conservation law.

The equations of motion of an ideal fluid (2.1) and (2.2) are the basic equations for deriving the equations of motion of the fluid surface. Hereafter, we assume that fluid is incompressible ($\rho = \text{const}$) and the fluid motion is potential (irrotational). The condition of potential flow rot $\boldsymbol{v} = 0$ means that one can introduce the velocity potential $\boldsymbol{\Phi}$ such as $\boldsymbol{v} = \nabla\boldsymbol{\Phi}$. Then the continuity equation (2.2) for the velocity $\boldsymbol{v}$ is transformed to the *Laplace equation*

$$\Delta\Phi = 0 \tag{2.3}$$

for the velocity potential. The Laplace equation describes the fluid motion in the entire flow domain. This equation should be supplemented with boundary conditions at each boundary, depending on the problem geometry:

$$\nabla F \cdot \boldsymbol{v} + \frac{\partial F}{\partial t} = 0, \tag{2.4}$$

where $F(\boldsymbol{r}, t) = 0$ is the equation of the boundary and $\boldsymbol{r} = \{x, y, z\}$ is the radius vector. Condition (2.4) implies that no particles cross the boundary. Such conditions are called kinematic.

We consider the flow domain that is unbounded in the horizontal xz-plane, whereas in the vertical xy-plane it is bounded by a flat bed from the bottom and a free surface from the top (Fig. 2.1). Then the kinematic boundary conditions at the moving free surface $y = \eta(x, t)$ and motionless bottom $y = -h$ are written as

$$\eta_t + \Phi_x \eta_x - \Phi_y = 0, \quad y = \eta(x, t); \tag{2.5}$$

$$\Phi_y = 0, \quad y = -h. \tag{2.6}$$

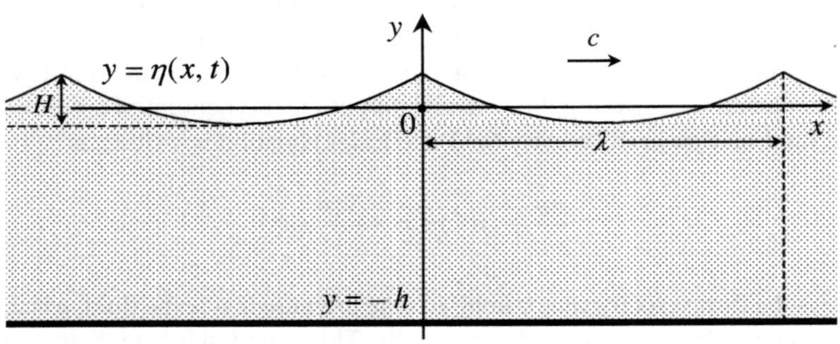

Fig. 2.1 Laboratory reference frame

Here, we assume the fluid motion to be planar (two-dimensional), i.e. such that wave parameters do not change along the wavefront parallel to the z-axis. The shape of the free surface is described by the unknown function $y = \eta(x, t)$ of the horizontal coordinate and time. It can be determined from the Euler equation that is reduced by integration to the Bernoulli equation in the case of incompressible fluid and irrotational motion, namely,

$$\Phi_t + \frac{1}{2}\left(\Phi_x^2 + \Phi_y^2\right) + \frac{p}{\rho} + gy = C(t), \tag{2.7}$$

where $C(t)$ is an integration constant. The pressure p is assumed to be constant and was therefore included in the integration constant. Since the velocity potential Φ is defined to within an arbitrary function of time, the integration constant $C(t)$ can be put equal to zero without loss of generality. This is equivalent to the redefinition of the velocity potential as

$$\widetilde{\Phi} = \Phi - \int C(t)\mathrm{d}t.$$

When Eq. (2.7) is written at the boundary $y = \eta(x, t)$, it is called the dynamic boundary condition. Laplace equation (2.3) and boundary conditions (2.5), (2.6) and (2.7) form the closed system of equations, which is usually referred to as the *canonical model* of hydrodynamics.

2.1.2 Stationary Fluid Motion

In the conservative system such as an ideal fluid, the fluid motion can become *stationary* after all the transient processes have finished. In this case, the parameters of surface waves and their shape do not change with time. The condition of stationary motion for progressive waves establishes an unambiguous correspondence between the coordinate x, which is parallel to the direction of wave propagation, and time t in terms of wave phase

$$\theta = kx - \omega t,$$

where k is the wave number, $\omega = c\,k$ is the wave frequency and c is the wave phase speed, so that

$$\frac{\partial}{\partial t} = -c\,\frac{\partial}{\partial x}\,. \tag{2.8}$$

In this case, any external initial perturbation affects only the wavelength λ and the wave energy (or amplitude), which serve as free parameters of the problem and can be determined from the relevant initial conditions.

Amongst a variety of stationary waves, we restrict our attention only to periodic waves. The periodicity condition implies that the following condition holds true for any function f describing the spatial and temporal parameters of the flow:

$$f(x,\ y,\ t) = f(x+\lambda,\ y,\ t) = f(x,\ y,\ t+T), \tag{2.9}$$

where $\lambda = 2\pi/k$ is the wavelength (spatial period) and T is the temporal period. In the case of condition (2.8), we have $\lambda = cT$ and $f(x,\ y,\ t) = f(\theta,\ y)$.

Depending on the ratio between the fluid depth and wavelength, the problem of propagation of two-dimensional stationary waves has two limiting cases: (i) $h/\lambda \to \infty$ (infinite depth and deep water waves) and (ii) $h/\lambda \to 0$ (solitary waves). In the latter case, the waves are not periodic (their wavelength is infinite), and they are beyond the scope of our consideration.

Let us determine the integration constant C in the Bernoulli equation (2.7) (in view of conditions (2.8) and (2.9), it is not arbitrary any longer). This constant does not depend on time when the flow is stationary, and it can be determined by averaging the Bernoulli equation at the bottom. To this end, let us find the mass of a fluid column between the free surface and the bottom, the column base having dimensions equal to the wavelength λ along the x-axis and length L along the z-axis parallel to the wavefront:

$$M = L\int_0^{\lambda}\int_{-h}^{\eta(\theta)} \rho\,\mathrm{d}x\,\mathrm{d}y = L\int_0^{\lambda} \rho\big(h+\eta(\theta)\big)\mathrm{d}x = \rho S h + \rho S\overline{\eta}, \tag{2.10}$$

where $S = L\lambda$ is the column base area and

$$\overline{\star} = \frac{1}{\lambda}\int_0^{\lambda} \star\ \mathrm{d}x \tag{2.11}$$

designates averaging over the spatial period. Note that $\overline{\eta} = 0$ if $y = 0$ is the wave *mean level* that separates the crests and the troughs into two domains of equal areas. Assuming that there is no external flow in and out of the domain filled with fluid, the column mass M remains constant in the course of wave propagation. Hence, the depth $h + \overline{\eta}$ corresponds to the *still fluid* depth, when there is no waves and flows. Since we have $\overline{\eta} = 0$ relative to the wave mean level, the mean level of a stationary wave coincides with the still fluid level, provided that the total mass of fluid remains constant.

Next we determine the fluid pressure at the bottom. In the case of infinite depth $h = \infty$, the motion of surface waves does not produce any flows at the bottom, so that $\Phi_x\,|_{y=-\infty} = 0$. From boundary condition (2.6) we have $\Phi_y\,|_{y=-\infty} = 0$, therefore it follows from the Bernoulli equation (2.7) that pressure is constant at infinite depth. In the case of finite depth, the motion of surface waves sets the entire fluid in motion, although it subsides with depth. The fluid velocity at the bottom (and at any other horizontal level located inside the domain filled with fluid) is a periodic function with spatial period λ, i.e.

$$\overline{\Phi_x}\,|_{y=\text{const}} = 0, \tag{2.12}$$

although the unaveraged velocity is not constant, namely, $\Phi_x\,|_{y=-h} \neq \text{const}$. Therefore, the pressure at the bottom is not constant but its average over the spatial period is constant at the entire bottom and is determined by the fluid mass above the bottom and the atmospheric pressure p_0 at the surface. Hence, taking into account relation (2.10) we have

$$\overline{p}\,|_{y=-h} = p_0 + \frac{Mg}{S} = p_0 + \rho g(h + \overline{\eta}). \tag{2.13}$$

Next, by averaging the Bernoulli equation (2.7) at the bottom over the spatial period and taking into account conditions (2.6), (2.12), and (2.13), we get the value of constant C:

$$C = \frac{p_0}{\rho} + \frac{1}{2}\,\overline{\Phi_x^2}\,|_{y=-h} + g\overline{\eta}. \tag{2.14}$$

Assuming the atmospheric pressure constant at the entire free surface and equal to the mean pressure p_0 and taking into account that $\Phi_t = -c\Phi_x$, the dynamic boundary condition can be rewritten as

$$\frac{1}{2}\left(\left(c - \Phi_x\right)^2 + \Phi_y^2\right) + gy = \frac{c^2}{2} + \frac{\overline{\Phi_x^2}\,|_{y=-h}}{2} + g\overline{\eta} = B, \quad y = \eta(x,\ t), \tag{2.15}$$

where the constant B is called the *Bernoulli constant.* If $y = 0$ is the wave mean level, we have $\overline{\eta} = 0$ and then $B = c^2/2$ in the case of infinite depth. This value of the Bernoulli constant is often used in the case of finite depth as well. To this end, one has to set $\overline{\eta} = -\frac{1}{2g}\,\overline{\Phi_x^2}\,|_{y=-h}$, which means that the coordinate origin $y = 0$ needs to be translated upwards by distance $\frac{1}{2g}\,\overline{\Phi_x^2}\,|_{y=-h}$ relative to the still fluid level.

2.1.3 Wave Energy, Momentum and Power

The energy of a unit fluid volume is a sum of volume densities of the kinetic and potential energies:

$$\mathscr{E} = \frac{\rho v^2}{2} + \rho\varepsilon,$$

where ε is the internal energy of the unit fluid mass [40]. For the external gravity field, we have $\varepsilon = gy$. Let us find the total energy of the fluid column between the free surface and the bottom, the column base having dimensions equal to the wavelength λ along the x-axis and length L along the z-axis:

$$\mathscr{H}_{\text{total}} = L \int_0^{\lambda} \int_{-h}^{\eta(\theta)} \left(\frac{\rho v^2}{2} + \rho g y \right) \mathrm{d}x\, \mathrm{d}y.$$

When there is no wave motion, the energy of this fluid column is

$$\mathscr{H}_{\text{still}} = L \int_0^{\lambda} \int_{-h}^{\overline{\eta}} \rho g y \, \mathrm{d}x\, \mathrm{d}y.$$

Hence, the contribution of the wave energy to the energy of the fluid column under consideration is

$$\mathscr{H}_{\text{wave}} = \mathscr{H}_{\text{total}} - \mathscr{H}_{\text{still}} = L \int_0^{\lambda} \int_{-h}^{\eta(\theta)} \frac{\rho v^2}{2} \mathrm{d}x\, \mathrm{d}y + L \int_0^{\lambda} \int_{\overline{\eta}}^{\eta(\theta)} \rho g y \, \mathrm{d}x\, \mathrm{d}y.$$

The ratio of this energy to the area $L\,\lambda$ of the column base is usually called the *wave energy density* (energy per unit horizontal area):

$$E = \frac{\mathscr{H}}{\lambda L} = K + U = \frac{\rho}{2\lambda} \int_0^{\lambda} \int_{-h}^{\eta(\theta)} \left(\Phi_x^2 + \Phi_y^2 \right) \mathrm{d}x\, \mathrm{d}y + \frac{\rho g}{2\lambda} \int_0^{\lambda} \left(\eta^2(\theta) - \overline{\eta}^2 \right) \mathrm{d}x, \quad (2.16)$$

where K and U are the densities of kinetic and potential energies, respectively.

The momentum of a unit fluid volume (fluid flow density) is $\rho \boldsymbol{v}$ [40]. Then the flow density of the fluid being transported along the direction of wave propagation is ρv_x. Similarly to the wave energy density, the *wave momentum density* (momentum per unit horizontal area) is defined as

$$I = \frac{\rho}{\lambda} \int_0^{\lambda} \int_{-h}^{\eta(\theta)} \Phi_x \, \mathrm{d}x\, \mathrm{d}y. \quad (2.17)$$

In what follows, we omit the words "per unit horizontal area" when referring to the wave energy and momentum.

The energy flux transferred by the wave is defined as the derivative of the total energy with respect to time:

$$\mathscr{P}(t) = \frac{\mathrm{d}}{\mathrm{d}t} \iiint_V \mathscr{E} \mathrm{d}V. \quad (2.18)$$

It can be shown that [74]

$$\mathscr{P}(t) = \rho c L \int_{-h}^{\eta(\theta)} \Phi_x^2 \, \mathrm{d}y. \tag{2.19}$$

The wave power is defined as the energy flux averaged over the wave period that we write normalised to the unit wavefront length:

$$P = \frac{1}{LT} \int_0^T \mathscr{P} \mathrm{d}t. \tag{2.20}$$

2.1.4 General Solution of the Laplace Equation

We search for a partial solution of the Laplace equation (2.3) by the method of separation of variables:

$$\Phi(x, y, t) = X(x - c\,t)\, Y(y) \Rightarrow \frac{X''}{X} = -\frac{Y''}{Y} = -\kappa^2.$$

Then we have

$$X(x - c\,t) = X_1 \mathrm{e}^{\mathrm{i}\kappa(x-c\,t)} + X_2 \mathrm{e}^{-\mathrm{i}\kappa(x-c\,t)},$$
$$Y(y) = Y_1 \mathrm{e}^{\kappa y} + Y_2 \mathrm{e}^{-\kappa y}.$$

Since the solution should be periodic with respect to coordinate x with period $\lambda = 2\pi/k$, we need to set $\kappa = n\,k$, n being a natural number. In addition, the following conditions should hold true.

1. Boundary condition (2.6) at the bottom implying that

$$Y_1 \mathrm{e}^{-n\,kh} - Y_2 \mathrm{e}^{n\,kh} = 0 \quad \Leftrightarrow \quad Y_1 = Y_2 \mathrm{e}^{2n\,kh}.$$

2. Transition to the limit at $h \to \infty$. To this end, the function $Y(y)$ should be bounded ($Y_2 \to 0$) and should not depend on depth h at $h \to \infty$. These conditions are satisfied if we introduce the scaling as follows

$$Y_2 = \frac{A_2 \mathrm{e}^{-n\,kh}}{\mathrm{e}^{n\,kh} + \mathrm{e}^{-n\,kh}} \Rightarrow Y_1 = \frac{A_2 \mathrm{e}^{n\,kh}}{\mathrm{e}^{n\,kh} + \mathrm{e}^{-n\,kh}}.$$

The constant A_2 can be included in the function X by redefinition $C_1 = X_1 \cdot A_2$ and $C_2 = X_2 \cdot A_2$.

3. The velocity potential Φ should be a real function:

$$\Phi^* = \Phi \quad \Rightarrow \quad C_1^* = C_2,$$

where * designates the complex conjugate. Thus,

$$\Phi(x, y, t) = X(x - c\,t)\,Y(y);$$
$$X(x) = C_n \mathrm{e}^{\mathrm{i}n\,kx} + C_n^* \mathrm{e}^{-\mathrm{i}n\,kx} = \left(\operatorname{Re} C_n\, \cos(n\,kx) - \operatorname{Im} C_n\, \sin(n\,kx)\right);$$
$$Y(y) = T_n\,\mathrm{e}^{n\,ky} + T_{-n}\,\mathrm{e}^{-n\,ky} = \frac{\cosh\left(n\,k(y+h)\right)}{\cosh(n\,kh)};$$
$$T_n = \frac{\mathrm{e}^{n\,kh}}{\mathrm{e}^{n\,kh} + \mathrm{e}^{-n\,kh}} = \frac{1}{2}\left(1 + \tanh(n\,kh)\right). \tag{2.21}$$

The general solution of the Laplace equation is a linear combination of all partial solutions ($n = 1, 2, 3, \ldots$):

$$\Phi(x, y, t) = \sum_{n=1}^{\infty} \left(C_n \mathrm{e}^{\mathrm{i}n\,k(x-ct)} + C_n^* \mathrm{e}^{-\mathrm{i}n\,k(x-ct)}\right)\left(T_n \mathrm{e}^{n\,ky} + T_{-n} \mathrm{e}^{-n\,ky}\right). \tag{2.22}$$

For symmetric waves we have $\Phi_y(\theta) = -\Phi_y(-\theta)$, so that

$$C_n^* = -C_n, \tag{2.23}$$

i.e. the coefficients C_n must be real in this case.

2.1.5 *Small-Amplitude Waves (Linear Approximation)*

Let us find the conditions when the nonlinearity of the boundary conditions can be neglected. To this end, we estimate the magnitudes of the terms present in the Euler equation (2.1):

$$|(\boldsymbol{v}\nabla)\boldsymbol{v}| \simeq \frac{v\delta v}{\delta r}, \qquad \left|\frac{\partial v}{\partial t}\right| \simeq \frac{\delta v}{\delta t},$$

where δv, δr and δt are typical variations of the velocity, distance and time, respectively. To get an estimate related to the wave motion, we can set δr equal to the wavelength λ and δt equal to the wave period T. Then we have

$$\frac{|(\boldsymbol{v}\nabla)\boldsymbol{v}|}{\left|\dfrac{\partial v}{\partial t}\right|} \simeq \frac{vT}{\lambda} = \frac{v}{c} \simeq \frac{H}{\lambda} \equiv A.$$

Therefore, the nonlinear term $|(\boldsymbol{v}\nabla)\boldsymbol{v}|$ is small as compared to the linear term $|\frac{\partial \boldsymbol{v}}{\partial t}|$ when $H \ll \lambda$ ($A \ll 1$), i.e. the wave amplitude is much smaller than the wavelength. The waves for which the contribution of nonlinearity is negligibly small (it happens at $A \to 0$) are called *linear waves* or *waves of infinitesimal amplitude*. Let us consider the main properties of linear waves.

The wave speed c and wave profile can be found from the boundary conditions at the free surface. In the linear approximation, all the derivatives at the free surface $y = \eta(x, t)$ are written at the zero level $y = 0$ (since the next terms of the corresponding Taylor series around the undisturbed level $y = 0$ have a higher order of smallness with respect to the parameter A). Then the kinematic and dynamic boundary conditions (2.5) and (2.15) take the following form:

$$\eta_x = -\frac{1}{c}\, \varPhi_y \,|_{y=0}, \tag{2.24}$$

$$\eta = \frac{c}{g}\, \varPhi_x \,|_{y=0}. \tag{2.25}$$

Eliminating the wave profile from these equations, we get

$$\left(\frac{c^2}{g}\, \varPhi_{xx} + \varPhi_y\right)\Big|_{y=0} = 0. \tag{2.26}$$

Substituting the solution (2.22) of the Laplace equation at $n = 1$ (linear approximation) in this equation, we obtain a relation that expresses the dependence of wave speed (or frequency) on wave number (or length):

$$c^2 = \frac{g}{k}\tanh(kh) \quad \Leftrightarrow \quad \omega^2 = gk\tanh(kh). \tag{2.27}$$

This expression specifies the relationship between the temporal and spatial parameters of the wave. Such a relationship is called the wave *dispersion* [40]. The function $\tanh(kh)$ rapidly tends to unity with increasing kh. For example, at $kh = 20$ we have $1 - \tanh(kh) \approx 8.5 \cdot 10^{-18}$ and at $kh = 100$ we have $1 - \tanh(kh) \approx 2.8 \cdot 10^{-87}$. Thus, even the case $kh = 20$ can be regarded as the infinite-depth case (deep fluid) to within high accuracy ($\approx 10^{-15}\%$). Needless to say the same conclusion holds true for larger kh.

The free surface profile in the linear approximation is found from boundary condition (2.25) with (2.21) taken into account:

$$\eta = \frac{\mathrm{i}ck}{g}\left(C_1 \mathrm{e}^{\mathrm{i}k(x-ct)} + C_1^* \mathrm{e}^{-\mathrm{i}k(x-ct)}\right) =$$

$$-\frac{2ck}{g}\left(\mathrm{Re}\, C_1 \sin\big(k(x-ct)\big) + \mathrm{Im}\, C_1 \cos\big(k(x-ct)\big)\right). \tag{2.28}$$

For *symmetric* waves we have $\eta(x - c\,t) = \eta(-x + c\,t)$, so that $\operatorname{Re} C_1 = 0$. Then we finally get

$$\eta = a\,\cos(kx - \omega\,t), \quad a = -\frac{2c\,k}{g}\,\operatorname{Im} C_1; \tag{2.29}$$

$$\Phi = \frac{ga}{\omega}\,\sin(kx - \omega\,t)\big(T_1^+ \mathrm{e}^{ky} + T_1^- \mathrm{e}^{-ky}\big) = ac\,\sin(kx - \omega\,t)\,\frac{\cosh\big(k(y+h)\big)}{\sinh(kh)}\,. \tag{2.30}$$

In the linear approximation, the parameter a is the height $\eta(0)$ of the wave above the still fluid level, which is equal to a half of the wave height:

$$a = \frac{H}{2} = \frac{\lambda A}{2} = \frac{\pi A}{k}. \tag{2.31}$$

For the velocity distribution we have

$$\begin{aligned} \Phi_x &= \frac{\mathrm{d}x}{\mathrm{d}\,t} = ack\,\cos(kx - \omega\,t)\,\frac{\cosh\big(k(y+h)\big)}{\sinh(kh)}\,; \\ \Phi_y &= \frac{\mathrm{d}y}{\mathrm{d}\,t} = ack\,\sin(kx - \omega\,t)\,\frac{\sinh\big(k(y+h)\big)}{\sinh(kh)}\,. \end{aligned} \tag{2.32}$$

Accordingly, the wave momentum density I calculated by formula (2.17) is equal to zero in the linear approximation with respect to the small parameter ak. Thus, *linear waves do not cause the mass transfer*.

The wave density energy calculated by formula (2.16) has the following form:

$$E = K + U = \frac{\rho g a^2}{2}, \quad K = U = \frac{\rho g a^2}{4}, \tag{2.33}$$

i.e. the kinetic and potential energies make the equal contribution to the total wave energy in the linear approximation.

The distribution of the kinetic energy over depth can be calculated by the formula

$$\begin{aligned} \overline{\mathscr{E}}_K &= \frac{1}{2\pi}\int_0^{2\pi} \mathscr{E}_K \mathrm{d}\theta = \frac{1}{2\pi}\int_0^{2\pi} \frac{\rho v^2}{2}\mathrm{d}\theta = \frac{\rho g}{k}\frac{a^2}{2}\,\frac{\cosh\big(2k(y+h)\big)}{\sinh(2kh)} \\ &= \Big|h \gg \lambda\Big| = \frac{\rho g}{k}\frac{a^2}{2}\,\exp(2ky), \end{aligned}$$

where $\mathscr{E}_K$ is the volume density of the kinetic energy. It can be seen that the major part of the wave's kinetic energy is concentrated in the near-surface layer of thickness $\lambda/2$. This fact is used as the basis for the operation of apparatuses that take the wave power off and convert it to the mechanical work or electric energy for subsequent use. Wave energy conversion methods and apparatuses are considered in detail in Ref. [59].

The wave power calculated by formula (2.20) is

$$P = \frac{\rho g a^2}{4}\, c \left(1 + \frac{2kh}{\sinh(2kh)}\right) = E\, c_g, \tag{2.34}$$

where

$$c_g = \frac{c}{2}\left(1 + \frac{2kh}{\sinh(2kh)}\right) = \frac{\mathrm{d}\omega}{\mathrm{d}k} \tag{2.35}$$

is the wave *group speed* by definition. Formula (2.34) means that the wave energy is transferred with group speed, which is always smaller than the phase speed ($c_g \to c$ only at $h \to 0$). Thus, the wave energy transfer always lags behind the wavefront motion. In the case of infinite depth, the group speed is twice as small as the phase speed:

$$c_g|_{h\to\infty} = \frac{c}{2}.$$

A more detailed insight into the theory of linear surface waves can be found in Refs. [39, 79, 92].

2.1.6 Dimensionless Equations and Parameters

For further analysis, we rewrite the equations of canonical model in dimensionless form. The units of length r, time t and mass m are selected such that in dimensionless form we could get

$$\widetilde{k} = \widetilde{g} = \widetilde{\rho} = 1,$$

where the tilde refers to dimensionless units. To this end, all the distances are normalised to the reciprocal of wave number k and time is normalised to the reciprocal of frequency $\omega_0 = \sqrt{g\,k}$, which is the linear wave frequency in the case of infinite depth. Then the relationship between the dimensionless and dimensional variables is given by the relations

$$\begin{gathered}\widetilde{r} = r\,k, \quad \widetilde{h} = h\,k, \quad \widetilde{t} = t\,\omega_0, \\ \widetilde{v} = \frac{v}{c_0}, \ \ \widetilde{\nabla} = \frac{\nabla}{k}, \ \ \widetilde{\Phi} = \frac{k}{c_0}\,\Phi, \ \ \widetilde{m} = \frac{k^3}{\rho}\,m, \ \ \widetilde{E} = \frac{k^2}{\rho g}\,E, \ \ \widetilde{I} = \frac{k^2}{\rho\omega_0}\,I,\end{gathered} \tag{2.36}$$

where $c_0 = \omega_0 / k$ is the linear wave phase speed in the case of infinite depth. In this case we have

$$\widetilde{\lambda} = 2\pi, \quad \widetilde{\omega} = \widetilde{c}, \quad \widetilde{\theta} = \widetilde{x} - \widetilde{c}\,\widetilde{t}, \quad \frac{\partial}{\partial\widetilde{x}} = \frac{\partial}{\partial\widetilde{\theta}}.$$

In terms of dimensionless variables (tildes are omitted henceforth), the equations of canonical model are

$$\Phi_{\theta\theta} + \Phi_{yy} = 0, \quad -h \leqslant y \leqslant \eta(\theta); \tag{2.37}$$

$$\frac{1}{2}\left(\left(c - \Phi_\theta\right)^2 + \Phi_y^2\right) + \eta = B, \quad y = \eta(\theta), \quad B = \frac{c^2}{2} + \frac{1}{2}\,\overline{\Phi_\theta^2}\,|_{y=-h} + \overline{\eta}\,; \tag{2.38}$$

$$\left(c - \Phi_\theta\right)\eta_\theta + \Phi_y = 0, \quad y = \eta(\theta); \tag{2.39}$$

$$\Phi_y = 0, \quad y = -h. \tag{2.40}$$

This system of equations at fixed depth h defines the unknown velocity potential $\Phi(\theta, y)$, free surface profile $\eta(\theta)$ and phase speed c as functions of only one dimensionless parameter—wave amplitude $A = H/(2\pi)$ (H being the wave height):

$$A = \frac{\eta(0) - \eta(\pi)}{2\pi}. \tag{2.41}$$

The Bernoulli equation (2.38) yields another expression:

$$A = \frac{q^2(\pi) - q^2(0)}{4\pi}, \tag{2.42}$$

where $q(0)$ and $q(\pi)$ are the velocities of fluid particles at the wave crest and trough, respectively, in the reference frame where the fluid is motionless. We refer to such a reference frame as the *intrinsic wave reference frame*.

The velocity potential and phase speed being determined, the particle trajectories can be found from a system of differential equations:

$$\frac{\mathrm{d}x}{\mathrm{d}t} = \Phi_\theta\,(\theta,\, y)\,, \quad \frac{\mathrm{d}y}{\mathrm{d}t} = \Phi_y\,(\theta,\, y)\,. \tag{2.43}$$

Wave energy (2.16) and momentum (2.17) have the following form in dimensionless variables:

$$E = K + U = \frac{1}{4\pi}\int_0^{2\pi}\int_{-h}^{\eta(\theta)} \left(\Phi_\theta^2 + \Phi_y^2\right) \mathrm{d}\theta\, \mathrm{d}y + \frac{1}{4\pi}\int_0^{2\pi} \left(\eta^2(\theta) - \overline{\eta}^2\right) \mathrm{d}\theta. \tag{2.44}$$

$$I = \frac{1}{2\pi}\int_0^{2\pi}\int_{-h}^{\eta(\theta)} \Phi_\theta \mathrm{d}\theta\, \mathrm{d}y. \tag{2.45}$$

2.1.7 Complex Potential

Kinematic boundary condition (2.4) means that the velocity of fluid particles at each boundary is directed along the tangent to the surface. The lines in the fluid flow at which the velocity vector is always directed along the tangent are called *streamlines* [42]. The equations for streamlines are found from the condition for the vectors $\boldsymbol{v}$ and $\mathrm{d}\boldsymbol{r}$ to be parallel:

$$\boldsymbol{v} \times \mathrm{d}\boldsymbol{r} = 0. \tag{2.46}$$

In the intrinsic wave reference frame and in the case of planar motion, the equations for streamlines can be rewritten as

$$\frac{\mathrm{d}\theta}{\phi_\theta} = \frac{\mathrm{d}y}{\phi_y} \quad \Leftrightarrow \quad -\phi_y\, \mathrm{d}\theta + \phi_\theta\, \mathrm{d}y = 0. \tag{2.47}$$

Here, the function $\phi(\theta, y)$ is the velocity potential in the intrinsic wave reference frame, its relationship to the velocity potential $\Phi(\theta, y)$ in the laboratory reference frame (which is motionless with respect to the bottom) is expressed as

$$\phi(\theta, y) = \Phi(\theta, y) - c\theta. \tag{2.48}$$

Then it follows from the Laplace equation that

$$\frac{\partial \phi_\theta}{\partial \theta} = \frac{\partial\left(-\phi_y\right)}{\partial y}.$$

This relation implies that there exists some function $\psi(\theta, y)$ such that

$$\phi_\theta = \psi_y, \quad \phi_y = -\psi_\theta. \tag{2.49}$$

The function $\psi(\theta, y)$ is called the *stream function* and equation (2.49) represent the well-known Cauchy–Riemann conditions for the complex function $w = \phi + \mathrm{i}\psi$ to be an analytic function of complex argument $\zeta = \theta + \mathrm{i}y$. The function $w(\zeta)$ is called the *complex potential* [40]. In this case, the expression $-\phi_y\, \mathrm{d}\theta + \phi_\theta\, \mathrm{d}y$ is the total differential of function ψ, so that the condition (2.47) at the streamline yields

$$\mathrm{d}\psi = 0 \quad \Leftrightarrow \quad \psi(\theta, y) = \mathrm{const}, \tag{2.50}$$

the constant in the right-hand side being different for different streamlines. The stream function $\psi(\theta, y)$ is constant along any streamline. In terms of velocity potential, the streamlines are determined from a system of differential equations:

$$\frac{\mathrm{d}\theta}{\mathrm{d}t} = \phi_\theta = \Phi_\theta(\theta, y) - c, \quad \frac{\mathrm{d}y}{\mathrm{d}t} = \phi_y = \Phi_y(\theta, y). \tag{2.51}$$

Note that streamlines (2.51) and particle trajectories (2.43) are different curves. Streamlines specify the velocity direction of different fluid particles in the successive points of the space at a *fixed time moment* (Euler variables), whereas trajectories specify the velocity direction of selected particles at *successive time moments* (Lagrange variables) [40].

The analytic function $w(\zeta)$ unambiguously maps the points of the complex plane $\zeta = \theta + \mathrm{i}y$ (which we call the *physical plane*) onto the complex plane $w = \phi + \mathrm{i}\psi$ (which we call the *inverse plane*). Such a mapping is called *conformal* [42]. The geometry of boundaries that enclose the flow domain are often much simpler in the inverse plane because the stream function ψ is constant at each boundary in the case of stationary velocity fields. Therefore, often it is more convenient to solve the planar problems in the inverse plane by taking the velocity potential ϕ and stream function ψ as independent variables and the spatial coordinates θ and y as functions of ϕ and ψ.

The squared velocity in the intrinsic wave reference frame is expressed in terms of the complex potential as

$$\left(c - \Phi_\theta\right)^2 + \Phi_y^2 = \left|\frac{\mathrm{d}w}{\mathrm{d}\zeta}\right|^2, \tag{2.52}$$

inasmuch as

$$\frac{\mathrm{d}w}{\mathrm{d}\zeta} = \frac{\partial\phi}{\partial\theta} + \mathrm{i}\frac{\partial\psi}{\partial\theta} = \phi_\theta - \mathrm{i}\phi_y \;\Rightarrow \phi_\theta = \mathrm{Re}\left(\frac{\mathrm{d}w}{\mathrm{d}\zeta}\right), \;\; \phi_y = -\,\mathrm{Im}\left(\frac{\mathrm{d}w}{\mathrm{d}\zeta}\right). \tag{2.53}$$

The quantity $\frac{\mathrm{d}w}{\mathrm{d}\zeta}$ is called the *complex velocity* [40]. In terms of complex variables, the equations for streamlines (2.51) are written as

$$\frac{\mathrm{d}\zeta}{\mathrm{d}t} = \left(\frac{\mathrm{d}w}{\mathrm{d}\zeta}\right)^*. \tag{2.54}$$

We define the stream function $\Psi(\theta,\ y)$ in the laboratory reference frame such that the Cauchy–Riemann conditions might hold true:

$$\Phi_\theta = \Psi_y, \quad \Phi_y = -\Psi_\theta. \tag{2.55}$$

To this end, it is sufficient to put $\Psi(\theta,\ y) = \psi(\theta,\ y) + cy$. The function

$$W = \Phi + \mathrm{i}\Psi = w + c\zeta$$

is the complex potential in the laboratory reference frame. Then

$$\begin{aligned}\Phi &= \frac{1}{2}\left(W + W^*\right) \equiv -\mathrm{i}c\left(R - R^*\right),\\ \Psi &= \frac{1}{2\mathrm{i}}\left(W - W^*\right) \equiv c\left(R + R^*\right),\end{aligned} \tag{2.56}$$

where we introduced the complex function

$$R = \frac{\mathrm{i}W^*}{2c},$$

for the sake of convenience. The Cauchy–Riemann conditions (2.55) imply that

$$R_\theta = \mathrm{i}R_y \quad \Leftrightarrow \quad R(\theta,\, y) = R(y + \mathrm{i}\theta). \tag{2.57}$$

Any complex function R that satisfies condition (2.57) also identically satisfies the Laplace equation $\Delta R = 0$. In this case, in view of relation (2.56), the Laplace equation for the velocity potential $\Delta\Phi = 0$ and stream function

$$\Delta\Psi = 0 \tag{2.58}$$

are also satisfied identically.

With relations (2.56) and (2.57) taken into account, the kinematic boundary condition (2.39) takes the form

$$\frac{\mathrm{d}}{\mathrm{d}\theta}\Big(R(\theta,\, \eta) + R^*(\theta,\, \eta) - \eta\Big) = 0$$

or

$$\Psi(\theta,\, \eta) - c\eta = \psi(\theta,\, \eta) = \text{const}\,. \tag{2.59}$$

The latter equation implies that the free surface $y = \eta(\theta)$ is a streamline. Since the stream function is defined to within an arbitrary constant, as it follows from the Cauchy–Riemann conditions (2.55), the integration constant in Eq. (2.59) is also arbitrary. However, this arbitrary constant is conventionally selected such that the stream function at the free surface might be equal to zero, $\psi = 0$, whence we have const $= 0$. Specifying const is this way, we unambiguously define the stream function.

The boundary condition (2.40) at the bottom implies that the bottom is a streamline, much like the free surface, i.e. $\psi(\theta,\, -h) = \mathfrak{const}$. The integration constant can be found by calculating the *fluid mass flux* Q through a curve connecting the arbitrary points 1 and 2 at the bottom and free surface, respectively. The orientation of the normal vector $\boldsymbol{n}$ to the curve at each point is selected such that the angle between the vector $\boldsymbol{n}$ and the direction of circulation around the curve $\mathrm{d}\boldsymbol{l}$ (by absolute value $\mathrm{d}l$ is the length element of the curve) might be $+\frac{\pi}{2}$, i.e. the shortest turn from the vector $\boldsymbol{n}$ to the vector $\mathrm{d}\boldsymbol{l}$ might be accomplished counterclockwise. Then we have

$$Q = \int_1^2 (\boldsymbol{v}\cdot\boldsymbol{n})\,\mathrm{d}l = \int_1^2 \big(-\phi_y\,\mathrm{d}\theta + \phi_\theta\,\mathrm{d}y\big) = \int_1^2 \mathrm{d}\psi = \big(\psi_2 - \psi_1\big). \tag{2.60}$$

In this case, the mass flux depends neither on the curve shape nor on the positions of points 1 and 2. Since we set $\psi = 0$ at the free surface, formula (2.60) implies that $\mathfrak{const} = -Q$. In terms of the velocity potential, the flux Q is defined as

$$Q = \int_{-h}^{\eta(\theta)} (\Phi_\theta - c)\mathrm{d}y. \tag{2.61}$$

The stream function in the intrinsic wave reference frame monotonically increases from zero at the free surface to $|Q|$ at the bottom, $\psi|_{y=-h} > 0$ ($Q < 0$), provided that $\Phi_\theta < c$ in the entire flow domain (wave propagates faster than fluid particles).

The mass flux Q and the stream function at the bottom can be expressed in terms of the wave momentum I. Taking into account the Cauchy–Riemann conditions (2.55), we obtain from formula (2.45) that

$$I = \frac{1}{2\pi}\int_0^{2\pi}\left(\int_{-h}^{\eta(\theta)} \Psi_y(\theta, y)\,\mathrm{d}y\right)\mathrm{d}\theta = \overline{\Psi}\,|_{y=\eta(\theta)} - \overline{\Psi}\,|_{y=-h}$$

$$= c\overline{\eta} - \left(\psi\,|_{y=-h} - c\,h\right).$$

Therefore,

$$\Psi(\theta, -h) = c\overline{\eta} - I \Leftrightarrow \psi(\theta, -h) = -Q = c(\overline{\eta} + h) - I. \tag{2.62}$$

By its absolute value, the quantity $Q_0 = c(\overline{\eta} + h)$ is the mass flux that originates in the intrinsic wave reference frame due to its motion with respect to the motionless laboratory reference frame (for the observer in the intrinsic reference frame fluid seems to move from right to left with speed c). The total mass flux Q differs from the "zero" flux Q_0 by the value equal to the wave momentum I. Thus, the wave momentum I is determined by the flow induced by the wave in the direction of its propagation. This flow is often called the *drift flow* because it is caused by the drift of fluid particles in the direction of wave propagation or the *near-surface flow* because the drift velocity of particles rapidly decreases with depth [39]. Sometimes it is called the *Stokes flow* in honour of Sir George Gabriel Stokes, who was the first to establish these properties of surface waves [80].

Since

$$-Q = c(\overline{\eta} + h) - I = c\left(\overline{\eta} + h - \frac{I}{c}\right) \equiv cd, \tag{2.63}$$

the parameter d is the depth of uniform flow with mass flux Q that moves with velocity c from right to left in the intrinsic wave reference frame. The depth d is called the *undisturbed depth* [10] and it determines the stream function at the bottom:

$$\frac{\psi\,|_{y=-h}}{c} = d. \tag{2.64}$$

In this case, the depth h measured with respect to the mean wave level ($\overline{\eta} = 0$) is larger than the undisturbed depth d by the value

$$h - d = \frac{I}{c} > 0. \tag{2.65}$$

The wave's kinetic energy K can be expressed in terms of the wave momentum I (or in terms of the Stokes flow) as

$$K = \frac{cI}{2}. \tag{2.66}$$

This relation was first derived by Levi-Cività in 1924 (see Ref. [43]).

The general solution to the Laplace equation (2.37) in the physical plane that satisfies the boundary condition (2.40) at the bottom is given by expansion (2.22) which has the following form in terms of dimensionless variables:

$$\Phi(\theta,\, y) = \sum_{n=1}^{\infty} \left(C_n \mathrm{e}^{\mathrm{i}n\theta} + C_n^* \mathrm{e}^{-\mathrm{i}n\theta} \right) \left(T_n \mathrm{e}^{ny} + T_{-n} \mathrm{e}^{-ny} \right), \tag{2.67}$$

$$T_n = \frac{\mathrm{e}^{nh}}{\mathrm{e}^{nh} + \mathrm{e}^{-nh}} = \frac{1}{2}\left(1 + \tanh(nh)\right). \tag{2.68}$$

The general solution (2.67) to the Laplace equation for the velocity potential can be rewritten for the stream function by taking into account the Cauchy–Riemann conditions (2.55):

$$\Psi(\theta,\, y) = (c\bar{\eta} - I) + \sum_{n=1}^{\infty} \left(\mathrm{i}C_n \mathrm{e}^{\mathrm{i}n\theta} - \mathrm{i}C_n^* \mathrm{e}^{-\mathrm{i}n\theta} \right) \left(T_n \mathrm{e}^{ny} - T_{-n} \mathrm{e}^{-ny} \right), \tag{2.69}$$

where the integration constant is determined from the value of the stream function at the bottom (2.62), inasmuch as $(T_n \mathrm{e}^{-nh} - T_{-n} \mathrm{e}^{nh}) = 0$ for any n. Then the complex potential $W = \Phi + \mathrm{i}\Psi$ in the *physical plane* can be written in the form of Fourier series for a periodic function of complex variable ζ with period $\lambda = 2\pi$:

$$\frac{W(\zeta)}{c} = 2\mathrm{i}\xi_0 + 2\mathrm{i} \sum_{n=1}^{\infty} \left(\xi_n^* T_n \mathrm{e}^{-\mathrm{i}n\zeta} - \xi_n T_{-n} \mathrm{e}^{\mathrm{i}n\zeta} \right),$$

$$\xi_0 = \frac{1}{2}\left(\bar{\eta} - \frac{I}{c} \right),\; \xi_n = \frac{\mathrm{i}C_n}{c}. \tag{2.70}$$

For the symmetric waves, the coefficients ξ_n are real in view of condition (2.23). The complex potential in the intrinsic wave reference frame is

$$\frac{w(\zeta)}{c} = -\zeta + B_0 + \mathrm{i} \sum_{n=1}^{\infty} \left(B_n^* \,\mathrm{e}^{-\mathrm{i}n\zeta} - \mathrm{e}^{-2nh} B_n \,\mathrm{e}^{\mathrm{i}n\zeta} \right),$$

$$B_0 = 2\mathrm{i}\xi_0,\; B_n = 2\xi_n T_n. \tag{2.71}$$

In the *inverse plane*, the complex potential is an independent variable and the solution is specified by the inverse expansion as follows [10]

$$\zeta(w) = -\frac{w}{c} + \mathrm{i}a_0 + \mathrm{i}\sum_{n=1}^{\infty}\left(a_n \mathrm{e}^{\mathrm{i}nw/c} - \mathrm{e}^{-2nd} a_n^* \mathrm{e}^{-\mathrm{i}nw/c}\right),\ a_0 = \left(\overline{\eta} - \frac{I}{c}\right), \quad (2.72)$$

where the undisturbed depth d appears from the condition (2.64) at the bottom. Note that the period $\lambda = 2\pi$ in the physical plane corresponds to the period $c\lambda$ in the inverse plane. In solving the problem in the inverse plane, the zero level $y = 0$ is often selected at the undisturbed depth level to set the free term a_0 equal to zero ($\overline{\eta} = I/c$). The coefficients a_n should be real for symmetric waves.

2.2 Stokes Waves and Methods of Their Calculation

2.2.1 Stokes Waves

Linear waves represent the solutions of Eqs. (2.37)–(2.40) at infinitesimal amplitudes ($A \to 0$). Their shape is described by a cosine curve and in terms of dimensionless variables their phase speed is $c = \tanh h$ ($c = 1$ in the case of infinite depth). When the wave amplitude can no longer be assumed infinitesimal, the nonlinearity of boundary conditions needs to be taken into account. In the case of finite but small amplitudes ($A \ll 1$), one can make use of typical perturbation techniques in terms of parameter A. For example, the coefficients of expansion (2.72) can be expressed as $a_n \sim a_1^n$. Stokes [80] (1847) was the first to construct the perturbation theory for gravity waves on deep water in the physical plane. Later he used expansions in the inverse plane to extend this theory to the case of finite depth [81]. Stokes established the following properties of finite-amplitude gravity waves [80–82].

1. The phase speed of nonlinear waves depends on their amplitude. In the case of deep water, this dependence is written as [39]

$$c^2 = 1 + a^2 + \frac{5}{4}\, a^4 + \dots, \quad (2.73)$$

where a is the first-harmonic amplitude of the wave profile:

$$\eta = a \cos\theta + \left(\frac{1}{2} + \frac{17}{24}\, a^2\right) a^2 \cos 2\theta + \frac{3}{8}\, a^3 \cos 3\theta + \frac{1}{3}\, a^4 \cos 4\theta + \dots\,. \quad (2.74)$$

For finite depth, the nonlinear dispersion relation (2.73) is much more cumbersome [92].

2. Finite-amplitude waves cause the mass transfer in the direction of wave propagation (particle trajectories are not closed curves in contrast to linear waves), i.e. the wave momentum $I \neq 0$.

3. Wave crests sharpen as the amplitude a increases and the troughs flatten.
4. There exists a limiting wave of the maximum amplitude (height), and the crests of this limiting wave form a 120° cusp. No stationary waves higher than the limiting wave can exist in the framework of the canonical model.

The one-parametric family of waves in amplitude a (or parameter A) that is approximately described by expansions (2.73) and (2.74) (or similar expansions in the case of finite depth) is usually referred to as *Stokes waves* [92]. It follows from formula (2.74) that the Stokes waves are symmetric both with respect to their crests ($\theta = 0$) and with respect to their troughs ($\theta = \pi$). This fact was first proved for small-amplitude waves by Levi-Cività [41] with the use of conformal transforms. Later Garabedian [27] used the variational approach and symmetrisation to show rigorously that the canonical model does not admit the existence of asymmetric waves with equal crests and equal troughs.

The Bernoulli equation (2.38) results in the following expression for the velocity of fluid particles at the surface in the intrinsic wave reference frame:

$$\Phi_\theta - c = \pm\sqrt{2(B-\eta) - \Phi_y^2}\,. \tag{2.75}$$

In particular, the particle velocity at the wave crest is

$$q(0) \equiv \left(\Phi_\theta\,|_{y=\eta(0)} - c\right) = \pm\sqrt{2\left(B-\eta(0)\right)}\,, \tag{2.76}$$

inasmuch as $\Phi_y\,|_{y=\eta(0)} = 0$. The minus sign corresponds to the family of Stokes waves. The plus sign corresponds to a family of irregular gravity waves described in Ref. [54].

Formula (2.76) implies that for the Stokes waves the fluid particle velocity at the wave crest increases from the value $q(0) = -c$ for the infinitesimal amplitude to the limiting value $q(0) = 0$. The wave with $q(0) = 0$ is called the *limiting Stokes wave*. The condition $(\Phi_\theta - c) < 0$ holds true in the entire flow domain, except for one point at the wave crest where $(\Phi_\theta - c) \equiv q(0) = 0$. This is a singular point at which the flow is motionless in the intrinsic wave reference frame. Such points are also referred to as *stagnation* or *critical* points [40]. It follows from the kinematic boundary condition (2.39) that the derivative η_θ is undetermined at the surface point where $(\Phi_\theta - c) = 0$ and $\Phi_y = 0$, i.e. it exhibits a discontinuity there. Therefore, the wave profile $y = \eta(\theta)$ forms a cusp at the point where $q(0) = 0$.

2.2.2 *Steep Stokes Waves and Their Main Properties*

Stokes performed calculations for the case of infinite depth with accuracy to within $O(A^5)$ and obtained results for the case of finite depth with accuracy to within $O(A^3)$. Stokes' results were confirmed by Lord Rayleigh [68]. Much later Wilton [94] could extend the infinite-depth expansions to the $O(A^{10})$ order (although with errors made in the eighth order) and De [16] published the results valid for an arbitrary depth with accuracy to within $O(A^5)$. This was a practical limit that could be achieved in the pre-computer era. The crests of Stokes waves sharpen as the wave amplitude increases and the wave profile becomes steeper. This makes the problem essentially nonlinear and much more terms are required in the perturbation expansions for the correct description of wave parameters.

Schwartz [72] was the first who used computer arithmetics to construct the Stokes expansions with accuracy to within $O(A^{48})$ for the case of finite depth and $O(A^{117})$ for the case of infinite depth. He used Fourier expansions in the inverse plane to derive a *nonlinear* system of algebraic equations for the unknown expansion coefficients a_n and wave speed c. Since solving those equations directly was not possible at that time, Schwartz used the perturbation techniques to expand the unknown parameters in power series in terms of small parameter ε:

$$a_n = \sum_{k=0}^{\infty} a_{nk}\, \varepsilon^{n+2k}, \quad c^2 = \sum_{k=0}^{\infty} \gamma_k\, \varepsilon^{2k}. \tag{2.77}$$

The coefficients a_{nk} and γ_k were calculated from the corresponding recurrence formulas. By setting $\varepsilon = a_1$ (a_1 being the first-harmonic amplitude), Schwartz found that the convergence of amplitude expansions (2.77) deteriorates at sufficiently large a_1, so that they eventually diverged. To increase the convergence radius of the perturbation series, Schwartz made use of Padé approximations. He showed that the divergence of perturbation expansions in terms of the first-harmonic amplitude a_1 was caused by the fact that a_1 is not a monotonic function of wave height A and attains a maximum before the maximum amplitude is reached. In the case of infinite depth, this maximum was found at $A \approx 0.13$. It is the position of this maximum that determines the convergence radius of amplitude expansions. Since the wave height A monotonically grows up to the limiting value, the convergence of the expansions can be improved by setting $\varepsilon \sim A$. Indeed, by making use of the expansions in terms of $\varepsilon = 2A$, Schwartz could reach nearly the limiting amplitude, although he had still to use Padé approximations to improve the convergence. For the case of infinite depth, he got the following estimate of the maximum wave amplitude: $A_{\text{max}} = 0.1412$.

Longuet-Higgins [43] recalculated Schwartz's expansions for the case of infinite depth in terms of parameter $\upsilon = \varepsilon^2 = 1 - \frac{q^2(0)q^2(\pi)}{c^2 c_0^2}$, where $q(0)$ and $q(\pi)$ are the fluid particle velocities at the wave crest and trough, respectively, and c_0 is the phase speed of linear waves. These expansions turned out to converge faster than Schwartz's expansions in terms of parameter $\varepsilon = 2A$. The new parameter υ also brings another advantage consisting in the fact that its variation range is known beforehand: it increases from $\upsilon = 0$ for linear waves to $\upsilon = 1$ for the limiting wave. Longuet-Higgins obtained expansions with accuracy to within $O(\upsilon^{40})$ and found that not only the first-harmonic amplitude but also the phase speed c, total energy E, potential energy U, kinetic energy K and momentum I of Stokes waves are not monotonic functions of their amplitude A and attain their maximum values before the limiting

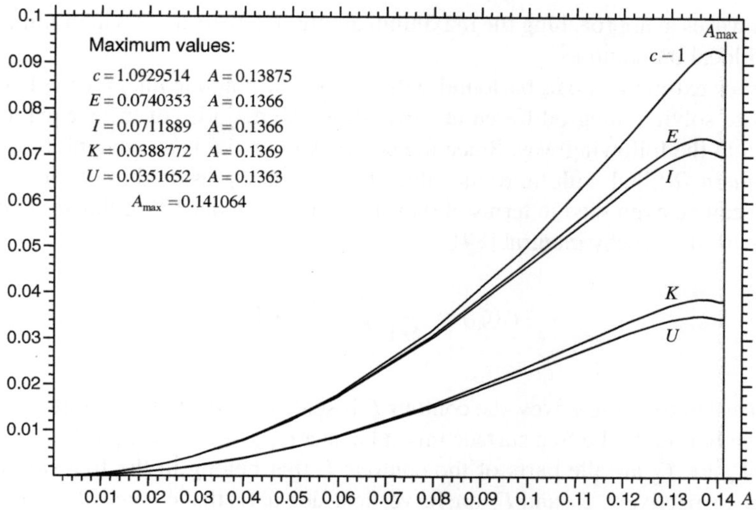

Fig. 2.2 Phase speed c, total energy E, potential energy U, kinetic energy K and momentum I of Stokes waves as functions of their amplitude A

amplitude is reached. The corresponding plots are shown in Fig. 2.2. Thus, *the wave of maximum height is not the fastest wave and does not possess the maximum energy and momentum*, at it had been assumed since the time of Stokes. Cokelet [10] proved that this result holds true for the case of other (finite) depths. To this end, he constructed Schwartz's expansions in terms of parameter $\varepsilon^2 = 1 - \frac{q^2(0)\,q^2(\pi)}{c^4}$ with accuracy to within $O(\varepsilon^{110})$. Cokelet found that many other integral wave parameters attain their maximums before the limiting amplitude is reached. Note that the maximum of the phase speed was first detected in Ref. [71], but the authors of that work did not make any definite conclusion in regard to the existence of the wave speed maximum because of insufficient accuracy of their calculations. Later Longuet-Higgins and Fox [49–51] constructed asymptotic expansions for waves close to the 120°-cusped wave (almost highest waves) and showed that these dependences oscillate infinitely as the limiting wave is approached. Longuet-Higgins [43] also rigorously proved the following relations that are valid for an arbitrary depth:

$$dE = d(K + U) = c\,dI, \quad d\mathscr{L} = d(K - U) = I\,dc, \tag{2.78}$$

where $\mathscr{L}$ is the Lagrange function. Thus, the extremums of the momentum and total energy coincide, as well as the extremums of the phase speed and Lagrangian.

Dallaston and McCue [13] used computer symbolic arithmetics to recalculate Schwartz's and Cokelet's power expansions in exact symbolic form to within $O(\varepsilon^{300})$ and numerically to within $O(\varepsilon^{846})$. Further, by using Padé approximations they could compute the positions of several next extremums in the phase speed and energy of

Stokes waves at approaching the maximum amplitude: their first local minimum and second local maximum.

Higher extremums can be found with the use of contour integral methods that deal with solving integrodifferential equations. The majority of these equations are derived in the following way. Since the complex potential is an analytic function in the domain Ω filled with fluid, its value at any point ζ_0 inside the selected domain $G \in \Omega$ can be expressed in terms of its value at the boundary Γ of that domain with the use of the Cauchy integral [84]:

$$w(\zeta_0) = \frac{1}{2\pi \mathrm{i}} \oint\limits_{\Gamma} \frac{w(\zeta)}{\zeta - \zeta_0}\, \mathrm{d}\zeta. \tag{2.79}$$

In the case of periodic waves, the contour Γ is selected such that the entire wave span from the bottom to the free surface might be encompassed: $\Gamma = \Gamma_s + \Gamma_- + \Gamma_b + \Gamma_+$. Here, Γ_s are Γ_b are the parts of the contour Γ that belong to the free surface and bottom, respectively, Γ_+ and Γ_- are its vertical side parts that connect the bottom and the surface and are spaced by the wave period from one another. The integral over the horizontal bottom Γ_b is always easy to calculate, and the integrals over Γ_+ and Γ_- cancel one another out due to periodicity condition. Thus, the Cauchy integral (2.79) expresses the complex potential at any point ζ_0 inside the fluid in terms of its value at the free surface Γ_s. The integral equation is obtained when the point ζ_0 is selected at the free surface. Although the Cauchy integral is divergent in conventional sense in this case, the quantity $w(\zeta_0)$ is equal to twice the principal value of the Cauchy integral [84].

Contour integral methods are the most accurate methods for calculating two-dimensional surface waves. As an example, let us consider one contour integral approach that was proposed by Tanaka in Ref. [86] for solitary waves and then extended to the case of periodic waves on finite depth [87]. Based on the equation

$$\frac{\mathrm{d}\left(\frac{1}{3}q^3\right)}{\mathrm{d}\phi} = \sin\vartheta, \quad \{\theta,\ y\} \in \Gamma_s, \tag{2.80}$$

that expresses the relationship between the absolute value of the fluid particle velocity q and the angle of inclination ϑ of the free surface to the horizontal axis, Tanaka derived an integral equation that defines $\ln q$ as a function of ϑ. This integral equation utilises one parameter defined as

$$\upsilon = 1 - \frac{q(0)}{q(\pi)}. \tag{2.81}$$

For linear waves, we have $\upsilon = 0$, and for the limiting wave we have $\upsilon = 1$. The obtained integral equation was solved by the method of successive approximations. To concentrate the computational mesh around the crest, Tanaka used the nonlinear transformation

Table 2.1 Extremums of the phase speed c of Stokes waves calculated by Tanaka's method in the case of infinite depth

Extremum c	v	A	c
First max	0.8637	0.138753	1.092951384
First min	0.9685	0.140920	1.092276839
Second max	0.9927	0.141056	1.092285150
Second min	0.9985	0.141063	1.092285047

$$\phi = \phi' - \gamma \sin \phi', \quad 0 \leqslant \gamma < 1, \tag{2.82}$$

that was earlier proposed in Ref. [6]. This transformation dramatically increased the computational accuracy, and the effect was the largest with parameter γ selected closer to unity. Tanaka's method allows the second local maximum of the wave phase speed to be located with high accuracy even at $N = 1000$ nodes, and the second local minimum can be computed at $N = 3500$ (Table 2.1).

In this context, we also need to mention the results obtained by Maklakov in Ref. [58]. He also calculated the second maximum and the second minimum of the wave phase speed and other integral wave parameters and reached the amplitude equal to 99.99997% of the limiting value. However, he did not tabulate those results. To this end, Maklakov used a modified contour integral method when no more than $N = 2000$ of mesh nodes were sufficient to achieve the required accuracy, which gives some advantage in practical computations. Later Dyachenko et al. [22] used another realisation of the conformal mapping and rational Padé approximations to calculate the parameters of Stokes waves on deep water with quadruple precision when approaching the limiting form with a 120° cusp. They could confidently tabulate the positions of the first local minimum and second local maximum of the phase speed. Lushnikov et al. [57] made an auxiliary conformal mapping in the numerical algorithm proposed in Ref. [22] that dramatically speeded up the numerical convergence by adapting the numerical grid for resolving singularities. As a result, they could compute the next two extremums of the wave phase speed: the second local minimum and the third local maximum.

2.2.3 Spectral Methods of Calculating the Stokes Waves

Expansions (2.71) and (2.72) represent the *exact* solution of the canonical problem in the form of Fourier series. In practice, these series are truncated to contain a finite number of terms. Such truncated series represent *approximate* solutions. After collecting the coefficients at the like powers of linearly independent functions (exp(i nf) in our case), this technique results in a fully nonlinear system of algebraic equations for the corresponding Fourier coefficients of expansion (2.70) in the physical plane (first Stokes method) or expansion (2.72) in the inverse plane (second Stokes method). The advantage of the latter expansions is that the free surface in the inverse plane is known and given by the horizontal line $\psi = 0$. Sometimes instead of solving the system of equations for the Fourier coefficients directly, the free surface is subdivided into a certain number of points (that are called *collocation* points) at which the boundary conditions need to be satisfied exactly. As a result, one gets a system of algebraic equations for the unknown coefficients at the collocation points. It is the approach that was used by Chappelear [5] to calculate the first nine terms in series (2.67). He computed several steep wave profiles and compared his results with the fifth-order Stokes expansions.

Longuet-Higgins [44] established simple quadratic relations between the coefficients a_n of expansion (2.72) and used them instead of solving a more cumbersome system of cubic equations that is obtained when Fourier series (2.72) are substituted directly in the Bernoulli equation. This approach, which is usually referred to as the *Longuet-Higgins method*, is described in detail in Refs. [45, 46, 70] for infinite depth and in Refs. [47, 60] for finite depth.

The methods of calculating the Stokes waves in the physical plane include the modification of the first Stokes method performed by Rienecker and Fenton in Ref. [69] in which the coefficients of the Fourier series for the velocity potential are calculated by the collocation method and the Hamiltonian method implemented by Zufiria in Ref. [104]. Rienecker and Fenton carried out the calculations with up to $N = 64$ terms in their expansions, and Zufiria took into account up to $N = 72$ terms. The accuracy of their calculations was sufficient to evaluate only the first local maximum of the wave phase speed. The results of Rienecker and Fenton were extended by Zhao et al. [103] who improved the iterative stability of their numerical scheme by a relaxation technique. Tao et al. [88] developed another method of calculating the progressive waves in the case of finite depth with the use of operator expansions employing homotopy analysis and Padé approximations.

Lukomsky et al. [53, 54] constructed a general realisation of the method of Fourier expansions in the physical plane with the use of explicit relations between the Fourier coefficients of the velocity potential and free surface profile. In this approach, the truncated Fourier series for the velocity potential in the physical plane is written as

$$R(\theta, y) = \xi_0 + \sum_{n=1}^{N} \left(\xi_n T_n e^{n(y+i\theta)} - \xi_n^* T_{-n} e^{-n(y+i\theta)} \right),$$

$$\xi_0 = \frac{1}{2}\left(\overline{\eta} - \frac{I}{c}\right), \quad \Phi = -ic(R - R^*), \tag{2.83}$$

where the coefficients T_n are defined by relations (2.68). The truncated Fourier series for the free surface profile is given by

$$\eta(\theta) = \sum_{n=-M}^{M} \eta_n \exp(in\theta), \quad \eta_{-n} = \eta_n^*. \tag{2.84}$$

Expansions (2.83) and (2.84) are substituted in the dynamic and kinematic boundary conditions (2.38) and (2.39), which, after collecting the coefficients at the like powers of exponential functions, reduce to a system of nonlinear algebraic equations for the unknown coefficients ξ_n, η_n and the phase speed c. This system of equations is solved numerically by Newton's method with the use of Fast Fourier Transform for calculating the Fourier harmonics of exponential functions. For the convergence of the method, M should be several times larger than N. (Employing different truncation numbers in the Fourier series of the velocity potential and free surface profile was originally proposed in Ref. [104].) The coefficients ξ_n and η_n are complex-valued in the general case, but for the symmetric waves they are real. This method could provide accurate data on the first local maximum of the wave phase speed and locate its first minimum [53]. The calculations were performed up to $N = 250$ and $M = 4N$.

The use of ordinary Fourier expansions (2.83) and (2.84) is limited by the fact that the expansion coefficients decrease with their number more and more slowly as the wave crests sharpen with increasing of the wave amplitude. Hence, the number of coefficients required for accurate calculating of the waves approaching the limiting amplitude grows up enormously beyond any practical limits. In view of this, Lukomsky and Gandzha [55] proposed a modified spectral method optimised for the efficient calculation of steep waves on deep water. The idea of this method was to select a more suitable set of functions in the expansion of the velocity potential. To this end, Lukomsky and Gandzha used the Euler method for the summation of series described in Ref. [30]. The resulting expansion of the velocity potential is written as $(h = \infty)$

$$R(\theta,\, y;\, y_0) = \sum_{n=0}^{N} \frac{\alpha_n}{\left(\exp(-y_0) - \exp\left(-y - i\theta\right)\right)^n}\,, \tag{2.85}$$

where y_0 is a free parameter used to accelerate the convergence of the series for the velocity potential. Fractional expansions (2.85) reduce to ordinary Fourier expansions (2.83) at $y_0 = \infty$. Expansions (2.85) and (2.83) are equivalent for all y_0 at $N = \infty$, but fractional expansion (2.85) converges mush faster at $y_0 \simeq 1$ and, therefore, needs less terms to maintain the same computational accuracy. The convergence of the series for the free surface profile was accelerated by the nonlinear transformation of the horizontal scale that was earlier proposed in Ref. [6]:

$$\theta(\chi;\, \gamma) = \chi - \gamma\, \sin \chi\,, \quad 0 < \gamma < 1. \tag{2.86}$$

Then the corresponding expansion in the transformed space is written as

$$\eta\left(\chi;\, \gamma\right) = \sum_{n=-M}^{M} \eta_n^{(\gamma)} \exp(in\chi), \quad \eta_{-n}^{(\gamma)} = \eta_n^{(\gamma)}. \tag{2.87}$$

As compared to expansion (2.84), much smaller number of terms is required at $\gamma \to 1$ to keep the same computational accuracy. Further steps are similar to the method of ordinary Fourier expansions. In practical calculations, the free parameters y_0 and γ

of the method as well as the ratio between the numbers N and M are selected such that the overall numerical mismatch would be the smallest. The advantage over the ordinary Fourier expansions was from one to ten orders of magnitude, depending on the wave amplitude [55].

Expansions (2.85) and (2.83) or any other analytic function

$$R(\theta,\ y) = R(y + \mathrm{i}\theta)$$

satisfy the Laplace equation (2.37) identically. The last fact was used by Clamond [8, 9] to introduce the so-called renormalisation principle which allows the velocity potential in the whole flow domain to be reconstructed from the form of the velocity potential at the bottom (or any other fluid level). Applying the renormalisation to the first approximation of cnoidal waves on shallow water, Clamond [9] obtained the velocity potential that identically coincided with the first term ($N = 1$) of fractional expansion (2.85). This first term proved to yield more accurate estimates than the fifth-order Stokes amplitude approximation.

2.3 The Limiting Stokes Wave with a Corner at the Crest and Its Calculation

The limiting wave with $q(0) = 0$ was first considered by Stokes is his pioneer work [82]. He showed that when the fluid flow at the wave crest is motionless with respect to the wave crest forms a 120° angle. Let us reproduce the proof of this statement. It follows from Eq. (2.76) that for the wave with $q(0) = 0$ the crest elevation above the wave mean level is

$$\eta(0) = B = \frac{c^2}{2} + \frac{1}{2}\,\overline{\Phi_x^2}\,|_{y=-h}. \tag{2.88}$$

The origin of the intrinsic wave reference frame is selected to lie at the wave crest, i.e. the level $y = 0$ is translated by $\eta(0)$ upwards. In this reference frame, the Bernoulli equation (2.38) is written in terms of the complex potential with the use of Eq. (2.52) as follows

$$\left|\frac{\mathrm{d}w}{\mathrm{d}\zeta}\right|^2 + 2\,\mathrm{Im}\,\zeta = 0, \quad \zeta \in \Gamma, \tag{2.89}$$

where $\zeta = \theta + \mathrm{i}\big(y - \eta(0)\big)$ and Γ is the surface. Consider the complex potential of the flow described by a power function:

$$w(\zeta) = \phi + \mathrm{i}\psi = \mathscr{A}\,\zeta^n. \tag{2.90}$$

Such a complex potential describes the flow into a $\frac{\pi}{n}$ corner. Substituting this expression in Eq. (2.89) yields the unknown n and $\mathscr{A}$. In particular, for the exponent n the only admissible value is

$$n = \frac{3}{2},$$

which indeed corresponds to a 120° angle. Accordingly, for the complex potential of the flow into a 120° corner we get

$$w(\zeta) = \mathrm{i}\frac{2}{3}\,(\mathrm{i}\zeta)^{\frac{3}{2}}, \tag{2.91}$$

the inverse relation being

$$\zeta(w) = -\mathrm{i}\Big(\frac{3}{2\mathrm{i}}\,w\Big)^{\frac{2}{3}}. \tag{2.92}$$

Thus, the analytic function $\zeta(w)$ has a singularity of order $\frac{2}{3}$. The corresponding complex velocity is written as

$$\frac{\mathrm{d}w}{\mathrm{d}\zeta} = -\big(\mathrm{i}\,\zeta\big)^{\frac{1}{2}} = -\Big(\frac{3}{2\mathrm{i}}\,w\Big)^{\frac{1}{3}}. \tag{2.93}$$

Stokes' solution (2.91) and (2.92) is valid only in the infinitesimal vicinity of wave crest. Therefore, the obtained corner flow is usually referred to as *local*. It represents only the first term of the expansion about the corner. The second term of such an expansion about the singular point was found by Grant [29]:

$$\zeta(w) = -\mathrm{i}\Big(\frac{3}{2\mathrm{i}}\,w\Big)^{\frac{2}{3}} + \mathrm{i}\gamma\,(-\mathrm{i}w)^{2\mu} + \ldots, \tag{2.94}$$

and higher terms were calculated by Norman [64] in the form of a power series in terms of the exponent μ that satisfies the transcendental equation

$$\tan \pi\mu = -\frac{2+3\mu}{3\sqrt{3}\mu}. \tag{2.95}$$

The first root of this equation is $\mu = 0.73467287\ldots$. Hence, the expansion about the singular point $q(0) = 0$ is not a regular power expansion but includes irrational exponents, i.e. the singularity formed by the stagnation point at the crest has an irregular nature.

Stokes' local solution (2.92) and expansions (2.94) satisfy the Laplace and Bernoulli equations but do not satisfy the asymptotic at the bottom, which has the following form:

$$\frac{\mathrm{d}w}{\mathrm{d}\zeta} = -c, \quad \psi = cd. \tag{2.96}$$

To make this condition hold, one needs to employ linear combinations of the solutions about the corner or to construct approximate expansions that take into account the singularity at the crest in one way or another. This task turns out to be more simple when the layer occupied by fluid in the complex plane w is transformed into a unit circle:

$$w = -\mathrm{i}c \ln u, \quad u = \rho \exp(\mathrm{i}s). \tag{2.97}$$

In this case, the velocity potential and stream function are written as

$$\phi = c\, s, \quad \psi = -c \ln \rho. \tag{2.98}$$

In the transformed space, the fluid occupies the circular domain bounded by the circles $\rho = 1$ (surface) and $\rho = \rho_0 = \mathrm{e}^{-d}$ (bottom). The wave crest and trough have the phases $s = 0$ and $s = \pm\pi$, respectively (Fig. 2.3). In the case of infinite depth, we have $\rho_0 = 0$ and the bottom is transformed into the point $u = 0$.

The singularity $\zeta = 0$ ($w = 0$) is located at the point $u = 1$ in the complex plane u, and Stokes' local solution (2.92), (2.93) has the following form:

$$\zeta = -\mathrm{i}\left(\frac{3}{2}\, c\,(1-u)\right)^{\frac{2}{3}}, \quad \frac{\mathrm{d}w}{\mathrm{d}\zeta} = -\left(\frac{3}{2}\, c\,(1-u)\right)^{\frac{1}{3}}. \tag{2.99}$$

With these asymptotics taken into account, Michell [61] proposed the following approximation for the wave with a singularity of order $\frac{2}{3}$ in the case of infinite depth:

$$\frac{\mathrm{d}w}{\mathrm{d}\zeta} = -c\,(1-u)^{\frac{1}{3}} \sum_{n=0}^{\infty} b_n\, u^n. \tag{2.100}$$

The asymptotic (2.96) at the bottom $u = 0$ is satisfied identically by setting $b_0 = 1$. All other unknown coefficients b_n and the wave speed c can be found from a system of algebraic equations that is obtained by substituting expansion (2.100) in the Bernoulli equation (2.89).

Michell calculated the first three coefficients (assuming that $b_n \sim b_1^n$) and obtained the estimates $c^2 \approx 1.2$ and $A_{\max} \approx 0.142$, i.e. the speed of the wave of limiting height is about 1.09 times higher than the linear wave speed. Later Havelock [32] calculated one additional (the fourth) coefficient and obtained $A_{\max} \approx 0.1418$. Nekrasov [63] confirmed Michell's results with the use of the inverse expansion

$$\frac{\mathrm{d}\zeta}{\mathrm{d}w} = -c^{-1}(1-u)^{-\frac{1}{3}} \sum_{n=0}^{\infty} \alpha_n\, u^n, \quad \alpha_0 = 1. \tag{2.101}$$

Davies [15] used a different approach and obtained $A_{\max} \approx 0.1443$. Michell's calculations could essentially be refined after the advent of the computer era. Yamada [95] could compute the first twelve coefficients of Nekrasov's expansion (2.101) and got $A_{\max} \approx 0.1412$. These calculations were also performed in the case of finite depth in Ref. [96].

Olfe and Rottman [65] formalised Michell's method and derived a general system of nonlinear algebraic equations for the coefficients b_n of expansion (2.100) and the phase speed c:

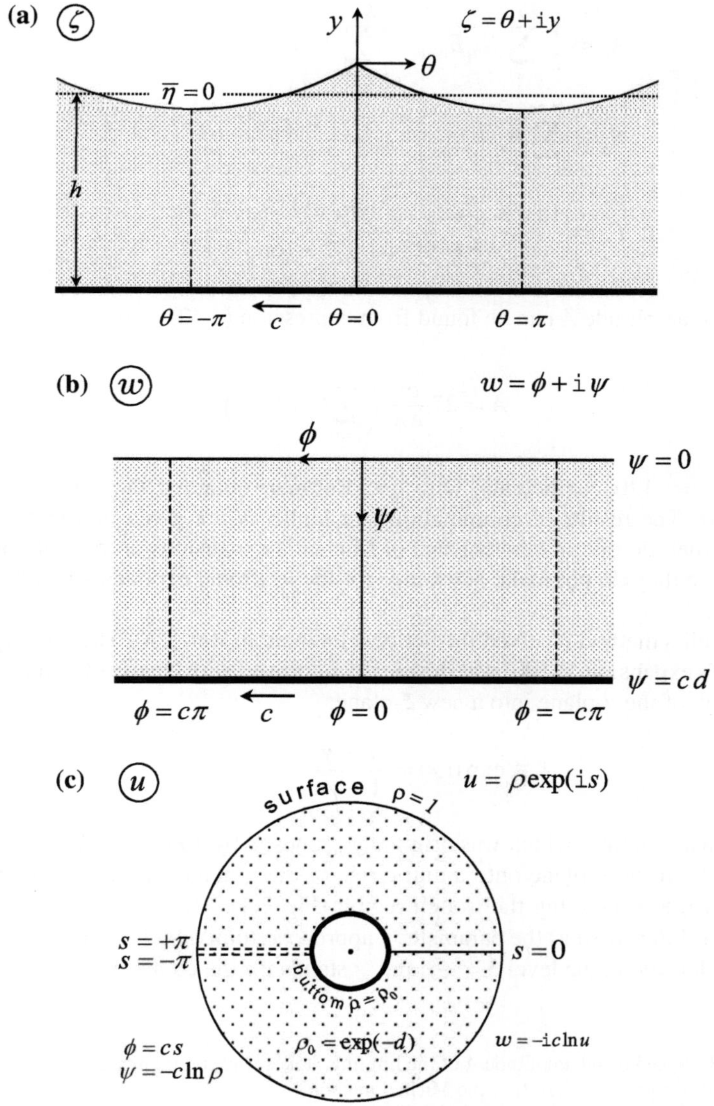

Fig. 2.3 Direct and inverse planes: **a** physical plane, **b** inverse plane of the complex potential, **c** inverse plane transformed into a unit circle

$$(3n+2)A_n - (3n+1)A_{n+1} - c^{-2}F_n = 0, \; n = \overline{0, N}, \tag{2.102}$$

where N is the number of the coefficients that were taken into account and

$$A_n = \frac{1}{2}\sum_{n_1=0}^{n} B_{n_1} B_{n-n_1} + \sum_{n_1=1}^{N-n} B_{n_1} B_{n+n_1},$$

$$B_n = \sum_{n_1=0}^{N-n} b_{n_1}\, b_{n+n_1},$$

$$F_n = \frac{18\sqrt{3}}{\pi}\sum_{n_1=0}^{N} \frac{(6n_1+1)\, b_{n_1}}{9(2n+1)2-(6n_1+1)^2},\ b_0 = 1.$$

The wave amplitude A can be found from expression (2.42) using relation (2.53):

$$A = 2^{\frac{2}{3}}\frac{c^2}{4\pi}\left(\sum_{n=0}^{N}(-1)^n b_n\right)^2. \tag{2.103}$$

With $N = 120$ coefficients, Olfe and Rottman obtained the estimate $A_{\max} \approx 0.141061$. The results of computations for higher N are given in Table 2.2. It can be seen that the rate of convergence of Michell's expansions drops with increasing N because they do not make allowance for the irrational exponents found by Grant in Ref. [29].

Michell's method received further development in Ref. [24]. The convergence of Michell's expansion (2.100) was drastically improved with the use of nonlinear transformation of the u-plane into a new ξ-plane:

$$\xi \equiv \exp(\mathrm{i}\chi) = \frac{u-\gamma}{1-\gamma u}, \quad 0 \leqslant \gamma < 1. \tag{2.104}$$

This transformation, which was originally proposed by Tanaka in Ref. [85], maps a unit circle of the u-plane onto a unit circle of the ξ-plane with stretching the crest region while contracting the trough region (Fig. 2.4). The rate of stretch and contraction is intensified as the parameter γ approaches unity. In the ξ-plane, the infinite depth is located at the level $\xi = -\gamma$ and χ stands for the complex potential.

Table 2.2 Speed c and amplitude A of the limiting Stokes wave on infinite depth calculated from the system of equations (2.102) of the Michell method

N	A	c	
100	0.14106002	1.09227679	$b_1 = 0.04119$
200	0.14106250	1.09228173	$b_2 = 0.01251$
300	0.14106307	1.09228335	$b_3 = 0.00605$
400	0.14106327	1.09228405	$b_4 = 0.00360$
500	0.14106337	1.09228441	$b_5 = 0.00240$

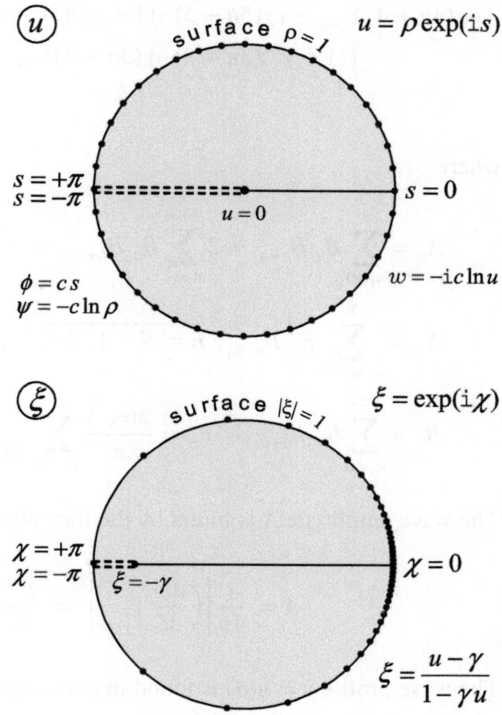

Fig. 2.4 Tanaka's transformation of the inverse plane. Reprinted from [24]

One can easily prove that asymptotic (2.99) remains unchanged in the ξ-plane:

$$\frac{\mathrm{d}w}{\mathrm{d}\zeta} \sim (1-\xi)^{1/3}. \tag{2.105}$$

So, the Michell expansion in the ξ-plane is written as

$$\frac{\mathrm{d}w}{\mathrm{d}\zeta} = \big(1-\exp(\mathrm{i}\chi)\big)^{\nu} \sum_{n=0}^{N} b_n \exp(\mathrm{i}n\chi), \quad \nu \equiv 1/3, \tag{2.106}$$

where N is the number of coefficients taken into account in the approximate solution. The boundary condition at infinite depth (2.96) takes the form

$$(1+\gamma)^{\nu} \sum_{n=0}^{N} b_n (-\gamma)^n = -c. \tag{2.107}$$

Further procedure is not much more awkward than that elaborated in Refs. [61, 65]. The corresponding system of nonlinear algebraic equations for the coefficients b_n of expansion (2.106) and phase speed c is

$$-\gamma(3n+4)A_{n+2} + \big(\gamma(3n+5)-(1+\gamma^2)(3n+1)\big)A_{n+1}+ \\ \big((1+\gamma^2)(3n+2)-\gamma(3n-2)\big)A_n + \gamma(3n-1)A_{n-1}+ \\ c(1-\gamma^2)F_n = 0,\ n=\overline{0,N}, \quad (2.108)$$

where

$$A_n = \sum_{n_1=0}^{n} B_{n_1}B_{n-n_1} + 2\sum_{n_1=1}^{N-n} B_{n_1}B_{n+n_1},\ n=\overline{0,N},$$

$$A_n = \sum_{n_1=n-N}^{N} B_{n_1}B_{n-n_1},\ n=\overline{N+1,\,2N}, \quad A_n \equiv 0,\ |n|>2N, \quad A_{-n}=A_n,$$

$$B_n = \sum_{n_1=0}^{N-n} b_{n_1}\,b_{n+n_1}, \quad F_n = \frac{36\sqrt{3}}{\pi}\sum_{n_1=0}^{N}\frac{(6n_1+1)\,b_{n_1}}{9(2n+1)^2-(6n_1+1)^2}.$$

The wave amplitude A is found by the formula

$$A = \frac{1}{4\pi}\left|\left(\frac{\mathrm{d}w}{\mathrm{d}\zeta}\right)_{\chi=\pi}\right|^2 = \frac{2^{2\nu}}{4\pi}\left(\sum_{n=0}^{N}(-1)^n\,b_n\right)^2. \quad (2.109)$$

The wave profile $y=\eta(\theta)$ is found in parametric form from the following integrals:

$$\theta(\chi) = \int_0^{\chi}\frac{\partial\theta}{\partial\chi'}\mathrm{d}\chi' = \int_0^{\chi}\frac{\partial\theta}{\partial\phi}\frac{\mathrm{d}\phi}{\mathrm{d}\chi'}\mathrm{d}\chi' = \int_0^{\chi}\mathrm{Re}\left(\left(\frac{\mathrm{d}w}{\mathrm{d}\zeta}\right)^{-1}\right)\frac{\mathrm{d}\phi}{\mathrm{d}\chi'}\mathrm{d}\chi', \quad (2.110)$$

$$y(\chi) = \int_0^{\chi}\frac{\partial y}{\partial\chi'}\mathrm{d}\chi' = \int_0^{\chi}\frac{\partial y}{\partial\phi}\frac{\mathrm{d}\phi}{\mathrm{d}\chi'}\mathrm{d}\chi' = \int_0^{\chi}\mathrm{Im}\left(\left(\frac{\mathrm{d}w}{\mathrm{d}\zeta}\right)^{-1}\right)\frac{\mathrm{d}\phi}{\mathrm{d}\chi'}\mathrm{d}\chi', \quad (2.111)$$

where χ ranges from 0 (crest) to π (trough) and

$$\frac{\mathrm{d}\phi}{\mathrm{d}\chi} = \frac{c(1-\gamma^2)\xi}{(\xi+\gamma)(1+\gamma\xi)}, \quad \xi = \exp(\mathrm{i}\chi). \quad (2.112)$$

The kinetic and potential energies of the wave can be calculated as

$$T = \frac{cI}{2} = \frac{c}{2}\frac{1}{2\pi}\int_0^{2\pi}\int_{-\infty}^{\eta(\theta)}(\phi_\theta + c)\,\mathrm{d}y\,\mathrm{d}\theta \\ = -\frac{c}{2\pi}\int_0^{\pi c}(y|_{\psi=0} + \overline{\eta})\,\mathrm{d}\phi = -\frac{c^4}{4} + \frac{c}{4\pi}\int_0^{\pi}\left|\frac{\mathrm{d}w}{\mathrm{d}\zeta}\right|^2\frac{\mathrm{d}\phi}{\mathrm{d}\chi}\,\mathrm{d}\chi, \quad (2.113)$$

$$U = \frac{1}{4\pi}\int_0^{2\pi}\left(y^2\,|_{\psi=0} - \overline{\eta}^2\right)\mathrm{d}\theta$$
$$= \frac{1}{8\pi}\int_0^{\pi}\left(c^4 - \left|\frac{\mathrm{d}w}{\mathrm{d}\zeta}\right|^4\right)\mathrm{Re}\left(\left(\frac{\mathrm{d}w}{\mathrm{d}\zeta}\right)^{-1}\right)\frac{\mathrm{d}\phi}{\mathrm{d}\chi}\,\mathrm{d}\chi, \quad (2.114)$$

where I is the wave momentum.

Based on the above-described approach, Gandzha and Lukomsky [24] obtained the most accurate values of the amplitude, phase speed and energy of the limiting Stokes wave:

$$c = 1.09228504859, \quad (2.115a)$$
$$A = 0.14106348398, \quad (2.115b)$$
$$U = 0.03456832591, \quad (2.115c)$$
$$T = 0.03829219689. \quad (2.115d)$$

These results are in agreement with the asymptotics derived earlier in Ref. [58] with the use of the theory of almost highest waves.

The method of calculating the limiting Stokes wave in the case of finite depth was developed by Willians [93] with Grant's asymptotic taken into account explicitly. Among approximate approaches to describing the limiting Stokes waves, we can mention a simple analytic approximation for the wave profile proposed by Longuet-Higgins [48] and more complex approximations for a broad range of depths presented by Karabut in Ref. [38].

2.4 High-Order Nonlinear Schrödinger Equation and Split-Step Fourier Technique

In this section, we briefly consider the evolution of narrow-band wave trains on the surface of an ideal fluid. In the case of finite depth, the evolution of surface gravity waves and associated fluid flows is governed by the following set of Euler equations derived in Sect. 2.1:

$$\Phi_{xx} + \Phi_{yy} = 0, \quad -h \leqslant y \leqslant \eta(x, t); \quad (2.116a)$$
$$\Phi_t + \frac{1}{2}\left(\Phi_x^2 + \Phi_y^2\right) + g\eta = 0, \quad y = \eta(x, t); \quad (2.116b)$$
$$\eta_t - \Phi_y + \eta_x\Phi_x = 0, \quad y = \eta(x, t); \quad (2.116c)$$
$$\Phi_y = 0, \quad y = -h. \quad (2.116d)$$

Here, the indices designate the partial derivatives with respect to the corresponding variables. In the case of slowly modulated wave trains and small-amplitude approximation, the unknown free-surface displacement and velocity potential can be expressed in the form of Fourier expansions around the carrier frequency and wave number:

$$\begin{pmatrix} \eta(x, t) \\ \Phi(x, y, t) \end{pmatrix} = \sum_{n=-\infty}^{\infty} \begin{pmatrix} \eta_n(x, t) \\ \Phi_n(x, y, t) \end{pmatrix} \exp\big(in(kx - \omega t)\big), \quad (2.117)$$
$$\eta_{-n} \equiv \eta_n^*, \quad \Phi_{-n} \equiv \Phi_n^*,$$

with $\eta_0 \sim \varepsilon^2$, $\Phi_0 \sim \varepsilon$, $\eta_n \sim \varepsilon^n$, $\Phi_n \sim \varepsilon^n$, $n \geqslant 1$, ε being a formal small parameter related to the smallness of wave amplitude as compared to the carrier wavelength $\lambda \equiv \frac{2\pi}{k}$. The carrier frequency ω and wave number k are connected by the linear dispersion relation derived in Sect. 2.1.5:

$$\omega^2 = gk \tanh(kh) \equiv gk\sigma, \quad \sigma \equiv \tanh(kh). \quad (2.118)$$

Further, by exploiting various perturbation techniques, the initial set of equations (2.116) can be reduced to two coupled equations for the first-harmonic amplitude η_1 of the wave profile and the mean flow expressed by the zero harmonic of the velocity potential: $\overline{\Phi} = \Phi_0$. In particular, it was done by Brinch-Nielsen and Jonsson [4] and then independently by Gramstad and Trulsen [28] and Craig et al. [11] in the Hamiltonian (canonical) form preserving momentum. Further simplification can be obtained when the mean flow is expanded into power series in terms of formal small parameter ε:

$$\Phi_0 = \varepsilon \Phi_0^{(1)} + \varepsilon^2 \Phi_0^{(2)} + O(\varepsilon^3). \quad (2.119)$$

This expansion is valid only in the case of finite depth, and its limitations are well discussed in Ref. [4]. The use of multi-scale expansions (see Sect. 1.6) allows the functions $\Phi_0^{(n)}$ to be expressed in terms of the first-harmonic amplitude η_1, so that two coupled equations for η_1 and Φ_0 might be reduced to a single evolution equation for η_1. It is the approach that was first utilised by Hasimoto and Ono [31] to derive the nonlinear Schrödinger equation (NLSE) for the first-harmonic amplitude η_1 of the wave profile:

$$\mathrm{i}\left(\frac{\partial \eta_1}{\partial t} + c_g \frac{\partial \eta_1}{\partial x}\right) + \left(\frac{\omega''}{2}\frac{\partial^2 \eta_1}{\partial x^2} + 4\omega k^2 a_{0,0,0}\, |\eta_1|^2 \eta_1\right) = 0, \quad (2.120)$$

where c_g is the wave group speed defined by formula (2.35) and ω'' is the second derivative of the carrier frequency with respect to k. The coefficient $a_{0,0,0}$ is defined in terms of dimensionless water depth kh:

$$a_{0,0,0} = -\frac{1}{16\sigma^4 \nu}\Big(\left(\sigma^2 - 1\right)^2 \left(9\sigma^4 - 10\sigma^2 + 9\right) k^2 h^2$$
$$+ 2\sigma\left(3\sigma^6 - 23\sigma^4 + 13\sigma^2 - 9\right) kh - \sigma^2\left(7\sigma^4 - 38\sigma^2 - 9\right)\Big), \quad (2.121)$$

where

$$\nu = \left(\sigma^2 - 1\right)^2 k^2 h^2 - 2\sigma\left(\sigma^2 + 1\right) kh + \sigma^2. \tag{2.122}$$

NLSE arises in describing nonlinear waves in various physical contexts, such as nonlinear optics [91], plasma physics [34], nanosized electronics [12], ferromagnetics [7], Bose–Einstein condensates [100], and hydrodynamics [18, 66, 98, 102]. In the general context of weakly nonlinear dispersive waves, this equation was first discussed by Benney and Newell [1]. In the case of gravity waves propagating on the surface of infinite-depth irrotational, inviscid and incompressible fluid, NLSE was first derived by Zakharov [99] using the Hamiltonian formalism and then by Yuen and Lake [98] using the averaged Lagrangian method.

Under certain relationship between the parameters, when

$$-\omega'' a_{0,0,0} < 0, \tag{2.123}$$

NLSE admits exact solutions in the form of *solitons* that exist due to the balance of dispersion and nonlinearity and propagate without changing their shape and keeping their energy [19]. In this case, the uniform carrier wave is unstable with respect to long-wave modulations allowing for the formation of envelope solitons. This type of instability is known as the *modulational* or *Benjamin–Feir instability* [101] (it was discovered for the first time in optics by Bespalov and Talanov [3]). In the case of surface gravity waves, condition (2.123) holds at $kh \gtrsim 1.363$. At the bifurcation point $a_{0,0,0} = 0$ ($kh \approx 1.363$), when the modulational instability changes to stability, NLSE of form (2.120) is not sufficient to describe the wave train evolution since the leading nonlinear term vanishes. In this case, high-order nonlinear and nonlinear dispersive terms should be taken into account.

In the case of infinite depth, such a high-order NLSE (HONLSE) was first derived by Dysthe [23]. It includes the third-order dispersion and cubic nonlinear dispersive terms as well as an additional nonlinear dispersive term describing the input of the wave-induced mean flow. This equation is usually referred to as the fourth-order HONLSE to emphasise the contrast with the third-order NLSE. Janssen [35] rederived Dysthe's equation and corrected the sign at one of the nonlinear dispersive terms. Worthy of mention is also the paper by Lukomsky [52] who derived Dysthe's equation in a different way. Later Trulsen and Dysthe [89] extended the equation derived by Dysthe to broader bandwidth by including the forth- and fifth-order linear dispersion. Debsarma and Das [17] derived a yet more general HONLSE that is one order higher than the equation derived by Trulsen and Dysthe. Original Dysthe's equation was written for the first-harmonic envelope of velocity potential rather than of surface profile. In the case of standard NLSE, this difference is not essential because in that order the first-harmonic amplitudes of the velocity potential and surface displacement differ by a dimensional factor only, which is not true any longer in the HONLSE case, as was discussed by Hogan [33]. Keeping this in mind, Trulsen et al. [90] rewrote Dysthe's equation in terms of the first-harmonic envelope of surface profile while taking into account the linear dispersion to an arbitrary order. Dyachenko and Zakharov [20, 21] made a conformal mapping of the fluid domain into the lower half-plane to derive a counterpart of Dysthe's equation in new canonical variables.

In the case of finite depth, the effect of induced mean flow manifests itself in the third order, so that the NLSE is generally coupled to the equation for the induced mean flow [2]. However, Davey and Stewartson [14] showed that these coupled equations are equivalent to the single NLSE (2.120) derived by Hasimoto and Ono [31] (see also Ref. [77] for more details). On the other hand, such an equivalence is not preserved for high-order equations. The first attempt to derive a HONLSE in the case of finite depth was made by Johnson [36], but only for $kh \approx 1.363$, when the cubic NLSE term vanishes. The similar attempt was made by Kakutani and Michihiro [37] (see also a more formal derivation made later by Parkes [67]). Sedletsky [75, 76] used the multiple-scale technique to derive a single fourth-order HONLSE for the first-harmonic envelope of surface profile by using the additional power expansion (2.119) of the induced mean flow. This equation is the direct counterpart of Dysthe's equation written in terms of the first-harmonic envelope of surface profile [90] but for the case of finite depth. Gandzha et al. [26] reduced this HONLSE to dimensionless form:

$$\begin{aligned} \mathrm{i}u_\tau = -\mathrm{i}a_1 u_\chi + a_2 u_{\chi\chi} - a_{0,0,0}\,|u|^2 u \\ +\mathrm{i}\Big(a_3 u_{\chi\chi\chi} - a_{1,0,0}\,u_\chi |u|^2 - a_{0,0,1}\,u^2 u^*_\chi\Big), \end{aligned} \tag{2.124}$$

where * stands for complex conjugate, the subscripts next to u denote the partial derivatives, and

$$\tau = \frac{\omega}{c}\,t, \quad \chi = kx \tag{2.125}$$

are the dimensionless time and coordinate. The free surface displacement η is related to the amplitude u as follows

$$\begin{aligned} k\eta = \alpha_0 |u|^2 + \alpha_1 \operatorname{Re}\big(u \exp(\mathrm{i}\theta)\big) + 2\alpha_2 \operatorname{Re}\big(u^2 \exp(2\mathrm{i}\theta)\big), \\ \alpha_0 = \tfrac{2(1-\sigma^2)kh+\sigma}{c\nu}, \quad \alpha_1 = \tfrac{1}{\sqrt{c}}, \quad \alpha_2 = \tfrac{3-\sigma^2}{8c\sigma^3}, \\ \nu = \left(\sigma^2 - 1\right)^2 k^2h^2 - 2\sigma\left(\sigma^2+1\right)kh + \sigma^2, \end{aligned} \tag{2.126}$$

where $\theta = kx - \omega t = \chi - c\tau$ is the wave phase; c and a_1 are the dimensionless phase and group speeds, respectively,

$$c = -\frac{\omega}{k}\left(k\frac{\partial^2\omega}{\partial k^2}\right)^{-1} = -\frac{4\sigma^2}{\upsilon} > 0, \tag{2.127}$$

$$\begin{aligned} a_1 = \frac{ck}{\omega}c_g = -\frac{2\sigma}{\upsilon}\Big(\left(1-\sigma^2\right)kh + \sigma\Big) > 0, \\ \upsilon = \left(\sigma^2 - 1\right)\left(3\sigma^2 + 1\right)k^2h^2 - 2\sigma\left(\sigma^2 - 1\right)kh - \sigma^2. \end{aligned} \tag{2.128}$$

Equation (2.124) describes the evolution of the first-harmonic envelope of the surface profile with taking into account the third-order dispersion and cubic nonlinear dispersive terms. The contribution of the mean flow that is "hidden" in the values of the coefficients $a_{0,0,0}$, $a_{1,0,0}$, and $a_{0,0,1}$ that depend only on dimensionless depth kh. The notation of coefficients $a_n \ldots$ was selected such that the number of their indices corresponds to the order of nonlinearity in the corresponding term, and the index values correspond to the orders of derivatives present in that term (with complex conjugate derivatives having a separate index). The second- and third-order dispersion coefficients are expressed as follows

$$\begin{aligned} a_2 &= \frac{1}{2}, \quad a_3 = \frac{1}{12\sigma v}\sum_{p=0}^{3} a_3^{(p)}(kh)^p, \\ a_3^{(0)} &= -3\sigma^3, \quad a_3^{(1)} = -3\sigma^2\left(\sigma^2-1\right), \\ a_3^{(2)} &= -3\sigma\left(\sigma^2-1\right)\left(3\sigma^2+1\right), \\ a_3^{(3)} &= \left(\sigma^2-1\right)\left(15\sigma^4-2\sigma^2+3\right). \end{aligned} \tag{2.129}$$

The cubic nonlinear dispersion coefficients are

$$\begin{aligned} \begin{pmatrix} a_{1,0,0} \\ a_{0,0,1} \end{pmatrix} &= \frac{1}{32\sigma^5 v^2}\sum_{p=0}^{5}\begin{pmatrix} 2a_{1,0,0}^{(p)} \\ a_{0,0,1}^{(p)} \end{pmatrix}(kh)^p, \\ a_{1,0,0}^{(0)} &= -\sigma^5\left(17\sigma^4-94\sigma^2-27\right), \\ a_{0,0,1}^{(0)} &= \sigma^5\left(\sigma^4-2\sigma^2+9\right), \\ a_{1,0,0}^{(1)} &= \sigma^4\left(55\sigma^6-269\sigma^4-99\sigma^2-135\right), \\ a_{0,0,1}^{(1)} &= \sigma^4\left(19\sigma^6-137\sigma^4-47\sigma^2-27\right), \\ a_{1,0,0}^{(2)} &= -2\sigma^3\left(19\sigma^8-82\sigma^6-92\sigma^4+98\sigma^2-135\right), \\ a_{0,0,1}^{(2)} &= -2\sigma^3\left(43\sigma^8-296\sigma^6+230\sigma^4-32\sigma^2-9\right), \\ a_{1,0,0}^{(3)} &= 2\sigma^2\left(1-\sigma^2\right)\left(15\sigma^8-14\sigma^6-24\sigma^4+94\sigma^2-135\right), \\ a_{0,0,1}^{(3)} &= 2\sigma^2\left(\sigma^2-1\right)\left(51\sigma^8-272\sigma^6+158\sigma^4+8\sigma^2-9\right), \\ a_{1,0,0}^{(4)} &= \sigma\left(\sigma^2-1\right)^3\left(39\sigma^6+35\sigma^4-3\sigma^2-135\right), \\ a_{0,0,1}^{(4)} &= -\sigma\left(\sigma^2-1\right)^3\left(27\sigma^6-57\sigma^4-7\sigma^2-27\right), \\ a_{1,0,0}^{(5)} &= -\left(\sigma^2-1\right)^5\left(9\sigma^4+10\sigma^2-27\right), \\ a_{0,0,1}^{(5)} &= -\left(\sigma^2-1\right)^5\left(9\sigma^4-10\sigma^2+9\right). \end{aligned} \tag{2.130}$$

The latter coefficients were first derived in Refs. [75, 76] and then rederived in Ref. [78] with keeping one additional term in the multi-scale expansion of the mean flow (this term introduces a small correction to the expressions presented in Ref. [76]). Here, we used the expressions with this additional term taken into account.

Equation (2.124) has exact solutions in the form of solitons [25] and numerical solutions in the form of quasi-solitons [26]. Quasi-solitons are the solutions that have slowly varying amplitude, propagate with nearly constant speed, and possess the unique property of solitons to exist over long periods of time without breaking. Their speed was found to be higher than the speed of the exact NLSE solitons taken as initial conditions in computations [26].

Numerical solutions to Eq. (2.124) can be found by the *split-step Fourier technique*. For the sake of convenience, we proceed to the reference frame moving with speed a_1 (dimensionless group speed):

$$\xi = \chi - a_1\tau, \quad T = \tau. \tag{2.131}$$

The relationship between the derivatives in new and old variables is given by the formulas

$$\frac{\partial}{\partial\chi} = \frac{\partial\xi}{\partial\chi}\frac{\partial}{\partial\xi} + \frac{\partial T}{\partial\chi}\frac{\partial}{\partial T} = \frac{\partial}{\partial\xi},$$
$$\frac{\partial}{\partial\tau} = \frac{\partial\xi}{\partial\tau}\frac{\partial}{\partial\xi} + \frac{\partial T}{\partial\tau}\frac{\partial}{\partial T} = -a_1\frac{\partial}{\partial\xi} + \frac{\partial}{\partial T},$$

so that

$$u_\tau = -\mathrm{i}a_2u_{\xi\xi} + \mathrm{i}a_{0,0,0}|u|^2u + \left(a_3u_{\xi\xi\xi} - a_{1,0,0}u_\xi|u|^2 - a_{0,0,1}u^2u_\xi^*\right). \tag{2.132}$$

Consider the linear part of HONLSE (2.132):

$$u_\tau = -\mathrm{i}a_2u_{\xi\xi} + a_3u_{\xi\xi\xi}, \quad u = u(\xi, \tau). \tag{2.133}$$

Apply the Fourier transform to the function $u(\xi, \tau)$:

$$\widehat{u}(\kappa, \tau) = \frac{1}{2\pi}\int_{-\infty}^{\infty} u(\xi, \tau)\exp(-\mathrm{i}\kappa\xi)\mathrm{d}\xi \equiv \mathscr{F}_\kappa[u(\xi, \tau)]. \tag{2.134}$$

The inverse Fourier transform is written as

$$u(\xi, \tau) = \int_{-\infty}^{\infty} \widehat{u}(\kappa, \tau)\exp(\mathrm{i}\kappa\xi)\mathrm{d}\kappa \equiv \mathscr{F}_\xi^{-1}[\widehat{u}(\kappa, \tau)]. \tag{2.135}$$

The Fourier transforms of the derivatives of function $u(\xi, \tau)$ are expressed as

$$\widehat{(u_\xi)} = \mathrm{i}\kappa\widehat{u}, \quad \widehat{(u_{\xi\xi})} = -\kappa^2\widehat{u}, \quad \ldots, \quad \widehat{(u_{n\xi})} = (\mathrm{i}\kappa)^n\widehat{u}. \tag{2.136}$$

Hence, linear equation (2.133) takes the following form in the Fourier space:

$$\widehat{u}_\tau = \left(-\mathrm{i}a_2(\mathrm{i}\kappa)^2 + a_3(\mathrm{i}\kappa)^3\right)\widehat{u}, \quad \widehat{u}(0) \equiv \widehat{u}_0. \tag{2.137}$$

This ordinary differential equation can easily be integrated,

$$\widehat{u} = \widehat{u}_0 \exp\big((\mathrm{i}a_2\kappa^2 - \mathrm{i}a_3\kappa^3)\tau\big), \tag{2.138}$$

and the following solution for $u(\xi, \tau)$ is obtained:

$$u_L = \int_{-\infty}^{\infty} \widehat{u}_0 \exp\big((\mathrm{i}a_2\kappa^2 - \mathrm{i}a_3\kappa^3)\tau\big) \exp(\mathrm{i}\kappa\xi)\mathrm{d}\kappa. \tag{2.139}$$

Nonlinear equation (2.132) can be split into the linear and nonlinear parts:

$$u_\tau = -\mathrm{i}a_2 u_{\xi\xi} + \mathrm{i}a_{0,0,0}u|u|^2 + \left(a_3 u_{\xi\xi\xi} - a_{1,0,0}u_\xi|u|^2 - a_{0,0,1}u^2 u_\xi^*\right) \equiv (\mathscr{L} + \mathscr{N})\, u,$$

where

$$\mathscr{L} \equiv -\mathrm{i}a_2\partial_{\xi\xi} + a_3\partial_{\xi\xi\xi}, \tag{2.140}$$

$$\mathscr{N} \equiv \mathrm{i}a_{0,0,0}|u|^2 - a_{1,0,0}u_\xi u^* - a_{0,0,1}uu_\xi^* \tag{2.141}$$

are the linear and nonlinear operators, respectively. The semi-discretisation in time is performed as follows

$$\left.\frac{u(\xi, \tau + \Delta\tau) - u(\xi, \tau)}{\Delta\tau}\right|_{\Delta\tau\to 0} = (\mathscr{L} + \mathscr{N})\, u(\xi, \tau) \quad \Rightarrow$$
$$u(\xi, \tau + \Delta\tau) \approx u(\xi, \tau) + \Delta\tau(\mathscr{L} + \mathscr{N})u(\xi, \tau) \approx \mathrm{e}^{\Delta\tau(\mathscr{L}+\mathscr{N})}u(\xi, \tau),$$

and then the second-order Strang formula for noncommuting operators [83] is used:

$$\mathrm{e}^{\Delta\tau(\mathscr{L}+\mathscr{N})} \equiv S^{(2)}(\Delta\tau)$$
$$= \exp\left(\frac{\Delta\tau}{2}\mathscr{N}\right) \exp\left(\Delta\tau\mathscr{L}\right) \exp\left(\frac{\Delta\tau}{2}\mathscr{N}\right), \tag{2.142}$$
$$= \exp\left(\frac{\Delta\tau}{2}\mathscr{L}\right) \exp\left(\Delta\tau\mathscr{N}\right) \exp\left(\frac{\Delta\tau}{2}\mathscr{L}\right). \tag{2.143}$$

In our computations splitting (2.143) proved to be more accurate than (2.142). The linear part is integrated exactly using relation (2.139)

$$\mathrm{e}^{\Delta\tau\mathscr{L}}u(\xi, \tau) = \mathscr{F}_\xi^{-1}\left[\mathrm{e}^{\Delta\tau(-\mathrm{i}a_2(\mathrm{i}\kappa)^2 + a_3(\mathrm{i}\kappa)^3)}\mathscr{F}_\kappa[u(\xi, \tau)]\right], \tag{2.144}$$

and the nonlinear part is corrected at each step as follows

$$\mathrm{e}^{\Delta\tau\mathscr{N}}u(\xi, \tau) = \mathrm{e}^{\Delta\tau\left(\mathrm{i}a_{0,0,0}|u|^2 - a_{1,0,0}u_\xi u^* - a_{0,0,1}uu_\xi^*\right)}u(\xi, \tau). \tag{2.145}$$

Following Yoshida [97], a more accurate fourth-order splitting can be introduced as well:

$$S^{(4)}(\Delta\tau) = S^{(2)}\left(p_1\Delta\tau\right) S^{(2)}\left(p_0\Delta\tau\right) S^{(2)}\left(p_1\Delta\tau\right), \\ p_0 = -\frac{2^{1/3}}{2-2^{1/3}} \approx -1.70, \quad p_1 = \frac{1}{2-2^{1/3}} \approx 1.35. \tag{2.146}$$

For a more detailed description of the split-step Fourier technique, the reader can also refer to Ref. [62]. The results of computations of the solutions to Eq. (2.132) are presented in Refs. [25, 26].

2.5 Two-parameter Method for Describing the Nonlinear Evolution of Narrow-Band Wave Trains

In this section, we consider the evolution of narrow-band wave trains of finite amplitude in a nonlinear dispersive system which is described by a (1 + 1) Klein–Gordon equation with arbitrary polynomial nonlinearity:

$$u_{tt} - c^2 u_{xx} + \sum_{p=1}^{P} \alpha_p u^p = 0. \tag{2.147}$$

Here, u is an unknown twice differentiable function of the wave process, $0 < t < \infty$ is time, $-\infty < x < \infty$ is coordinate, c and α_p are arbitrary real constants ($\alpha_1 \neq 0$), and P is an arbitrary positive integer. Let the initial condition at $t = 0$ have the form

$$u(x, 0) = Q(x)\big(\exp(\mathrm{i}kx) + \exp(-\mathrm{i}kx)\big), \\ u_t(x, 0) = P(x)\big(\exp(\mathrm{i}kx) + \exp(-\mathrm{i}kx)\big),$$

where k is the carrier wave number.

In the multi-scale method, the unknown function $u(x, t)$ of coordinate and time is looked for in the form of asymptotic expansion in powers of a small nonlinearity parameter ε:

$$u(x, t) = \sum_{n=1}^{\infty} \varepsilon^n u^{(n)}(x, t). \tag{2.148}$$

The wave motion is classified into slow one and fast one by introducing different time scales and different spatial scales:

$$T_n \equiv \mu^n t, \quad X_n \equiv \mu^n x.$$

The derivatives with respect to time and coordinate are expanded into the following series:

$$\frac{\partial}{\partial t} = \sum_{n=0}^{\infty} \mu^n \frac{\partial}{\partial T_n}, \qquad \frac{\partial}{\partial x} = \sum_{n=0}^{\infty} \mu^n \frac{\partial}{\partial X_n}, \tag{2.149}$$

the times T_n and coordinates X_n being assumed to be independent variables. A principal drawback of this method lies in the fact that the parameters ε and μ with different physical meanings (the former characterising the smallness of nonlinearity, and the latter describing the slowness of temporal and spatial variations) are tentatively taken equal: $\varepsilon = \mu$. This admission produces the so-called secular terms in the equations for $u^{(n)}(x, t)$. Such terms, which indefinitely grow with time, are eliminated in each new order of ε by an appropriate choice of free parameters emerging in the solutions of the linear inhomogeneous wave equations derived from the original nonlinear equations for the function $u(x, t)$. This procedure is very awkward, and it is difficult to formulate in algorithmic form.

From here on we follow Ref. [56] to introduce a perturbative technique that was first proposed in Ref. [52] to reduce the original wave equation to a model equation for the wave train envelope (high-order nonlinear Schrödinger equation). The time derivative is expanded into an asymptotic series in two independent parameters that characterise the smallness of amplitudes (ε) and the slowness of their spatial variations (μ). In contrast to the multi-scale method and other perturbative methods in which these two parameters are taken equal, the two-parameter method produces no secular terms.

We look for a solution to Eq. (2.147) in the form of truncated Fourier series with variable coefficients:

$$u(x, t) = \sum_{n=-N_u}^{N_u} u_n(x, t) e^{in(\omega t - kx)}, \quad u_{-n} \equiv u_n^*, \tag{2.150}$$

where ω is the wave train carrier frequency, $N_u + 1$ is the number of harmonics taken into consideration, and * stands for complex conjugate. The same series can be written for all integer powers of the function u:

$$u^p(x, t) = \sum_{n=-pN_u}^{pN_u} (u^p)_n(x, t) e^{in(\omega t - kx)}, \quad p = \overline{2, P}, \tag{2.151}$$

where $(u^p)_{-n} \equiv (u^p)_n^*$. The coefficients $(u^p)_n$ can be expressed recurrently in terms of the coefficients u_n:

$$(u^p)_n = \sum_{n_1=\max(-N_u, n-(p-1)N_u)}^{\min(N_u, n+(p-1)N_u)} u_{n_1} (u^{p-1})_{n-n_1}.$$

The corresponding expansions of the derivatives are

$$u_{tt}(x,\,t) = \sum_{n=-N_u}^{N_u} \left((u_n)_{tt} + 2\mathrm{i}n\omega(u_n)_t - n^2\omega^2 u_n\right)\mathrm{e}^{\mathrm{i}n(\omega t - kx)}, \tag{2.152}$$

$$u_{xx}(x,\,t) = \sum_{n=-N_u}^{N_u} \left((u_n)_{xx} - 2\mathrm{i}nk(u_n)_x - n^2k^2 u_n\right)\mathrm{e}^{\mathrm{i}n(\omega t - kx)}. \tag{2.153}$$

Substituting (2.150)–(2.153) in (2.147) and equating the coefficients at the like powers of the function $\exp(\mathrm{i}(\omega t - kx))$, we obtain a system of nonlinear differential equations for the coefficients $u_n(x,\,t)$ ($n = \overline{0,\,N_u}$):

$$(u_n)_{tt} - c^2(u_n)_{xx} + 2\mathrm{i}n\left(\omega(u_n)_t + c^2k(u_n)_x\right) + \\ (n^2c^2k^2 - n^2\omega^2 + \alpha_1)u_n + \sum_{p=2}^{P} \alpha_p(u^p)_n = 0. \tag{2.154}$$

Linearisation of these equations at $n = 1$ gives the dispersion relation in the linear approximation:

$$\omega^2 = \alpha_1 + c^2k^2. \tag{2.155}$$

Generally, the system of equations (2.154) is by no means more simple than original equation (2.147). It can be simplified if solutions are looked for in a class of functions with narrow spectrum, $|\Delta k| \ll k$ (*quasi-monochromaticity* condition). In this case, the problem has a formal small parameter $\mu \sim |\Delta k|/k$, and the coefficients $u_n(x,\,t)$ can be regarded as slow functions of x and t. Let us introduce a slow coordinate $\xi = \mu x$ and go over to the variables $u_n = u_n(\mu x,\,t)$.

When there are no resonances between higher harmonics, the amplitudes of Fourier coefficients decrease with increasing number (*quasi-harmonicity* condition):

$$u_n \sim \varepsilon^n A, \quad n \geqslant 1, \quad u_0 \sim \varepsilon^2 A, \quad \varepsilon < 1, \tag{2.156}$$

where $u_1 \equiv \varepsilon A$. The parameter ε can be chosen as the second formal parameter, which is independent of the dispersion parameter μ in the general case. The use of two independent formal parameters is a distinctive feature of this approach as compared to the multi-scale method and other perturbative methods, where these parameters are not distinguished ($\varepsilon = \mu$). When these incomparable parameters are set equal, a perturbative procedure produces nonphysical secular terms.

In contrast to perturbative methods which use the expansions of form (2.148) and (2.149) to reduces Eq. (2.154) to evolution equations of NLSE type, we start from the most general explicit form of such an evolution equation. To this end, the time derivative $(u_1)_t \equiv \varepsilon A_t$ should be expressed in terms of the derivatives $(u_1)_{nx} \equiv \varepsilon\mu^n A_{n\xi}$ with respect to coordinate (designation $A_{n\xi}$ means the n-th derivative with respect to ξ) and all possible combinations of nonlinear terms $\varepsilon^{2n+1}A^{(n+1)}(A^*)^n$. Hence, the derivative A_t can be written as the following asymptotic expansion in terms of parameters ε and μ:

$$A_t = \mathrm{i}\sum_{n_0=0}^{\infty}(\mathrm{i}\mu)^{n_0}\Big(a_{n_0}A_{n_0\xi}+ \\ +\varepsilon^2\sum_{n_1=0}^{n_0}\sum_{n_2=0}^{n_1}a_{n_0-n_1,n_1-n_2,n_2}A_{(n_0-n_1)\xi}A_{(n_1-n_2)\xi}A^*_{n_2\xi}+O(\varepsilon^4)\Big). \tag{2.157}$$

Equation (2.157) is the general form of the evolution equation for the complex first-harmonic amplitude A. The unknown coefficients $a_{n\_}$ can be determined from Eqs. (2.154). To this end, the amplitudes of all other harmonics (u_0, u_2, u_3, ...) are expanded in terms of the amplitude of the first harmonic A in the same way as it is done in expansion (2.157):

$$u_0 = \varepsilon^2\sum_{n_0=0}^{\infty}(\mathrm{i}\mu)^{n_0}\Big(\sum_{n_1=0}^{n_0}b^{(0)}_{n_0-n_1,n_1}A_{(n_0-n_1)\xi}A^*_{n_1\xi} \\ +\varepsilon^2\sum_{n_1=0}^{n_0}\sum_{n_2=0}^{n_1}\sum_{n_3=0}^{n_2}b^{(0)}_{n_0-n_1,n_1-n_2,n_2-n_3,n_3} \\ \times A_{(n_0-n_1)\xi}A_{(n_1-n_2)\xi}A^*_{(n_2-n_3)\xi}A^*_{n_3\xi}+O(\varepsilon^4)\Big), \tag{2.158}$$

$$u_2 = \varepsilon^2\sum_{n_0=0}^{\infty}(\mathrm{i}\mu)^{n_0}\Big(\sum_{n_1=0}^{n_0}b^{(2)}_{n_0-n_1,n_1}A_{(n_0-n_1)\xi}A_{n_1\xi} \\ +\varepsilon^2\sum_{n_1=0}^{n_0}\sum_{n_2=0}^{n_1}\sum_{n_3=0}^{n_2}b^{(2)}_{n_0-n_1,n_1-n_2,n_2-n_3,n_3} \\ \times A_{(n_0-n_1)\xi}A_{(n_1-n_2)\xi}A_{(n_2-n_3)\xi}A^*_{n_3\xi}+O(\varepsilon^4)\Big), \tag{2.159}$$

$$u_3 = \varepsilon^3\sum_{n_0=0}^{\infty}(\mathrm{i}\mu)^{n_0}\Big(\sum_{n_1=0}^{n_0}\sum_{n_2=0}^{n_1}b^{(3)}_{n_0-n_1,n_1-n_2,n_2}A_{(n_0-n_1)\xi}A_{(n_1-n_2)\xi}A_{n_2\xi} \\ +\varepsilon^2\sum_{n_1=0}^{n_0}\sum_{n_2=0}^{n_1}\sum_{n_3=0}^{n_2}\sum_{n_4=0}^{n_3}b^{(3)}_{n_0-n_1,n_1-n_2,n_2-n_3\,n_3-n_4,n_4} \\ \times A_{(n_0-n_1)\xi}A_{(n_1-n_2)\xi}A_{(n_2-n_3)\xi}A_{(n_3-n_4)\xi}A^*_{n_4\xi}+O(\varepsilon^4)\Big), \quad \ldots . \tag{2.160}$$

The unknown coefficients $b^{(n)}_{n\_}$ are found along with the coefficients $a_{n\_}$ from the system of equations (2.154) by substituting expansions (2.157)–(2.160) and equating to zero the coefficients at the like powers of the products $\varepsilon^k\mu^m$ in different combinations ($A\_\ldots A^*\_\ldots$). In its essence, this procedure is similar to the method of undetermined coefficients. The coefficient calculation order and the general form of the expansions for A_t and u_n in arbitrary order of ε are given in Ref. [56]. The use of two parameters in anzats (2.157) is of key importance for the coefficient calculation procedure, since the expansions could not be split into linear-independent terms at $\varepsilon = \mu$.

It should be noted that the reduction of Eq. (2.147) with the second time derivative to Eq. (2.157) with the first time derivative puts a constraint on the initial condition for u_t. In this case, $u_t(x, 0)$ is a function of $u(x, 0)$ defined by formula (2.157).

The two-parameter expansions were programmed in symbolic form for an arbitrary order of μ and ε. The evolution equation for the complex amplitude of the first harmonic is

$$\begin{aligned}A_t = \mathrm{i}\Big(&(\mathrm{i}\mu)a_1 A_\xi + (\mathrm{i}\mu)^2 a_2 A_{\xi\xi} + (\mathrm{i}\mu)^3 a_3 A_{\xi\xi\xi} + (\mathrm{i}\mu)^4 a_4 A_{\xi\xi\xi\xi} + O(\mu^5)\\ &+ \varepsilon^2\Big[a_{0,0,0}A|A|^2 + (\mathrm{i}\mu)\ \left(a_{1,0,0}A_\xi|A|^2 + a_{0,0,1}A^2A_\xi^*\right)\\ &+ (\mathrm{i}\mu)^2\left(a_{2,0,0}A_{\xi\xi}|A|^2 + a_{1,1,0}A_\xi^2A^* + a_{1,0,1}|A_\xi|^2A + a_{0,0,2}A^2A_{\xi\xi}^*\right) + O(\mu^3)\Big]\\ &+ \varepsilon^4\Big[a_{0,0,0,0,0}A|A|^4 + (\mathrm{i}\mu)\ \left(a_{1,0,0,0,0}A_\xi|A|^4 + a_{0,0,0,1,0}A^2|A|^2A_\xi^*\right)\\ &+ (\mathrm{i}\mu)^2\left(a_{2,0,0,0,0}A_{\xi\xi}|A|^4 + a_{1,1,0,0,0}A_\xi^2|A|^2A^* + a_{1,0,0,1,0}|A_\xi|^2A|A|^2\right.\\ &\left.+ a_{0,0,0,2,0}A^2|A|^2A_{\xi\xi}^* + a_{0,0,0,1,1}A^3A_\xi^2\right) + O(\mu^3)\Big] + O(\varepsilon^6)\Big).\end{aligned} \tag{2.161}$$

In each term of this equation, the power of the formal parameter μ points to the overall order of the derivatives with respect to ξ, and the power of the formal parameter ε points to the nonlinearity order. These parameters disappear after going back to the original variables $u_1 = \varepsilon A$ and $x = \xi/\mu$. Taking into account dispersion relation (2.155), the coefficients $a\_$ can be written as (at $P = 5$)

$$a_0 = 0, \quad a_1 = \frac{c^2 k}{\omega}, \quad a_2 = \frac{c^2\alpha_1}{2\omega^3}, \quad a_3 = -\frac{c^4 k\alpha_1}{2\omega^5},$$

$$a_4 = \frac{\alpha_1 c^4(4c^2k^2 - \alpha_1)}{8\omega^7}, \quad a_n = \frac{1}{n!}\frac{\mathrm{d}^n\omega}{\mathrm{d}k^n};$$

$$a_{0,0,0} = \frac{3\alpha_3}{2\omega} - \frac{5\alpha_2^2}{3\omega\alpha_1}, \quad a_{1,0,0} = 2a_{0,0,1} = \frac{c^2k}{\omega^3}\left(\frac{10\alpha_2^2}{3\alpha_1} - 3\alpha_3\right);$$

$$a_{2,0,0} = \frac{c^2}{18\alpha_1\omega^5}\left(2c^2k^2\left(27\alpha_1\alpha_3 - 14\alpha_2^2\right) - \alpha_1\left(27\alpha_1\alpha_3 - 62\alpha_2^2\right)\right),$$

$$a_{1,1,0} = \frac{c^2}{36\alpha_1\omega^5}\left(4c^2k^2\left(27\alpha_1\alpha_3 - 28\alpha_2^2\right) - \alpha_1\left(27\alpha_1\alpha_3 - 38\alpha_2^2\right)\right),$$

$$a_{1,0,1} = \frac{c^2}{6\alpha_1\omega^5}\Big(4c^2k^2(9\alpha_1\alpha_3 - 4\alpha_2^2) - \alpha_1(9\alpha_1\alpha_3 - 34\alpha_2^2)\Big),$$

$$a_{0,0,2} = \frac{c^2}{6\alpha_1\omega^5}\Big(c^2k^2(9\alpha_1\alpha_3 + 2\alpha_2^2) + 12\alpha_1\alpha_2^2\Big);$$

$$a_{0,0,0,0,0} = \frac{1}{\omega}\Big(-\frac{335\alpha_2^4}{108\alpha_1^3} - \frac{25\alpha_2^4}{18\omega^2\alpha_1^2} + \frac{143\alpha_2^2\alpha_3}{12\alpha_1^2} + \frac{5\alpha_2^2\alpha_3}{2\omega^2\alpha_1} - \frac{9\alpha_3^2}{8\omega^2} + \frac{3\alpha_3^2}{16\alpha_1} - \frac{14\alpha_2\alpha_4}{\alpha_1} + 5\alpha_5\Big),$$

$$a_{1,0,0,0,0} = \frac{c^2k}{\omega^3}\Big(\frac{925\alpha_2^4}{108\alpha_1^3} + \frac{275\alpha_2^4}{18\omega^2\alpha_1^2} - \frac{421\alpha_2^2\alpha_3}{12\alpha_1^2} - \frac{55\alpha_2^2\alpha_3}{2\omega^2\alpha_1} + \frac{99\alpha_3^2}{8\omega^2} - \frac{9\alpha_3^2}{16\alpha_1} + \frac{42\alpha_2\alpha_4}{\alpha_1} - 15\alpha_5\Big),$$

$$a_{0,0,0,1,0} = \frac{c^2k}{\omega^3}\Big(\frac{295\alpha_2^4}{54\alpha_1^3} + \frac{100\alpha_2^4}{9\omega^2\alpha_1^2} - \frac{139\alpha_2^2\alpha_3}{6\alpha_1^2} - \frac{20\alpha_2^2\alpha_3}{\omega^2\alpha_1} + \frac{9\alpha_3^2}{\omega^2} - \frac{3\alpha_3^2}{8\alpha_1} + \frac{28\alpha_2\alpha_4}{\alpha_1} - 10\alpha_5\Big).$$

The expressions for the subsequent coefficients $a\_$ are too cumbersome to be presented in explicit form. The expressions for the complex amplitudes of other harmonics are given in Ref. [56].

Let us illustrate the evolution of a wave train envelope described by the equation of form (2.157). To this end, we rewrite original equation (2.147) in terms of dimensionless variables $\widetilde{x} \equiv kx$ and $\widetilde{t} \equiv ckt$:

$$\widetilde{u}_{\widetilde{t}\widetilde{t}} - \widetilde{u}_{\widetilde{x}\widetilde{x}} + \sum_{p=1}^{P} \widetilde{\alpha}_p \widetilde{u}^p = 0, \quad \widetilde{u} = \frac{u}{u_A}, \quad \widetilde{\alpha}_p = \frac{\alpha_p\, u_A^{p-1}}{(ck)^2}. \tag{2.162}$$

In this case, we have $\widetilde{c} = 1$, $\widetilde{k} = 1$, $\widetilde{\omega}^2 = \widetilde{\alpha}_1 + 1$, and u_A is a typical amplitude of the function u. As an example, let us consider the case $P = 3$ with $\widetilde{\alpha}_1 = 1$, $\widetilde{\alpha}_2 = 0$, and $\widetilde{\alpha}_3 = -1/6$. The values of these parameters correspond to the first two terms in the Taylor expansion of the function $\sin u$. Hereafter, the tildes over the dimensionless variables are omitted. The corresponding coefficients of evolution equation (2.161) are

$$a_1 = \frac{1}{\sqrt{2}}, \quad a_2 = \frac{1}{4\sqrt{2}}, \quad a_3 = -\frac{1}{8\sqrt{2}}, \quad a_4 = \frac{3}{64\sqrt{2}},$$

$$a_{0,0,0} = -\frac{1}{4\sqrt{2}}, \quad a_{1,0,0} = \frac{1}{4\sqrt{2}}, \quad a_{0,0,1} = \frac{1}{8\sqrt{2}},$$

$$a_{2,0,0} = a_{0,0,2} = -\frac{1}{16\sqrt{2}}, \quad a_{1,1,0} = a_{1,0,1} = -\frac{3}{16\sqrt{2}},$$

$$a_{0,0,0,0,0} = -\frac{1}{96\sqrt{2}}, \quad \dots . \tag{2.163}$$

Initially, we retain only those terms in Eq. (2.161) whose overall order of smallness with respect to the parameters ε and μ is no more than two. In this case we obtain a classical NLSE:

$$(u_1)_t = -a_1(u_1)_x - \mathrm{i}a_2(u_1)_{xx} + \mathrm{i}a_{0,0,0}\, u_1|u_1|^2. \tag{2.164}$$

It has an exact one-soliton solution at $a_2\, a_{0,0,0} < 0$:

$$u_1(x,\, t) = U_0\, \mathrm{sech}\big(K(x - x_0 - Vt)\big)\, \mathrm{e}^{\mathrm{i}\kappa x - \mathrm{i}\Omega t}. \tag{2.165}$$

Here, U_0, κ, Ω, and V are the soliton's complex amplitude, wave number, frequency and speed, respectively; x_0 is the soliton's arbitrary initial position, which is usually set equal to zero. By selecting $x_0 = \pm\mathrm{i}\frac{\pi}{2}$, the sech profile can be transformed to csch. Solutions of form (2.165) are called *bright solitons*. They were first derived in Ref. [102]. The following relationships between the soliton parameters can be established in this case:

$$K = |U_0|\sqrt{-\frac{a_{0,0,0}}{2a_2}}, \quad \Omega = \kappa a_1 + \left(K^2 - \kappa^2\right) a_2, \quad V = a_1 - 2\kappa a_2. \tag{2.166}$$

The parameters U_0 and κ are free parameters of the problem. The corresponding approximate solution of Eq. (2.162) is

$$u(x,\, t) = u_1(x,\, t)\exp\big(\mathrm{i}(\sqrt{2}\,t - x)\big) + u_1^*(x,\, t)\exp\big(-\mathrm{i}(\sqrt{2}\,t - x)\big). \tag{2.167}$$

To analyse the effect of high-order dispersive terms on the shape of one-soliton solution (2.165), we consider HONLSE

$$\begin{aligned}(u_1)_t = &- a_1(u_1)_x - \mathrm{i}a_2(u_1)_{xx} + a_3(u_1)_{xxx} + \mathrm{i}a_4(u_1)_{xxxx} + \mathrm{i}a_{0,0,0}\, u_1|u_1|^2 \\ &- a_{1,0,0}(u_1)_x|u_1|^2 - a_{0,0,1}\, u_1^2(u_1^*)_x - \mathrm{i}a_{2,0,0}(u_1)_{xx}|u_1|^2 - \mathrm{i}a_{1,1,0}(u_1)_x^2\, u_1^* \\ &- \mathrm{i}a_{1,0,1}|(u_1)_x|^2 u_1 - \mathrm{i}a_{0,0,2}\, u_1^2(u_1^*)_{xx} + \mathrm{i}a_{0,0,0,0,0}\, u_1|u_1|^4\end{aligned} \tag{2.168}$$

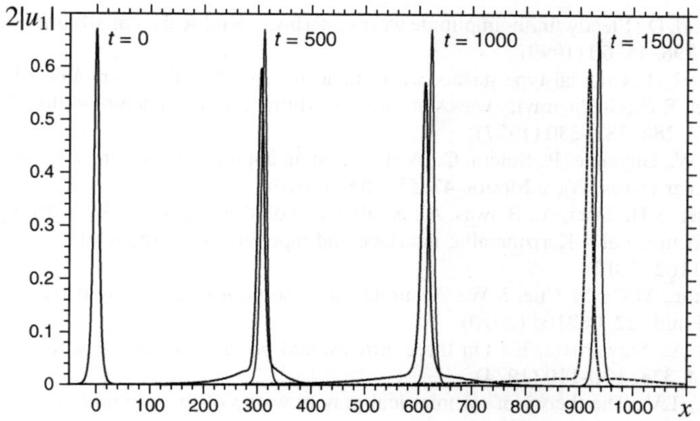

Fig. 2.5 Evolution of the wave train envelope which at the initial moment $t = 0$ is given by function (2.165). (*Solid curve*) exact one-soliton NLSE solution (2.164), (*dashed curve*) numerical solution of HONLSE (2.168)

with the coefficients defined by formulas (2.163). The initial condition is chosen in the form of function (2.165) with

$$U_0 = \frac{1}{10}\sqrt{\frac{2}{|a_{0,0,0}|}}, \quad \kappa = \frac{1}{10\sqrt{|a_2|}},$$

and $x_0 = 0$. Figure 2.5 shows the evolution of such an envelope. To solve Eq. (2.168) numerically, we used the split-step Fourier method described in the previous section. High-order dispersive terms are seen to affect the amplitude, shape and velocity of the soliton solution. This solution shows a typical quasi-soliton behaviour discussed in more detail in Ref. [26].

References

1. Benney, D.J., Newell, A.C.: The propagation of nonlinear wave envelopes. J. Math. Phys. **46**, 133–139 (1967)
2. Benney, D.J., Roskes, G.J.: Wave instabilities. Stud. Appl. Math. **48**(4), 377–385 (1969)
3. Bespalov, V., Talanov, V.: Filamentary structure of light beams in nonlinear liquids. JETP Lett. **3**, 307–309 (1966)
4. Brinch-Nielsen, U., Jonsson, I.G.: Fourth-order evolution equations and stability analysis for Stokes waves on arbitrary water depth. Wave Motion **8**, 455–472 (1986)
5. Chappelear, J.E.: Direct numerical calculation of wave properties. J. Geophys. Res. **66**(2), 501–508 (1961)
6. Chen, B., Saffman, P.G.: Numerical evidence for the existence of new types of gravity waves of permanent form on deep water. Stud. Appl. Math. **62**, 1–21 (1980)
7. Chen, M., Nash, J.M., Patton, C.E.: A numerical study of nonlinear Schrödinger equation solutions for microwave solitons in magnetic thin films. J. Appl. Phys. **73**, 3906–3909 (1993)

8. Clamond, D.: Steady finite-amplitude waves on a horizontal seabed of arbitrary depth. J. Fluid Mech. **398**, 45–60 (1999)
9. Clamond, D.: Cnoidal-type surface waves in deep water. J. Fluid Mech. **489**, 101–120 (2003)
10. Cokelet, E.D.: Steep gravity waves in water of arbitrary uniform depth. Philos. Trans. R. Soc. Lond. A **286**, 183–230 (1977)
11. Craig, W., Guyenne, P., Sulem, C.: A Hamiltonian approach to nonlinear modulation of surface water waves. Wave Motion **47**, 552–563 (2010)
12. Crutcher, S.H., Osei, A., Biswas, A.: Nonlinear evolution equations for surface plasmons for nano-focusing at a Kerr/metallic interface and tapered waveguide. Opt. Laser Technol. **44**, 1156–1162 (2012)
13. Dallaston, M.C., McCue, S.W.: Accurate series solutions for gravity-driven Stokes waves. Phys. Fluids **22**, 082104 (2010)
14. Davey, A., Stewartson, K.: On three-dimensional packets of surface waves. Proc. R. Soc. Lond. A **338**, 101–110 (1974)
15. Davies, T.V.: The theory of symmetrical gravity waves of finite amplitude. I. Proc. R. Soc. Lond. A **208**, 475–486 (1951)
16. De, S.C.: Contributions to the theory of Stokes waves. Proc. Camb. Philos. Soc. **51**, 713–736 (1955)
17. Debsarma, S., Das, K.P.: A higher-order nonlinear evolution equation for broader bandwidth gravity waves in deep water. Phys. Fluids **17**, 104101 (2005)
18. Dias, F., Kharif, C.: Nonlinear gravity and capillary-gravity waves. Annu. Rev. Fluid Mech. **31**, 301–346 (1999)
19. Dodd, R.K., Eilbeck, J.C., Gibbon, J.D., Morris, H.C.: Solitons and Nonlinear Wave Equations. Academic Press, London (1984)
20. Dyachenko, A.I., Zakharov, V.E.: Compact equation for gravity waves on deep water. JETP Lett. **93**(12), 701–705 (2011)
21. Dyachenko, A.I., Zakharov, V.E.: A dynamic equation for water waves in one horizontal dimension. Eur. J. Mech. B Fluids **32**, 17–21 (2012)
22. Dyachenko, S.A., Lushnikov, P.M., Korotkevich, A.O.: Branch cuts of Stokes wave on deep water. Part I: numerical solution and Padé approximation. Stud. Appl. Math. **137**(4), 419–472 (2016)
23. Dysthe, K.B.: Note on a modification to the nonlinear Schrödinger equation for application to deep water waves. Proc. R. Soc. Lond. A **369**, 105–114 (1979)
24. Gandzha, I.S., Lukomsky, V.P.: On water waves with a corner at the crest. Proc. R. Soc. Lond. A **463**, 1597–1614 (2007)
25. Gandzha, I.S., Sedletsky, Y.V.: Bright and dark solitons on the surface of finite-depth fluid below the modulation instability threshold. Phys. Lett. A **381**, 1784–1790 (2017)
26. Gandzha, I.S., Sedletsky, Yu.V., Dutykh, D.S.: High-order nonlinear Schrödinger equation for the envelope of slowly modulated gravity waves on the surface of finite-depth fluid and its quasi-soliton solutions. Ukr. J. Phys. **59**(12), 1201–1215 (2014)
27. Garabedian, P.R.: Surface waves of finite depth. J. d'Analyse Mathématique **14**, 161–169 (1965)
28. Gramstad, O., Trulsen, K.: Hamiltonian form of the modified nonlinear Schrödinger equation for gravity waves on arbitrary depth. J. Fluid Mech. **670**, 404–426 (2011)
29. Grant, M.A.: The singularity at the crest of a finite amplitude progressive Stokes wave. J. Fluid Mech. **59**, 257–262 (1973)
30. Hamming, R.W.: Numerical Methods for Scientists and Engineers. McGraw-Hill, New York (1973)
31. Hasimoto, H., Ono, H.: Nonlinear modulation of gravity waves. J. Phys. Soc. Jpn. **33**, 805–811 (1972)
32. Havelock, T.H.: Periodic irrotational waves of finite height. Proc. R. Soc. Lond. A **95**, 38–51 (1918)
33. Hogan, S.J.: The potential form of the fourth-order evolution equation for deep-water gravity-capillary waves. Phys. Fluids **29**, 3479–3480 (1986)

34. Infeld, E., Rowlands, G.: Nonlinear Waves. Solitons and Chaos. Cambridge University Press, Cambridge (2000)
35. Janssen, P.A.E.M.: On a fourth-order envelope equation for deep-water waves. J. Fluid Mech. **126**, 1–11 (1983)
36. Johnson, R.S.: On the modulation of water waves in the neighbourhood of $kh \approx 1.363$. Proc. R. Soc. Lond. A **357**, 131–141 (1977)
37. Kakutani, T., Michihiro, K.: Marginal state of modulational instability—Note on Benjamin-Feir instability. J. Phys. Soc. Jpn. **52**, 4129–4137 (1983)
38. Karabut, E.A.: An approximation for the highest gravity waves on water of finite depth. J. Fluid Mech. **372**, 45–70 (1989)
39. Lamb, H.: Hydrodynamics. Dover Publications, New York (1945)
40. Landau, L.D., Lifshitz, E.M.: Course of theoretical physics. Fluid Mechanics, vol. 6. Pergamon Press, Oxford (1959)
41. Levi-Cività, T.: Détermination rigoureuse des ondes permanentes d'ampleur finie. Math. Ann. **93**, 264–314 (1925)
42. Loitsyanskii, L.G.: Mekhanika zhidkosti i gaza (Fluid and Gas Mechanics). Nauka, Moscow (1987)
43. Longuet-Higgins, M.S.: Integral properties of periodic gravity waves of finite amplitude. Proc. R. Soc. Lond. A **342**, 157–174 (1975)
44. Longuet-Higgins, M.S.: Some new relations between Stokes's coefficients in the theory of gravity waves. J. Inst. Math. Appl. **22**, 261–273 (1978)
45. Longuet-Higgins, M.S.: New integral relations for gravity waves of finite amplitude. J. Fluid Mech. **149**, 205–215 (1984)
46. Longuet-Higgins, M.S.: Bifurcation in gravity waves. J. Fluid Mech. **151**, 457–475 (1985)
47. Longuet-Higgins, M.S.: Lagrangian moments and mass transport in Stokes waves Part 2. Water of finite depth. J. Fluid Mech. **186**, 321–336 (1988)
48. Longuet-Higgins, M.S.: On an approximation to the limiting Stokes wave in deep water. Wave Motion **45**, 770–775 (2008)
49. Longuet-Higgins, M.S., Fox, M.J.H.: Theory of the almost-highest wave: the inner solution. J. Fluid Mech. **80**(4), 721–741 (1977)
50. Longuet-Higgins, M.S., Fox, M.J.H.: Theory of the almost-highest wave. Part 2. Matching and analytic extension. J. Fluid Mech. **85**(4), 769–786 (1978)
51. Longuet-Higgins, M.S., Fox, M.J.H.: Asymptotic theory for the almost-highest solitary wave. J. Fluid Mech. **317**, 1–19 (1996)
52. Lukomskiĭ, V.P.: Modulational instability of gravity waves in deep water with allowance for nonlinear dispersion. JETP **81**(2), 306–310 (1995)
53. Lukomsky, V.P., Gandzha, I.S., Lukomsky, D.V.: Computational analysis of the almost-highest waves on deep water. Comput. Phys. Commun. **147**(1–2), 548–551 (2002)
54. Lukomsky, V.P., Gandzha, I.S., Lukomsky, D.V.: Steep sharp-crested gravity waves on deep water. Phys. Rev. Lett. **89**(16), 164502 (2002)
55. Lukomsky, V.P., Gandzha, I.S.: Fractional Fourier approximations for potential gravity waves on deep water. Nonlinear Processes in Geophysics **10**(6), 599–614 (2003)
56. Lukomsky, V.P., Gandzha, I.S.: Two-parameter method for describing the nonlinear evolution of narrow-band wave trains. Ukr. J. Phys. **54**(1–2), 207–215 (2009)
57. Lushnikov, P.M., Dyachenko, S.A., Silantyev, D.A.: New Conformal Mapping for Adaptive Resolving of the Complex Singularities of Stokes Wave (2017). arXiv:1703.06343v1
58. Maklakov, D.V.: Almost-highest gravity waves on water of finite depth. Euro. J. Appl. Math. **13**, 67–93 (2002)
59. McCormick, M.: Ocean Wave Energy Conversion. Wiley, New York (1981)
60. McLean, J.W.: Instabilities of finite-amplitude gravity waves on water of finite depth. J. Fluid Mech. **114**, 331–341 (1982)
61. Michell, J.H.: The highest waves in water. Philos. Mag. **5**(36), 430–437 (1893)
62. Muslu, G.M., Erbay, H.A.: Higher-order split-step Fourier schemes for the generalized nonlinear Schrödinger equation. Math. Comput. Simul. **67**, 581–595 (2005)

63. Nekrasov, A.I.: O volne Stoksa (On the Stokes wave). Izv. Ivanovo-Voznesensk. Politekhn. Inst. **2**, 81–89 (1919)
64. Norman, A.C.: Expansions for the shape of maximum amplitude Stokes waves. J. Fluid Mech. **66**, 261–265 (1974)
65. Olfe, D.B., Rottman, J.W.: Some new highest-wave solutions for deep-water waves of permanent form. J. Fluid Mech. **100**(4), 801–810 (1980)
66. Onorato, M., Residori, S., Bortolozzo, U., Montina, A., Arecchi, F.T.: Rogue waves and their generating mechanisms in different physical contexts. Phys. Rep. **528**(2), 47–89 (2013)
67. Parkes, E.J.: The modulation of wealkly non-linear dispersive waves near the marginal state of instability. J. Phys. A: Math. Gen. **20**, 2025–2036 (1987)
68. Rayleigh, Lord: On waves. Philos. Mag. **5**(1), 257–279 (1876)
69. Rienecker, M.M., Fenton, J.D.: A Fourier approximation method for steady water waves. J. Fluid Mech. **104**, 119–137 (1981)
70. Saffman, P.G.: Long wavelength bifurcation of gravity waves on deep water. J. Fluid Mech. **101**(3), 567–581 (1980)
71. Sasaki, K., Murakami, T.: Irrotational progressive surface gravity waves near the limiting height. J. Ocean. Soc. Jpn. **29**, 94–105 (1973)
72. Schwartz, L.W.: Computer extension and analytic continuation of Stokes' expansion for gravity waves. J. Fluid Mech. **62**(3), 553–578 (1974)
73. Schwartz, L.W., Fenton, J.D.: Strongly nonlinear waves. Ann. Rev. Fluid Mech. **14**, 39–60 (1982)
74. Sclavounos, P.D.: Wave energy density and flux. In: MIT Lecture Open Course in Mechanical Engineering No. 2.24 Ocean wave interaction with ships and offshore energy systems. http://ocw.mit.edu/courses/mechanical-engineering/2-24-ocean-wave-interaction-with-ships-and-offshore-energy-systems-13-022-spring-2002/lecture-notes/lecture4.pdf (2002). Assessed 27 Mar 2017
75. Sedletsky, Yu.V.: Higher order nonlinear Schrödinger equation for Stokes waves on the surface of a water layer of arbitrary depth. Ukr. J. Phys. **48**(1), 82–95 (2003)
76. Sedletsky, Yu.V.: The fourth-order nonlinear Schrödinger equation for the envelope of Stokes waves on the surface of a finite-depth fluid. JETP **97**(1), 180–193 (2003)
77. Sedletsky, Yu.V.: Variational approach to the derivation of the Davey–Stewartson system. Fluid Dyn. Res. **48**, 015506 (2016)
78. Slunyaev, A.V.: A high-order nonlinear envelope equation for gravity waves in finite-depth water. JETP **101**, 926–941 (2005)
79. Stoker, J.J.: Water Waves: the Mathematical Theory with Applications. Wiley, New York (1992)
80. Stokes, G.G.: On the theory of oscillatory waves. Camb. Phil. Soc. Trans. **8**, 441–455 (1847)
81. Stokes, G.G.: Supplement to a paper on the theory of oscillatory waves. Math. Phys. Pap. **1**, 314–326 (1880)
82. Stokes, G.G.: Considerations relative to the greatest height of oscillatory irrotational waves which can be propagated whithout change of form. Math. Phys. Pap. **1**, 225–228 (1880)
83. Strang, G.: On the construction and comparison of difference schemes. SIAM J. Numer. Anal. **5**(3), 506–517 (1968)
84. Sveshnikov, A.G., Tikhonov, A.N.: Teoriya funktsii kompleksnoi peremennoi (The Theory of Functions of a Complex Variable). Nauka, Moscow (1979)
85. Tanaka, M.: The stability of steep gravity waves. J. Phys. Soc. Jpn. **52**(9), 3047–3055 (1983)
86. Tanaka, M.: The stability of solitary waves. Phys. Fluids **29**(3), 650–655 (1986)
87. Tanaka, M.: A method of calculation of periodic steady waves in water of finite depth: the manual to the program `Stokes_finite.f` (1996)
88. Tao, L., Song, H., Chakrabarti, S.: Nonlinear progressive waves in water of finite depth–an analytic approximation. Coast. Eng. **54**, 825–834 (2007)
89. Trulsen, K., Dysthe, K.B.: A modified nonlinear Schrödinger equation for broader bandwidth gravity waves on deep water. Wave Motion **24**, 281–289 (1996)

90. Trulsen, K., Kliakhandler, I., Dysthe, K.B., Velarde, M.G.: On weakly nonlinear modulation of waves on deep water. Phys. Fluids **12**(10), 2432–2437 (2000)
91. Turitsyn, S.K., Bale, B.G., Fedoruk, M.P.: Dispersion-managed solitons in fibre systems and lasers. Phys. Rep. **521**, 135–203 (2012)
92. Whitham, G.B.: Linear and Nonlinear Waves. Wiley, New York (1974)
93. Williams, J.M.: Limiting gravity waves in water of finite depth. Philos. Trans. R. Soc. Lond. A **302**, 139–188 (1981)
94. Wilton, J.R.: On deep water waves. Philos. Mag. **6**(27), 385–394 (1914)
95. Yamada, H.: Highest waves of permanent type on the surface of deep water. Rep. Res. Inst. Appl. Mech. Kyushu Univ. **5**, 37–52 (1957)
96. Yamada, H., Shiotani, T.: On the highest water waves of permanent type. Bull. Disaster Prev. Res. Inst. Kyushu Univ. **18**, 1–22 (1968)
97. Yoshida, H.: Construction of high order symplectic operators. Phys. Lett. A **150**, 262–268 (1990)
98. Yuen, H.C., Lake, B.M.: Nonlinear deep water waves: theory and experiment. Phys. Fluids **18**(8), 956–960 (1975)
99. Zakharov, V.E.: Stability of periodic waves of finite amplitude on the surface of a deep fluid. J. Appl. Mech. and Tech. Phys. **9**(2), 190–194 (1968)
100. Zakharov, V.E., Kuznetsov, E.A.: Solitons and collapses: two evolution scenarios of nonlinear wave systems. Physics-Uspekhi **55**(6), 535–556 (2012)
101. Zakharov, V.E., Ostrovsky, L.A.: Modulation instability: the beginning. Physica D **238**, 540–548 (2009)
102. Zakharov, V.E., Shabat, A.B.: Exact theory of two-dimensional self-focusing and one-dimensional self-modulation of waves in nonlinear media. Sov. Phys. JETP **34**(1), 62–69 (1972)
103. Zhao, H., Song, Zh., Li, L., Kong, J.: On the Fourier approximation method for steady water waves. Acta Oceanol. Sin. **33**(5), 37–47 (2014)
104. Zufiria, J.A.: Non-symmetric gravity waves on water of infinite depth. J. Fluid Mech. **181**, 17–39 (1987)

[illegible] of light pulse. Physica D [illegible] Conf. [illegible]

Chapter 3
Ray Method of Investigating the Wave Evolution over Arbitrary Topography

3.1 On the Geometric Theory of Water Wave Refraction

It is well known in optics that the accuracy of the geometric approximation of wave refraction is limited by the phenomenon of wave diffraction [4]. The linearised equations describing the evolution of water waves can be written as [52]

$$\nabla^2 \Phi = 0 \quad \text{for} \quad -h(\boldsymbol{r}) < y < 0, \tag{3.1}$$

$$\Phi_{tt} + g\Phi_y = 0 \quad \text{at} \quad y = 0, \tag{3.2}$$

$$\Phi_y + h_x \Phi_x + h_z \Phi_z = 0 \quad \text{at} \quad y = -h(\boldsymbol{r}), \tag{3.3}$$

where g is the acceleration due to gravity, $\Phi(\boldsymbol{r}, y, t)$ is the velocity potential, $h(\boldsymbol{r})$ is the water depth, $\boldsymbol{r} = \{x, z\}$ is the position vector in the horizontal xz plane, and $y = 0$ is a free surface. The velocity potential can be presented as

$$\Phi = \Phi_0(\boldsymbol{r}, y) \exp\Big(-\mathrm{i}\big(\omega t - \Theta(\boldsymbol{r}, y)\big)\Big). \tag{3.4}$$

The phase function $\Theta(\boldsymbol{r}, y)$ is supposed to be real, but all the transformations remain valid when Θ has an imaginary part.

When the vertical variation Φ_0 over a slightly sloping bottom is assumed to be such as in the case of horizontal bottom, Eqs. (3.1)–(3.4) result in the following approximation:

$$\Phi_0(\boldsymbol{r}, y) = \frac{ga(\boldsymbol{r})}{\omega} \frac{\cosh\big(k(h + y)\big)}{\cosh(kh)}, \tag{3.5}$$

where $a(\boldsymbol{r})$ is the amplitude of the local wave elevation and the wave number k is determined from the dispersion relation

I.T. Selezov et al., *Wave Propagation and Diffraction*,
Foundations of Engineering Mechanics, DOI 10.1007/978-981-10-4923-1_3

$$\omega^2 = gk \tanh(kh). \tag{3.6}$$

Formulas (3.1)–(3.6) are valid in the framework of linear theory for the case of fluid with constant depth independently of amplitude variations. These results still remain approximately valid for the case of variable depth, provided that the bottom slope is small.

It follows from expression (3.5) that in the domain occupied by fluid the following relation holds true in the linear approximation:

$$\left(\nabla_{\parallel}\Theta\right)^2 = k^2 + \frac{a_{xx} + a_{zz}}{a}, \tag{3.7}$$

where $\nabla_{\parallel} = \frac{\partial}{\partial r}$ is the planar Hamilton operator. Since the wave number is the absolute value of the wave vector whose components in any direction characterise the rate of phase variation in that direction, relation (3.7) characterises the contribution of the wave number k with the correction for the amplitude variations. Introducing the parameter

$$\delta = \frac{a_{xx} + a_{zz}}{k^2 a}, \tag{3.8}$$

we can rewrite relation (3.7) as

$$\left(\nabla_{\parallel}\Theta\right)^2 = k^2\left(1 + \delta\right). \tag{3.9}$$

Then an expression for the phase speed c follows from (3.6)–(3.9):

$$c = \left(\frac{g}{k}\tanh(kh)\right)^{1/2}(1 + \delta)^{1/2}. \tag{3.10}$$

When $\delta = 0$, Eq. (3.9) is reduced to the eikonal equation from geometrical optics that forms the basic set of equations together with Eq. (3.6) [43]. The parameter δ characterises the effect of diffraction. It provides an estimate as to how the geometrical optics approximation deviates from the linearised theory. The effect of parameter δ on the wave refraction was estimated in Ref. [30] based on the shallow water theory and from the experiments. The accuracy of the theory in the geometrical optics approximation was shown to deteriorate with increasing δ.

The above analysis can be presented in a more general form [22]. Consider the case when the scales of spatial inhomogeneities $h(\boldsymbol{r})$ and temporal velocity variations are large as compared to the wavelength and wave period. In this case, the wave can be assumed to be *locally planar*, and its profile can be written as

$$\eta(\boldsymbol{r}, t) = a(\boldsymbol{r}, t)\exp\left(\mathrm{i}\theta(\boldsymbol{r}, t)\right),$$

where the amplitude a is a slowly varying function and the phase θ (or *eikonal*) is an approximately linear function of $\boldsymbol{r}$ and t (the velocity potential and other wave parameters can be written in the same way). In such cases, the wave vector $\boldsymbol{k}$ and the wave frequency ω can be defined at each point $\{\boldsymbol{r}, t\}$ such that

$$\boldsymbol{k} = \nabla_{\parallel} \theta, \quad \omega = -\frac{\partial \theta}{\partial t}. \tag{3.11}$$

Thus, the wave vector field $\boldsymbol{k}(\boldsymbol{r}, t)$ proves to be potential and the following *equation of wave number conservation* holds true:

$$\frac{\partial \boldsymbol{k}}{\partial t} + \nabla_{\parallel} \omega = 0. \tag{3.12}$$

Since there is one wave with $\theta = \boldsymbol{k}(\boldsymbol{r} - \boldsymbol{c}\, t)$, $\boldsymbol{c}$ being the wave phase velocity vector aligned with the vector $\boldsymbol{k}$, the number of waves per unit length is equal to $|\boldsymbol{k}|$, to within a $\pi/2$ factor. In this case, the parameter $\omega = \boldsymbol{k} \cdot \boldsymbol{c}$ is proportional to the wave number flux. The frequency ω is determined by dispersion relation (3.6) at $h = h(\boldsymbol{r})$. Substituting the dispersion relation

$$\omega(\boldsymbol{r}, t) = \omega\big(\boldsymbol{k}(\boldsymbol{r}, t)\big)$$

into Eq. (3.12), we get

$$\frac{\partial k_i}{\partial t} + \frac{\partial \omega}{\partial k_j} \frac{\partial k_j}{\partial r_i} = 0, \quad i = 1, 2,$$

with the summation being held over the paired index. Taking into account that

$$\nabla_{\parallel} \times \boldsymbol{k} = 0 \quad \Leftrightarrow \quad \frac{\partial k_j}{\partial r_i} = \frac{\partial k_i}{\partial r_j},$$

the equation of wave number conservation can be rewritten as [20]

$$\frac{\partial \boldsymbol{k}}{\partial t} + \boldsymbol{c}_g \cdot \nabla_{\parallel} \boldsymbol{k} = 0, \tag{3.13}$$

where

$$\boldsymbol{c}_g = \frac{\partial \omega}{\partial \boldsymbol{k}} = \frac{\omega}{2k} \left(1 + \frac{2kh}{\sinh(2kh)}\right) \frac{\boldsymbol{k}}{k} \tag{3.14}$$

is the wave group velocity by definition. In fact, Eq. (3.13) is a version of a more general conservation law

$$\frac{\mathrm{d}\boldsymbol{k}}{\mathrm{d}t} = 0, \tag{3.15}$$

where the total derivative operator is written as

$$\frac{\mathrm{d}}{\mathrm{d}t} = \frac{\partial}{\partial t} + \boldsymbol{c}_g \cdot \nabla_{\parallel}. \tag{3.16}$$

This operator can be regarded as translation along the group velocity vector [20]. The corresponding path is called the *wave ray*, and the approximation of the geometric refraction theory is called the *ray theory*.

The ray theory has much in common with the Hamiltonian mechanics of a point particle: the eikonal θ is similar to the *action* S, the wave vector $\boldsymbol{k}$ to the momentum $\boldsymbol{p}$, the frequency ω to the *Hamiltonian* $\mathscr{H}$ and rays to particle trajectories. Therefore, the ray equations are the analogues of the Hamilton equations [20], namely,

$$\frac{\partial \boldsymbol{k}}{\partial t} = -\frac{\partial \omega}{\partial \boldsymbol{r}}, \quad \frac{\partial \boldsymbol{r}}{\partial t} = \frac{\partial \omega}{\partial \boldsymbol{k}} \equiv \boldsymbol{c}_g. \tag{3.17}$$

Substituting ω into Eq. (3.17) leads to the following system of ordinary differential equations that define the ray coordinates and wave vector components:

$$\frac{\mathrm{d}\boldsymbol{r}}{\mathrm{d}s} = \frac{\boldsymbol{k}}{k}, \quad \frac{\mathrm{d}\boldsymbol{k}}{\mathrm{d}s} = -\frac{gk^2 \nabla_{\parallel} h}{2c_g\, \omega \cosh^2(kh)}, \tag{3.18}$$

where we used the notation $s = \int_{t_0}^{t} c_g \mathrm{d}t$.

The ray method approximation is equivalent to the geometrical optics approximation or to the quasi-classical approximation (WKB method) [20]. The ray theory application to the geometrical optics of waveguides is considered in Ref. [13].

3.2 Wave Refraction over Inhomogeneous Bottom

The development of refraction theory stems from the works of I. Newton and A.N. Krylov on astronomical refraction [18]. An extensive approximate theory of water wave refraction was developed by V.V. Shuleikin [44, 45]. These studies were based on the Fermat principle stating that the wave ray in an inhomogeneous medium always selects the path corresponding to the minimum consumption of wave energy.

Consider a problem of the refraction of harmonic waves coming from a deep water to a shallow coastal zone that is characterised by a variable depth. Solving such problems in the framework of the equations pertaining to the theory of waves in a finite-depth fluid poses significant challenges even in the shallow-water approximation. Therefore, a simplified approach is often used that is based on the ray method [2, 14, 17, 23, 24, 28, 46–48]. Such simplified techniques were also considered in Refs. [25, 37–40, 50, 60]. To this end, small-amplitude waves are considered under the assumption that there is no energy dissipation and shear flows and that the bottom surface is characterised by a small curvature. The effects of wave reflection and diffraction are negligible in this case, with the formula for the phase velocity pertaining to the constant-depth fluid being approximately valid. This approach allows an approximate pattern of the refracted field to be constructed everywhere in the domain under consideration prior to the stage of wave breaking.

3.2.1 *Equations of the Ray Method*

The equations of the ray method are based on the Fermat principle, according to which the wavefront travels the path $\delta s = c\,\delta t$ along the wave ray over the minimum time δt. Written in parametric form $x = x(t)$, $z = z(t)$, it satisfies the equations

$$\begin{aligned} \frac{\mathrm{d}x}{\mathrm{d}t} &= c\cos\vartheta, \\ \frac{\mathrm{d}z}{\mathrm{d}t} &= c\sin\vartheta, \end{aligned} \tag{3.19}$$

where the phase speed of wave propagation is set in the flat bottom approximation:

$$c = \left(\frac{g}{k}\tanh\big(kh(x,z)\big)\right)^{1/2}, \tag{3.20}$$

$k = \frac{2\pi}{\lambda}$ being the wave number, $h(x, z)$ being the fluid depth.

Taking into account a relationship between the phase speed and wavelength λ,

$$c = \frac{\lambda}{T} = \frac{\omega}{k}, \tag{3.21}$$

expression (3.20) can be rewritten as

$$c = \frac{g}{\omega}\tanh\big(kh(x,z)\big), \tag{3.22}$$

where $T = \frac{2\pi}{\omega}$ is the wave period.

Note that while writing equations (3.19), we use the term wave *ray* to describe the so-called wave *orthogonal*—a curve normal to the wave surface point. It has already been mentioned in the previous section that in a more general sense the wave ray is a curve whose direction in each point coincides with the wave group speed vector. However, in the stationary case when there are no shear flows or wind, the rays and orthogonals coincide [47]. In the presence of shear flows (or in a reference frame moving relative to the bottom), the wave ray and orthogonal are two different curves that should be distinguished [16]. Keeping this remark in mind, we will identify wave rays with wave orthogonals in this chapter.

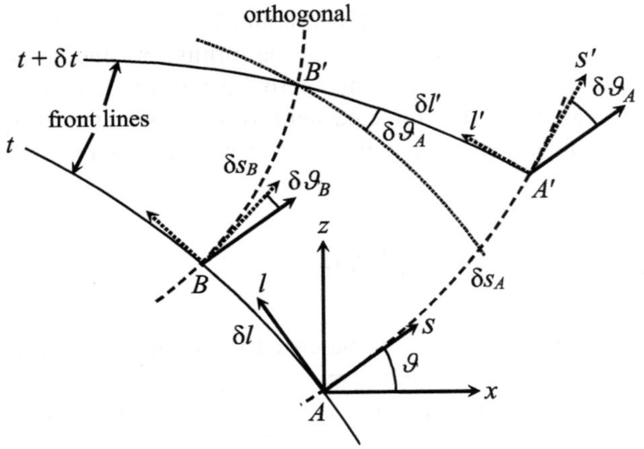

Fig. 3.1 Refraction: a wavefront section δl that travels a distance δs in the horizontal plane along the wave ray undergoes rotation by an angle $\delta\vartheta$

The equation for the angle ϑ between the ray and x-axis can be found from the following considerations [8, 14]. The deviation of angle $\vartheta(s, l)$ in any point of the wavefront is determined by the following relation for the total differential:

$$\delta\vartheta = \frac{\partial\vartheta}{\partial x}\,\mathrm{d}x + \frac{\partial\vartheta}{\partial y}\,\mathrm{d}y = \frac{\partial\vartheta}{\partial s}\,\delta s + \frac{\partial\vartheta}{\partial l}\,\delta l, \tag{3.23}$$

where δs is the translation along the wave ray and δl is the translation along the wavefront. Consider the translation of an infinitesimal element of the wavefront $\delta l = AB$ in a time equal to δt (Fig. 3.1). The distances travelled by the points A and B along the rays are

$$\delta s_A = AA' = c\ \delta t, \tag{3.24}$$

$$\delta s_B = BB' = (c + \delta c)\ \delta t = \left(c + \frac{\partial c}{\partial l}\,\delta l\right)\delta t. \tag{3.25}$$

Then the angle deviation at the point A' can be determined as

$$\delta\vartheta_A = \frac{\delta s_A - \delta s_B}{\delta l'} = -\frac{\partial c}{\partial l}\frac{\delta l}{\delta l'}\,\delta t.$$

By adopting $\delta l = \delta l'$ in the first order of smallness, we finally get

$$\frac{\delta\vartheta_A}{\delta t} = -\frac{\partial c}{\partial l} \tag{3.26}$$

or, taking into account (3.25) and (3.23) at $\delta l = 0$,

$$\frac{\partial\vartheta}{\partial s} = -\frac{1}{c}\frac{\partial c}{\partial l}. \tag{3.27}$$

By going over to the limit $\delta t \to 0$, equation (3.26) can be rewritten as

$$\frac{d\vartheta}{dt} = -\frac{\partial c}{\partial l} = -\left(\frac{\partial c}{\partial x}\frac{\partial x}{\partial l} + \frac{\partial c}{\partial z}\frac{\partial z}{\partial l}\right) = \sin\vartheta\,\frac{\partial c}{\partial x} - \cos\vartheta\,\frac{\partial c}{\partial z}. \tag{3.28}$$

Equations (3.19), (3.28) and dispersion relation (3.20) form the closed system of equations of the ray method [53]. The derivatives along the wavefront and ray are expressed in terms of the rotation matrix by the angle ϑ:

$$\frac{\partial}{\partial l} = -\sin\vartheta\frac{\partial}{\partial x} + \cos\vartheta\frac{\partial}{\partial z}, \tag{3.29}$$

$$\frac{\partial}{\partial s} = \cos\vartheta\frac{\partial}{\partial x} + \sin\vartheta\frac{\partial}{\partial z}. \tag{3.30}$$

Note that the coordinates l and s are curvilinear, so that the cross derivative difference is not equal to zero:

$$\begin{aligned}\frac{\partial^2}{\partial l\,\partial s} - \frac{\partial^2}{\partial s\,\partial l} &= \frac{\partial}{\partial l}\left(\cos\vartheta\frac{\partial}{\partial x} + \sin\vartheta\frac{\partial}{\partial z}\right)\\ &- \frac{\partial}{\partial s}\left(-\sin\vartheta\frac{\partial}{\partial x} + \cos\vartheta\frac{\partial}{\partial z}\right) = \frac{\partial\vartheta}{\partial x}\frac{\partial}{\partial x} + \frac{\partial\vartheta}{\partial z}\frac{\partial}{\partial z}.\end{aligned} \tag{3.31}$$

Therefore, taking into account relation (3.23), we get a second-order differential equation for the angle ϑ:

$$\frac{\partial^2\vartheta}{\partial l\,\partial s} - \frac{\partial^2\vartheta}{\partial s\,\partial l} = \left(\frac{\partial\vartheta}{\partial x}\right)^2 + \left(\frac{\partial\vartheta}{\partial z}\right)^2 = \left(\frac{\partial\vartheta}{\partial l}\right)^2 + \left(\frac{\partial\vartheta}{\partial s}\right)^2. \tag{3.32}$$

3.2.2 Ray Separation Factor

The degree of refraction is determined by the degree of ray separation. Let us derive an equation for calculating the deviation of the arc distance δl between the rays. To this end, we define the deviation of angle ϑ at the point B (Fig. 3.1):

$$\delta\vartheta_B = \frac{\delta l' - \delta l}{\delta s_B} = \frac{\delta\beta}{\delta s_B}\,\delta l_\infty, \tag{3.33}$$

where the parameter

$$\beta = \frac{\delta l}{\delta l_\infty} \tag{3.34}$$

is called the ray separation factor [47], δl_∞ is the distance between the rays at infinite depth, and the parameter

$$K_p = \frac{1}{\sqrt{\beta}}, \tag{3.35}$$

is called the refraction coefficient [24, 55].

For the ray separation factor β, one can derive a second-order differential equation, which was first obtained by Munk and Arthur [23] and then reproduced by Skovgaard and Jonsson [47] (see also [11]). Indeed, the first-order equation can be obtained from relation (3.33) with taking into account (3.23) at $\delta s = 0$:

$$\frac{1}{\beta}\frac{\partial \beta}{\partial s} = \frac{\partial \vartheta}{\partial l}. \tag{3.36}$$

Next using Eq. (3.32) with substituted derivatives (3.27) and (3.36), we get

$$\frac{1}{\beta}\frac{\partial^2 \beta}{\partial s^2} + \frac{1}{c}\frac{\partial^2 c}{\partial l^2} = 0. \tag{3.37}$$

The second term can be transformed using relations (3.29) and (3.36):

$$\frac{1}{c}\frac{\partial^2 c}{\partial l^2} = \frac{1}{c}\frac{\partial}{\partial l}\left(-\sin\vartheta\frac{\partial c}{\partial x} + \cos\vartheta\frac{\partial c}{\partial z}\right) =$$
$$\frac{1}{c}\left(-\sin\vartheta\frac{\partial}{\partial l}\frac{\partial c}{\partial x} + \cos\vartheta\frac{\partial}{\partial l}\frac{\partial c}{\partial z} - \cos\vartheta\frac{\partial c}{\partial x}\frac{\partial \vartheta}{\partial l} - \sin\vartheta\frac{\partial c}{\partial z}\frac{\partial \vartheta}{\partial l}\right) =$$
$$\frac{1}{c}\left(\sin^2\vartheta\frac{\partial^2 c}{\partial x^2} + \cos^2\vartheta\frac{\partial^2 c}{\partial z^2} - 2\sin\vartheta\cos\vartheta\frac{\partial^2 c}{\partial x \partial z} - \right.$$
$$\left.\left(\cos\vartheta\frac{\partial c}{\partial x} + \sin\vartheta\frac{\partial c}{\partial z}\right)\frac{1}{\beta}\frac{\partial \beta}{\partial s}\right).$$

Then Eq. (3.37) takes the final form:

$$\frac{\partial^2 \beta}{\partial s^2} + P_1(s)\frac{\partial \beta}{\partial s} + Q_1(s)\beta = 0, \tag{3.38}$$

where

$$P_1(s) = -\frac{1}{c}\left(\cos\vartheta\frac{\partial c}{\partial x} + \sin\vartheta\frac{\partial c}{\partial z}\right),$$

$$Q_1(s) = \frac{1}{c}\left(\sin^2\vartheta\frac{\partial^2 c}{\partial x^2} - \sin(2\vartheta)\frac{\partial^2 c}{\partial x\,\partial z} + \cos^2\vartheta\frac{\partial^2 c}{\partial z^2}\right).$$

Using the relation $\mathrm{d}s = c\,\mathrm{d}t$, Eq. (3.38) can be reduced to an equation where the time t is an independent variable [46]:

$$\frac{\mathrm{d}^2\beta}{\mathrm{d}t^2} + P_1(t)\frac{\mathrm{d}\beta}{\mathrm{d}t} + Q_1(t)\,\beta = 0. \tag{3.39}$$

Here, the functions $P_1(t)$ and $Q_1(t)$ can be determined by the formulas

$$P_1(t) = -2\left(\cos\vartheta\frac{\partial c}{\partial x} + \sin\vartheta\frac{\partial c}{\partial z}\right),$$

$$Q_1(t) = c\left(\sin^2\vartheta\frac{\partial^2 c}{\partial x^2} - \sin(2\vartheta)\frac{\partial^2 c}{\partial x\,\partial z} + \cos^2\vartheta\frac{\partial^2 c}{\partial z^2}\right),$$

keeping in mind that $x = x(t)$ and $z = z(t)$. In deriving Eq. (3.39), one needs to take into account that

$$\frac{\partial^2\beta}{\partial s^2} = \frac{\partial}{\partial s}\left(\frac{\mathrm{d}\beta}{c\mathrm{d}t}\right) = \frac{1}{c^2}\frac{\partial^2\beta}{\partial s^2} - \frac{1}{c^2}\frac{\partial c}{\partial s}\frac{\mathrm{d}\beta}{\mathrm{d}t}$$

and to use relation (3.30).

By introducing two auxiliary variables $\beta = b_1$ and $\frac{d\beta}{dt} = b_2$, Eq. (3.39) can be reduced to a system of first-order equations:

$$\frac{\mathrm{d}b_1}{\mathrm{d}t} = b_2, \quad \frac{\mathrm{d}b_2}{\mathrm{d}t} = -P(t)\,b_2 - Q(t)\,b_1. \tag{3.40}$$

Since the wave speed $c = c(t)$ is a function of fluid depth and the fluid depth in its turn is a function of coordinates, $h = h(x, z)$, the wave speed derivatives can be written as

$$\frac{\partial c}{\partial x} = \frac{\mathrm{d}c}{\mathrm{d}h}\frac{\partial h}{\partial x}, \quad \frac{\partial c}{\partial z} = \frac{\mathrm{d}c}{\mathrm{d}h}\frac{\partial h}{\partial z},$$

$$\frac{\partial^2 c}{\partial x^2} = \frac{\mathrm{d}^2 c}{\mathrm{d}h^2}\left(\frac{\partial h}{\partial x}\right)^2 + \frac{\mathrm{d}c}{\mathrm{d}h}\frac{\partial^2 h}{\partial x^2}, \quad \frac{\partial^2 c}{\partial z^2} = \frac{\mathrm{d}^2 c}{\mathrm{d}h^2}\left(\frac{\partial h}{\partial z}\right)^2 + \frac{\mathrm{d}c}{\mathrm{d}h}\frac{\partial^2 h}{\partial z^2},$$

$$\frac{\partial^2 c}{\partial x\partial z} = \frac{\mathrm{d}^2 c}{\mathrm{d}h^2}\frac{\partial h}{\partial x}\frac{\partial h}{\partial z} + \frac{\mathrm{d}c}{\mathrm{d}h}\frac{\partial^2 h}{\partial x\partial z}.$$

The phase speed derivatives with respect to h can be found by differentiating relation (3.22). In doing so, one needs to keep in mind that in the ray method the wave period $T = \frac{2\pi}{\omega}$ is supposed to be independent of time t and depth h in all the instances except for those when the wave travels over considerable distances [21]. Therefore, the derivative with respect to h in relation (3.22) should be taken only from the factor with hyperbolic tangent [14]:

$$\frac{\mathrm{d}c}{\mathrm{d}h} = \frac{gT}{2\pi}\frac{1}{\cosh^2(kh)}\left(k + \frac{\mathrm{d}k}{\mathrm{d}h}h\right) = \frac{2c}{\sinh(2kh)}\left(k + \frac{\mathrm{d}k}{\mathrm{d}h}h\right).$$

The derivative $\frac{dk}{dh}$ can be found by differentiating expression (3.21):

$$\frac{\mathrm{d}k}{\mathrm{d}h} = -\frac{k}{c}\frac{\mathrm{d}c}{\mathrm{d}h},$$

whence, after some algebraic transformations, we finally get

$$\frac{\mathrm{d}c}{\mathrm{d}h} = \frac{c}{h}\frac{R}{(1+R)},$$

where

$$R = \frac{2kh}{\sinh(2kh)}.$$

The second derivative can be calculated in the same way:

$$\frac{\mathrm{d}^2c}{\mathrm{d}h^2} = \frac{c}{h(1+R)^2}\left(\frac{\mathrm{d}R}{\mathrm{d}h} - \frac{R}{h}\right).$$

Taking into account that

$$\frac{\mathrm{d}R}{\mathrm{d}h} = \frac{R}{h(1+R)}\big(1 - R\cosh(2kh)\big),$$

we finally obtain

$$\frac{\mathrm{d}^2c}{\mathrm{d}h^2} = -\frac{c}{h^2}\frac{R^2}{(1+R)^3}\big(1 + \cosh(2kh)\big).$$

Further, we introduce the following designations:

$$x_1 = x, \quad x_2 = z, \quad x_3 = \vartheta, \quad x_4 = b_1, \quad x_5 = b_2,$$

$$k_1 = \frac{\partial c}{\partial x} = \frac{\mathrm{d}c}{\mathrm{d}h}\frac{\partial h}{\partial x}, \quad k_2 = \frac{\partial c}{\partial z}, \quad k_3 = \frac{\partial^2 c}{\partial x \partial z},$$

$$k_4 = \frac{\partial^2 c}{\partial x^2} = \frac{\mathrm{d}^2c}{\mathrm{d}h^2}\left(\frac{\partial h}{\partial x}\right)^2 + \frac{\mathrm{d}c}{\mathrm{d}h}\frac{\partial^2 h}{\partial x^2}, \quad k_5 = \frac{\partial^2 c}{\partial z^2}.$$

Then relations (3.19) and (3.28) lead to a system of first-order differential equations:

$$
\begin{cases}
\dfrac{dx_1}{dt} = c\cos x_3, \quad \dfrac{dx_2}{dt} = c\sin x_3, \\
\dfrac{dx_3}{dt} = k_1 \sin x_3 - k_2 \cos x_3, \quad \dfrac{dx_4}{dt} = x_5, \\
\dfrac{dx_5}{dt} = -P(t)x_5 - Q(t)x_4,
\end{cases}
\tag{3.41}
$$

which is integrated at the given initial conditions

$$
x_i(0) = (x_i)_0, \quad i = 1, \ldots, 5. \tag{3.42}
$$

The solution to the initial-value problem (3.41) and (3.42) determines the ray direction and the ray separation factor.

Next we introduce the dimensionless variables as follows

$$
x_1^\star = \frac{x_1}{\lambda_\infty}, \quad x_2^\star = \frac{x_2}{\lambda_\infty}, \quad h^\star = \frac{h}{\lambda_\infty}, \quad t^\star = \frac{t}{T}, \quad c^\star = \frac{c}{c_\infty}, \quad \lambda^\star = \frac{\lambda}{\lambda_\infty},
$$

where λ_∞ and c_∞ are the wavelength and phase speed at infinite depth, $\lambda_\infty = \frac{g}{2\pi}T^2$, $c_\infty = \frac{\lambda_\infty}{T} = \frac{g}{2\pi}T$, T being the wave period. Asterisks will be omitted henceforth.

3.2.3 *Wave Amplitude*

The wave amplitude a can be defined from the equation of wave action conservation

$$
\frac{\partial}{\partial t}\left(\frac{E}{\omega}\right) + \nabla_\parallel \cdot \left(\mathbf{c}_g \frac{E}{\omega}\right) = 0, \tag{3.43}
$$

where E is the wave energy density, which in the framework of linear theory is given by the following relation:

$$
E = \frac{1}{2}\rho g a^2. \tag{3.44}
$$

Note that wave action is a quantity that is defined as a ratio of the spectral energy to the wave frequency [5]. The derivation of Eq. (3.43) can be found in [58] based on the variation principle in the linear approximation and also in [20]. Reference [1] provides the most general approach without assuming the wave perturbations to be small.

In the stationary case and in the case of constant frequency ω (when the wave period does not change in the horizontal plane), Eq. (3.43) can be rewritten as

$$
\nabla_\parallel \cdot (\mathbf{c}_g E) = 0. \tag{3.45}
$$

Let us select a ray tube that occupies the space between two adjacent rays and integrate this equation over the ray tube area:

$$\iint_\sigma \nabla_\| \cdot (\boldsymbol{c}_g E)\,\mathrm{d}\sigma = \iint_\sigma \frac{\partial}{\partial s}(c_g E)\,\mathrm{d}s\,\mathrm{d}l = 0.$$

Here, l is the direction along the wavefront connecting the points of equal phase and s is the direction along the ray (perpendicular to the wavefront line). When integrating, we made an assumption that the gradient direction coincides with the normal to the wavefront, although it is not always the case in a more general formulation (see Ref. [17]). After integrating and taking into account relation (3.44), we come to an expression for the wave amplitude transformation:

$$c_g\, a^2\, l = \left(c_g\, a^2\, l\right)_{s=0}, \tag{3.46}$$

where l is the distance between the ray points of equal phase in the ray tube. This relation then leads to an expression for the relative wave amplitude:

$$\frac{a}{a_{h=\infty}} = K_T K_p, \tag{3.47}$$

where the refraction coefficient K_p is determined by formula (3.35) and the transformation coefficient K_T is determined by the following relation with the expression (3.14) for the group speed taken into account:

$$K_T = \left(\frac{c_g|_{h=\infty}}{c_g}\right)^{1/2} = \left[\frac{\left(\tanh(kh)\right)_\infty (1+R)_\infty}{\tanh(kh)\,(1+R)}\right]^{1/2}. \tag{3.48}$$

This coefficient is normalised so that the relative wave height is equal to unity at the initial moment, i.e. at infinite depth. Thus, the wave height variation along one ray can be determined by solving the system of equations (3.41) and using formulas (3.47) and (3.48).

3.2.4 Specifying the Bottom Relief in Analytical Form. The Case of Linear Depth Variation

Here we investigate the wave refraction in the case of flat sloping bottom when the fluid depth varies according to the formula

$$h\left(x_1\right) = \kappa x_1 + h_0, \tag{3.49}$$

where $\kappa = -\frac{1}{m}$ and m is the slope ratio. The origin of the coordinate system Ox_1x_2 is set on deep water, the Ox_1 axis being perpendicular to the shore horizon, which forms a straight line in the undisturbed state.

The system of equations (3.41) with (3.49) taken into account can be simplified and takes the following form:

$$\begin{cases} \frac{\mathrm{d}x_1}{\mathrm{d}t} = c\cos x_3, & \frac{\mathrm{d}x_2}{\mathrm{d}t} = c\sin x_3, \\ \frac{\mathrm{d}x_3}{\mathrm{d}t} = k_1 \sin x_3, & \frac{\mathrm{d}x_4}{\mathrm{d}t} = x_5, \\ \frac{\mathrm{d}x_5}{\mathrm{d}t} = -P(t)\,x_5 - Q(t)\,x_4, \end{cases} \tag{3.50}$$

where $P(t) = -2k_1x_3$, $Q(t) = ck_4\sin^2 x_3$,

$$k_1 = \frac{c}{h}\frac{R}{(1+R)}\kappa, \quad k_4 = -\frac{c}{h^2}\frac{R^2}{(1+R)^3}\left(1+\cosh(2kh)\right)\kappa^2.$$

The initial conditions corresponding to the wavefront curvature at the moment $t = 0$ are as follows:

$$x_1(0) = x_2(0) = x_5(0) = 0, \ \ x_3(0) = \frac{\pi}{2a_1}, \ \ x_4(0) = 1, \tag{3.51}$$

where a_1 is some adjustable constant.

The system of equations (3.50) with initial conditions (3.51) can be integrated numerically by the Runge–Kutta method, so that the wave height might be found afterwards.

The calculations were performed at different slope ratios m equal to 5, 30, 50 and 70 and for the initial angles ϑ_0 equal to $\pi/8$, $\pi/4$, $3\pi/8$, and $\pi/2$. The corresponding a_1 values in (3.51) are 4, 2, 4/3 and 1. The parameter h_0 is selected according to the deep-water criterion $kh \geq 3$ [19], whence we have $h_0 = 3/(2\pi)$.

Figure 3.2 shows some results of calculations in the form of wave rays for shallow water, when $h/\lambda = 1/20$ and $c = \sqrt{gh}$. The dashed curves represent the rays corresponding to the exact solution found by Sager in Ref. [34]. The ray deviation towards the shore line is seen to slow down substantially with increasing the initial angle.

Figure 3.3 demonstrates the relative wave height as a function of depth for several initial angles ϑ_0 equal to $\pi/8$, $\pi/4$, $3\pi/8$ and for $m = 5$. The relative wave height a/a_∞, which is equal to unity at the initial moment, first decreases and then starts to increase, so that there is such a depth when the relative wave height is at its minimum. This result is in accord with the field observations and theoretical predictions described in Refs. [7, 31, 41, 56]. The dashed curve shows the variations of the relative wave height according to the data from Ref. [7].

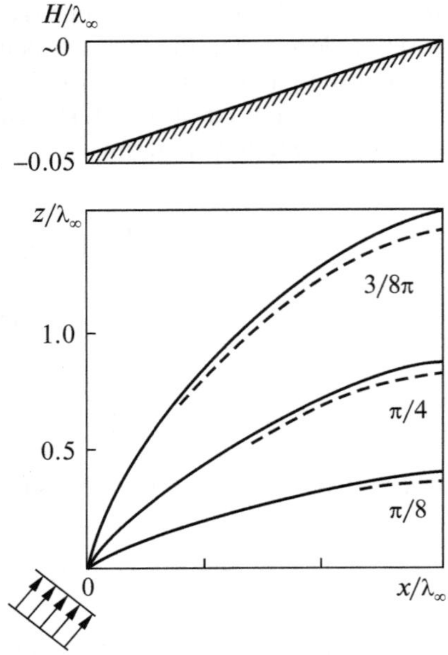

Fig. 3.2 Wave rays over a flat sloping bottom (*dashed curves* correspond to the exact solution)

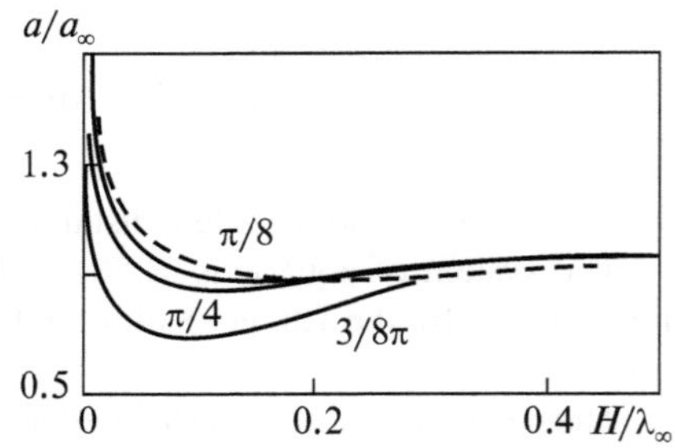

Fig. 3.3 Relative wave height as a function of depth for different initial ray angles and $m = 5$

3.2.5 *The Case of Parabolic Depth Variation*

Consider the wave refraction in the case of parabolic depth variation:

$$h(x_1) = \frac{\kappa^2}{h_0} x_1^2 + h_0,$$

where $\kappa = -1/m$.

In this case, the system of equations (3.41) takes form (3.50), with the coefficients determined by the following formulas:

$$P(t) = -2k_1 \cos x_3, \quad Q(t) = ck_4 \sin^2 x_3,$$

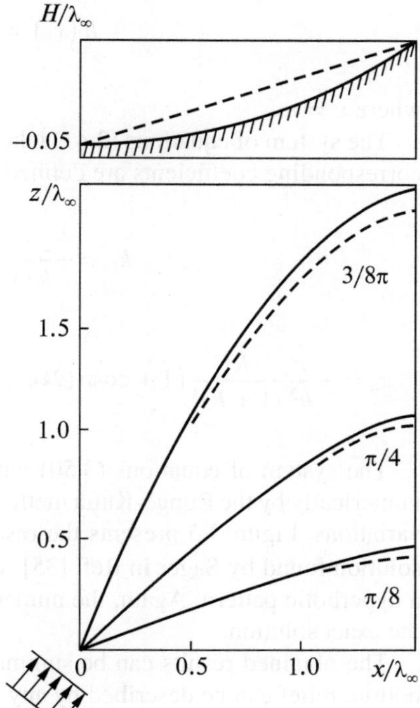

Fig. 3.4 Wave rays over a parabolic bottom (*dashed curves* correspond to the exact solution)

$$k_1 = \frac{c}{h}\frac{R}{(1+R)}\left(\frac{2x_1\kappa^2}{h}\right),$$

$$k_4 = -\frac{c}{h^2}\frac{R^2}{(1+R)^3}(1+\cosh(2kh))\left(\frac{2x_1\kappa^2}{h_0}\right)^2 + \frac{c}{h}\frac{R}{(1+R)}\frac{2\kappa^2}{h_0}.$$

The initial conditions at $t = 0$ are defined by formula (3.42). The system of equations (3.50) with initial conditions can be integrated numerically by the Runge-Kutta method, as in the case of linear depth variation. Figure 3.4 presents the results of calculations as compared to the exact solution found by Sager in Ref. [36] for the case of long waves with $c = \sqrt{gh}$, when the depth h varies according to a parabolic pattern. The numerical results provide a good approximation to the exact solution.

3.2.6 The Case of Hyperbolic Depth Variation

Consider the wave refraction when the fluid depth varies according to a hyperbolic pattern:

$$h\left(x_1\right) = \frac{h_0\left(\kappa h_0 - x_1\right)}{\kappa h_0 + x_1},$$

where $\kappa = m$.

The system of equations (3.41) takes form (3.50), as in the previous case, and the corresponding coefficients are defined as

$$k_1 = -\frac{c}{h}\frac{R}{(1+R)}\frac{2\kappa h_0^2}{\left(\kappa h_0 + x_1\right)^2},$$

$$k_4 = -\frac{c}{h^2}\frac{R^2}{(1+R)^3}\left(1+\cosh(2kh)\right)\left(\frac{2\kappa h_0^2}{(\kappa h_0 + x_1)^2}\right)^2 + \frac{c}{h}\frac{R}{(1+R)}\frac{4\kappa h_0^2}{\left(\kappa h_0 + x_1\right)^3}.$$

The system of equations (3.50) with initial conditions (3.42) can be integrated numerically by the Runge-Kutta method, as in the cases of linear and parabolic depth variations. Figure 3.5 presents the results of calculations as compared to the exact solution found by Sager in Ref. [35], when $c = \sqrt{gh}$ and depth varies according to a hyperbolic pattern. Again, the numerical results provide a good approximation to the exact solution.

The obtained results can be summarised in the following conclusion: when the bottom relief can be described by any functional dependence, the refraction pattern and wave height variations along each ray can be calculated based on the analytical formula for the depth variations.

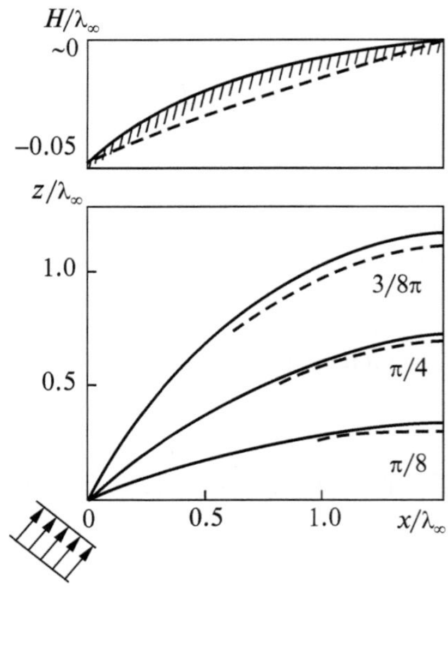

Fig. 3.5 Wave rays over a hyperbolic bottom (*dashed curves* correspond to the exact solution)

3.2.7 Specifying the Fluid Depth in Tabular Form

The bottom relief can be presented in analytical form only in some particular cases. It can hardly be described by a specific function in real conditions. In this case, the above-described numerical procedure can be used for an arbitrary relief when the fluid depth can be presented in tabular form [50].

To specify the depth h and to calculate its derivatives, the refraction zone under consideration is subdivided into squared areas (Fig. 3.6). The depth values are specified in the nodes of the mesh obtained in this way, and the corresponding derivatives are calculated by the formulas

$$\begin{aligned}
\frac{\partial h}{\partial x} &= \frac{1}{2d}\big(h\,(i+1,j) - h\,(i-1,j)\big), \quad \frac{\partial h}{\partial z} = \frac{1}{2d}\big(h\,(i,j+1) - h\,(i,j-1)\big),\\
\frac{\partial^2 h}{\partial x \partial z} &= \frac{1}{4d^2}\big(h\,(i+1,j+1) - h\,(i-1,j+1) - h\,(i+1,j-1) + h\,(i-1,j-1)\big),\\
\frac{\partial^2 h}{\partial x^2} &= \frac{1}{d^2}\big(h\,(i+1,j) - 2h\,(i,j) + h\,(i-1,j)\big),\\
\frac{\partial^2 h}{\partial z^2} &= \frac{1}{d^2}\big(h\,(i,j+1) - 2h\,(i,j) + h\,(i,j-1)\big),
\end{aligned} \tag{3.52}$$

where d is the mesh step $\left(d^\star = d/\lambda_\infty\right)$.

In calculations, we use the x, z coordinate system. The lines $x = 0$, $x = i_{max}$ and $z = 0$, $z = j_{max}$ specify the boundaries of the domain under consideration. Interpolation to an arbitrary point M inside the mesh is carried out by the formula

$$\begin{aligned}
f\,(i+r_1, j+r_2) &= (1-r_1)\,(1-r_2)f\,(i,j)\\
&+ r_1\,(1-r_2)f\,(i+1,j) + (1-r_1)\,r_2 f\,(i,j+1) + r_1 r_2 f\,(i+1,j+1),
\end{aligned} \tag{3.53}$$

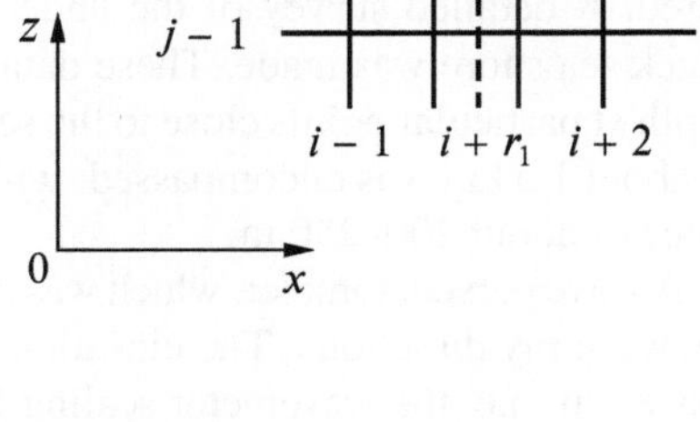

Fig. 3.6 Subdivision of the refraction zone into squared areas in the case when the fluid depth is specified in tabular form

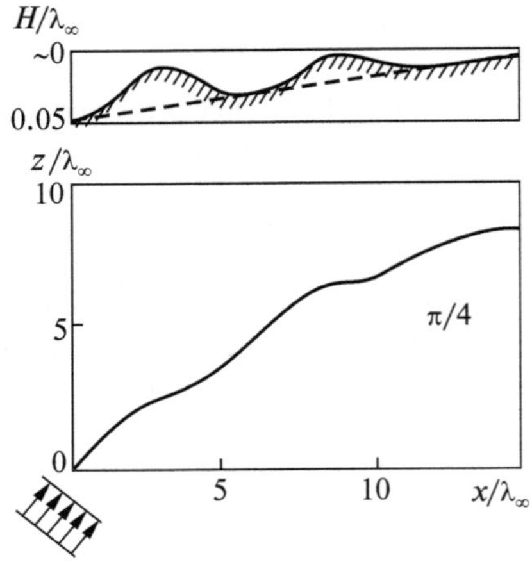

Fig. 3.7 Calculations of the wave refraction over the sloping bottom with two sinusoidal humps when the initial ray angle is equal to $\pi/4$

where f is the dimensionless depth or one of its derivatives. The dimensionless distances r_1, r_2 can be found based on the principle shown in Fig. 3.6. After calculating the depth and its derivatives, one can find the velocity and its derivatives. Next, the refraction pattern at the given depth distribution can be obtained by integrating the system of equations (3.50) at the given initial conditions (3.42).

The algorithm was tested on a reference problem for a flat sloping bottom at different initial ray angles equal to $\pi/8$, $\pi/4$ and $3\pi/8$. The dimensionless depth table was composed with the use of formula (3.49) with the mesh step $d = 1$ and the slope ratio $m = 30$. The results showed good agreement with the calculations made with the analytical depth pattern described by formula (3.49). Figure 3.7 demonstrates the results of calculating the wave refraction over the sloping bottom with two sinusoidal humps when the initial ray angle is equal to $\pi/4$.

3.2.8 Comparing the Theoretical Results and Field Observations

To substantiate the above-described numerical technique, the field observations of the wave refraction were performed and the measured and numerical data were compared. A detailed survey of the underwater relief in one accumulative area of the Black sea shore was made. These data served as a reference base for tabulating the depth at particular points close to the sea shore horizon. A coastal basin with a length of about 1.5 km was encompassed, with a mesh step of 50 m and a sea-to-shore distance of about 200–250 m.

A wave perspectometer, which was mounted on a special tower, was used to record the wave ray directions. The elevation of the perspectometer axis over the sea level was 8.5 m, i.e. the wavemeter scaling factor was equal to 0.85 m. The ray direction (azimuth) was recorded at particular points of the basin as a deviation (in polar coordinates) from the 3.5-m isobath and the shore horizon. Tracing one wave ray from the moment of its appearance in the instrument's field of view and to the moment when it reached the shore horizon could not be accomplished technically with one instrument. Therefore, the wave ray azimuths were recorded at 14 points of the basin with approaching the shore horizon successively. The last record point was located at the shore horizon. An average of ten accurate records with equal polar coordinates was used in further data processing to plot one wave ray recorded at the given wave conditions.

Table 3.1 gives the polar coordinates of the wave ray record points and main parameters of the wave disturbances recorded at depths of 3–3.5 m. These are the depths at which the visual and instrumental recording of the wave refraction could be practically initiated.

Each wave ray was reconstructed from the data acquired at 14 polar points by the successive tracing of the wave ray vector at each subsequent record point. The connection point was located at half a distance between the adjacent record points along the normal to the shore line. The wave ray vector of the farthest record point, which was usually located at a depth of 3–3.5 m, served as the initial ray vector, and the final point of the wave ray vector was located at the shore horizon. Figure 3.8 shows the ray plotted in this way for the wave period $T = 5.2$ s at the South wind direction. Here the points marked as 1 show the instrument's record points, the corresponding wave ray vector is marked as 2, and the wave ray is marked as 3.

Based on the survey of the underwater relief, a bathymetric map of the basin under investigation was plotted. The coordinate origin is located at the edge of this basin on the 3.5-m isobath, the Ox axis being perpendicular to the shore line and the Oz axis being parallel to that line. The basin was subdivided into sections with a step of 10 m. Two data arrays were built for each section that were used as the initial data in the computational programme. The first array contains the distances of the isobaths crossing the given section starting from the coordinate line $x = 0$, and the second array contains the depth values at the corresponding points of the section. The depth values at each 10-m mesh point along the section are calculated with the use of linear interpolation. Thus, the refraction zone under investigation is covered by a uniform squared mesh with a step of 10 m, with the corresponding depth values specified at its nodes.

The deep-water wavelength can be calculated from the period T by the formula $\lambda_\infty = gT^2/2\pi$. The wave period is found from observations and is a characteristic of wave conditions, it is assumed to be constant at the transition to shallow water. The depth derivatives $\partial h/\partial x$, $\partial h/\partial z$, $\partial^2 h/(\partial x \partial z)$, $\partial^2 h/\partial x^2$ and $\partial^2 h/\partial z^2$ are calculated by formulas (3.52), and the interpolation to any arbitrary mesh point is performed using formula (3.53). The refraction pattern of the domain under investigation can be plotted after solving the system of equations (3.41) at the given initial conditions. Figure 3.9 shows the rays obtained by calculations (solid curve) and from observations (dashed curve) at the given wave conditions ($T = 4.56$ s, $\vartheta_0 = 22°$, South wind

Table 3.1 Polar coordinates of the wave ray acquired at 14 polar points in a coastal basin of the Black sea

Period (s)		Wind		Wave ray azimuths at the record points located away from the wavemeter at the given distance in m		
By float	Speed (m/s)	By surf	Azimuth (°)	1190	1020	850
6.0	5.2	9.6	184.6	185.0	186.7	190.3
5.1	5.5	5.3	188.4	189.0	191.7	189.9
4.6	4.8	3.8	162.2	164.2	168.3	174.6
6.0	6.0	4.2	187.5	188.9	191.2	195.2
2.0	3.3	5.9	228.5	–	–	–
1.3	2.4	7.3	392.7	328.2	325.7	316.5

Wave ray azimuths at the record points located away from the wavemeter at the given distance in m										
680	510	425	383	340	297.5	255	212.5	170	127.5	85
193.4	197.0	201.1	207.6	206.2	207.8	212.2	216.1	222.3	225.3	230.4
197.3	205.7	220.3	223.9	227.9	230.1	232.3	232.9	229.8	230.8	234.7
178.5	181.6	185.7	189.3	192.7	196.1	199.8	205.4	199.2	213.2	219.2
198.5	201.9	205.6	208.0	211.6	215.1	219.4	213.1	208.1	222.2	231.0
235.9	239.0	242.3	244.8	246.8	249.6	250.2	245.9	241.2	249.5	246.6
310.5	304.2	320.1	316.9	301.9	294.9	292.6	284.4	281.4	278.3	271.1

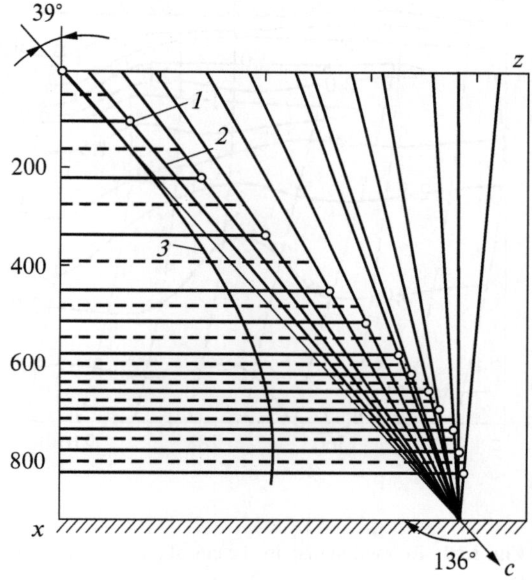

Fig. 3.8 Wave ray plotted on the basis of 14 polar points in a basin of the Black sea shore

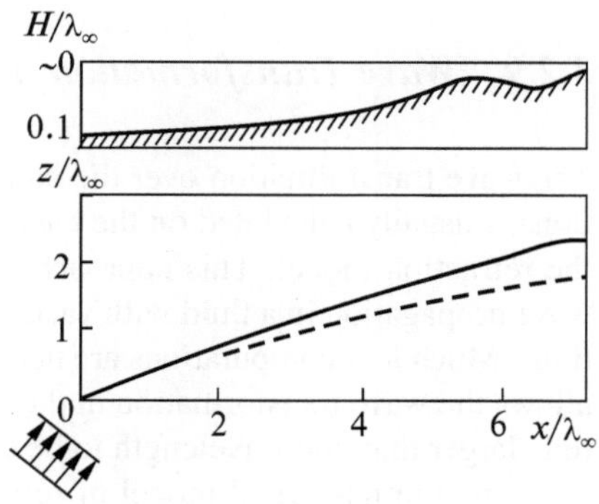

Fig. 3.9 Results of numerical modelling (*solid curve*) as compared to the field observations (*dashed curve*)

direction). The maximum deviation between these two sets of data is no more than 20%. Some disagreement between the calculations and observations can be attributed to the fact that the theory does not take into account friction at the bottom and a possible shear flow along the shore.

Figure 3.10 demonstrates the refraction pattern of the basin under consideration for $T = 5.2$ s, $\vartheta_0 = 12°$, and the South wind direction. Solid curves depict the isobaths, and the dashed curves depict the rays obtained by calculations.

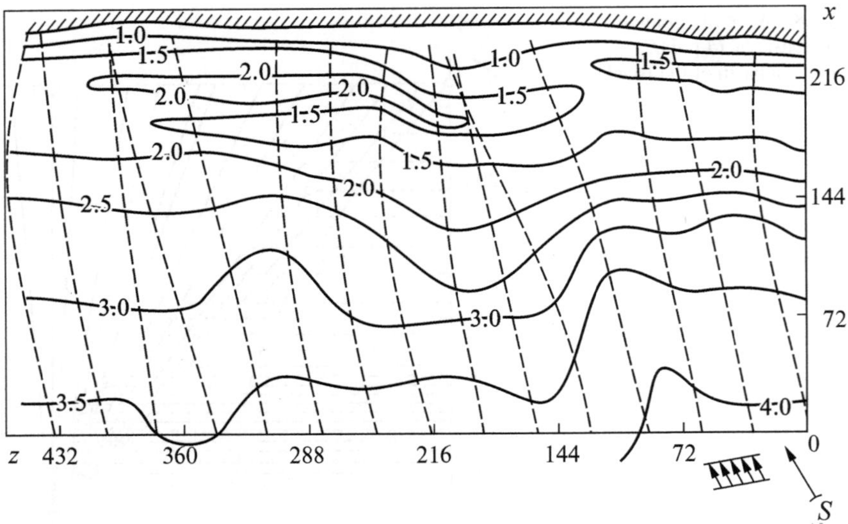

Fig. 3.10 Refraction map in one costal sea zone (*solid curves*: isobaths, *dashed curves*: calculated rays)

3.2.9 *Wave Transformation in the Zones of Caustics*

The wave transformation over the natural bottom relief and embankment constructions is usually calculated on the basis of the ray approach, i.e. in the framework of the refraction model. This approach has an advantage of reducing the problem of wave propagation in a fluid with variable depth to a system of four differential equations. Much less computations are needed to solve this system, so that this approach allows the wave transformation in the water basins that are several orders of magnitude larger than the wavelength to be calculated [38, 46].

Using the numerical model of refraction in application to real objects revealed some problems related to the interpretation of wave amplitude data in the regions of ray intersection, i.e. in the so-called zones of caustics [38].[1] Rays can intersect because of some inevitable errors arising when the bathymetry is specified on the basis of observation data (the so-called false caustics) or they can intersect in the regions where the slopes are large enough to cause the formation of caustics due to ray focusing. The intersections of the first kind can be avoided by an appropriate relief smoothing.

Calculations in the zones of real caustics cannot be performed beyond the bounds of the rigorous refraction theory. Nevertheless, the above-mentioned advantages of

[1]A caustic is a curve (or surface) that serves as a wave ray envelope. It can be defined as a surface where each point forms an intersection of two rays emanating from a wave source at an infinitesimal angle to each other and coming together after refraction. Thus, a caustic is the place of wave energy accumulation.

the ray method stimulated the development of the techniques of constructing the approximate solutions in the ray intersection zone. The method of averaging of the wave amplitudes over the computational mesh elements on the basis of ray density determination was used in Refs. [6, 46]. Hereafter, we consider a similar modified technique for calculating the wave amplitudes that was proposed in Ref. [60].

Let us recall that the wave amplitude a can be determined from the relation (3.46) for the wave amplitude transformation. In the absence of ray intersections, this relation allows the refraction pattern over a smooth bottom relief to be constructed by solving the system of equations of the refraction theory numerically. However, relation (3.46) is no longer applicable after the intersection of the rays. In this case, it can be modified using the following averaging procedure.

Consider the initial wavefront in the form of a line with the initial distance between the rays equal to l_0 and integrate equation (3.45) over the rectangle with sides d that encompasses a bunch of ray tubes. After approximating the integral by a finite sum, we get

$$\frac{\overline{a}}{a_{s=0}} = \left(\frac{l_0}{d} \left(c_g\right)_{s=0} \sum_{i=1}^{m} \frac{1}{\left(c_g\right)_i \cos\phi_i} \right)^{1/2}, \tag{3.54}$$

where $\overline{a}$ is the wave amplitude averaged over the segment with length d, m is the number of rays crossing d, $\left(c_g\right)_i$ is the group speed of the i-th ray crossing the segment d (which is calculated by formula (3.14)), and ϕ_i is the angle between the i-th ray and a normal to d. Next, system (3.18) is solved numerically by the Runge–Kutta method, the average amplitudes at the nodes of the computational mesh with step d being calculated by formula (3.54).

The efficiency of the proposed technique can be tested by considering a model problem of wave transformation over a submerged circular elevation whose shape is defined as [60]

$$h = \begin{cases} h_R + \dfrac{h_0 - h_R}{R^2} r^2, & r < R, \\ h_0, & r \geq R, \end{cases}$$

where $r^2 = (x - x_0)^2 + (z - z_0)^2$. The calculations were performed with $d/\lambda = 0.5$, $R/\lambda = 2$, $h_0/\lambda = 0.37$, $h_R/\lambda = 0.125$ and the initial distance between the rays equal to 0.1λ (λ being the wavelength). The same relief was considered in Ref. [15], where the problem of wave transformation in a finite-depth fluid was solved by the finite-difference method.

Figure 3.11 shows some calculated rays. Ray focusing by the bottom irregularity does not allow the wave amplitudes to be calculated from classical relation (3.46). Figure 3.12 demonstrates the results of calculations by formula (3.54), along with the solution of the problem obtained by the finite-difference method in Ref. [15]. It can be seen that the technique of wave amplitude averaging allows the wave transformation in the zones of caustics to be calculated with satisfactory accuracy keeping all the advantages of the ray method pertaining to the small volume of computations and

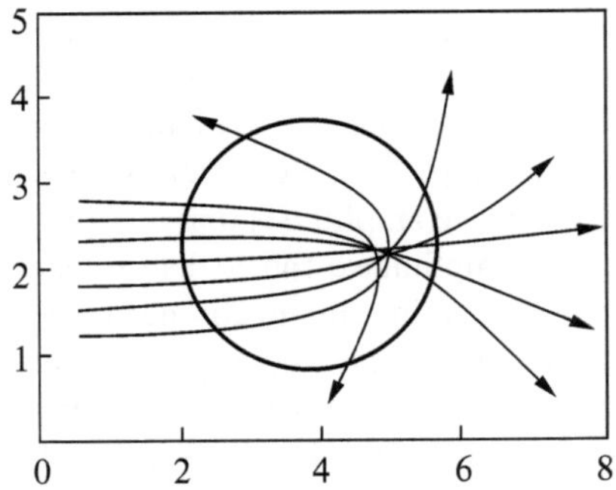

Fig. 3.11 Wave refraction pattern over a circular bottom elevation

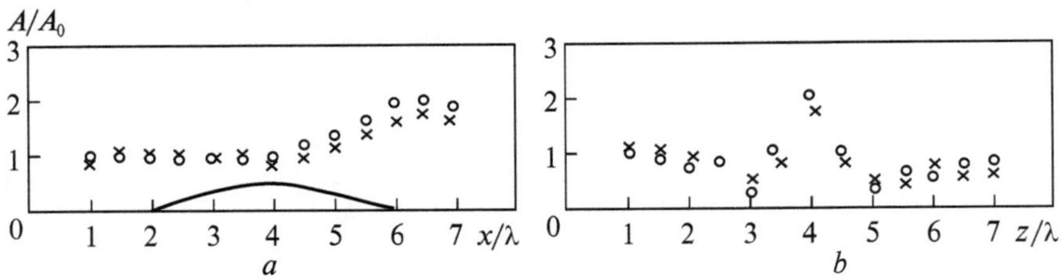

Fig. 3.12 Relative wave amplitudes over the bottom irregularity with $z/\lambda = 4$ (a) and $x/\lambda = 7$ (b). Circles mark the solution from Ref. [15] obtained by the finite-difference method and criss-crosses mark the results obtained by formula (3.54)

a possibility to deal with larger water basins. Yet more accurate formulation of the ray theory with allowance for caustics was considered in Ref. [57]. A generalisation of the ray theory was also made in Ref. [54].

3.3 Nonlinear Theory of Wave Refraction in a Fluid with Variable Depth

Nonlinear wave propagation in a fluid with free surface is a remarkable phenomenon that has been the subject of numerous studies by distinguished mathematicians and mechanicians. An elegant solution to this problem was given by M.A. Lavrentiev in 1946. Since then various approximate nonlinear-dispersive and refractive models have been developed to describe certain features of the complex process of wave propagation on water. The most widespread technique of investigating the refraction problems is based on the assumption that the effect of the wave amplitude on the ray pattern is small and that the ray equations and the energy relation can be considered separately. The Stokes waves are not compatible with such an assumption and represent a good object of investigating the effect of parameter α on the wave transformation.

Further development in the theory of refraction of surface gravity waves on the basis of the ray method was made in Refs. [39, 40, 60]. These works dealt with the wave refraction problems in the case of arbitrary bottom topography, some generalised models were derived with allowance for nonlinear-dispersive effects (as the development of the Stokes theory) and certain diffraction effects, with a possibility of taking into account the temporal transformations of the bottom surface. This made possible to investigate some new effects of nonstationary refraction and the effect of anomalous refraction when the ray bends towards the normal to the coastal line [60].

Let us outline some studies on the nonlinear theory of wave refraction. The wave speed is supposed to satisfy the formula $c = \sqrt{g\,(h+\eta)}$, with a correction for the deviation of the free surface η taken into account [52]. Such an approximation is made by drawing an analogy between the problem of wave propagation on the surface of finite-depth fluid and the problem of one-dimensional flow of a compressible gas in the adiabatic approximation [32, 49]. It was demonstrated in Ref. [3] that the self-focusing and defocusing of waves in a cubic nonlinear medium is possible due to the wave diffraction in the case of wave propagation on deep water. The equations describing the obliquely incident waves in the case of linear variation of the bottom surface were averaged over depth and solved numerically in Ref. [27]. In the case of anomalous refraction, the capture of waves can take place with the rays deflected in the opposite direction. In this regard, noteworthy are the results on the propagation of finite-amplitude waves in a fluid with slowly varying depth obtained in Refs. [9, 26, 42, 51]. The nonlinear ray theory was presented in Ref. [29].

Tsunamis can be regarded as wave trains at the distances from the source larger than the dispersion length. Therefore, the refraction techniques, in view of their simplicity and clarity, provide a useful tool for investigating the transformation of tsunami waves. Their application to nonlinear problems has long been limited to the case when the effect of nonlinear parameter on the ray path could be ignored. However, the propagation of waves from the tsunami source and their run-up on the coast cannot often be explained within the limits of the linear theory. The integral approach proposed in Ref. [27] led to the discovery of anomalous refraction, when at some initial angles larger than the critical angle the ray bent away from the coastline rather than to the coastline, as in the linear theory. The numerical results from Ref. [26] confirmed the fact of decreasing of the refraction coefficient for nonlinear waves. However, no anomalous effects were discovered, probably because of the use of constant-depth relations in the calculations.

It was shown both theoretically and experimentally that the surface waves running on the coast from deep water can generate the edge waves. The amplitude of edge waves is at maximum at the coastline, their wavelength is larger than that of the incident wave and the frequency is smaller. In the linear case, the formation of edge waves can be explained merely by the interaction with a current or with other edge waves [12]. However, some experiments show that edge waves can also be formed without external disturbing forces. Hence, there is another mechanism of the formation of edge waves that can be described in the framework of a more general model. The nonlinear interaction of incident and edge waves was investigated in Refs. [33, 59]. The formation of edge waves was proved to be possible merely owing to the nonlinearity of the incident wave.

Let us reproduce the derivation of the equations describing the refraction of Stokes waves in the fluid of variable depth. In contrast to the cumbersome integral approach from [27], which can provide solutions only in the one-dimensional case, the proposed model is two-dimensional and more simple. Our calculations showed the anomalous behaviour of wave parameters near the breaking point at any nonzero initial incidence angles. This effect can explain the onset of the capture of tsunami waves by the coastal slope.

We use the dimensionless variables

$$(\boldsymbol{r}, y) = \lambda_0(\boldsymbol{r}^\star, y^\star), \quad t = \sqrt{\frac{\lambda_0}{g}}\, t^\star, \quad \Phi = \frac{\lambda_0 a}{\sqrt{\lambda_0/g}} \Phi^\star, \quad \eta = a\eta^\star$$

to write the system of equations for an ideal incompressible fluid (asterisks are hereafter omitted):

$$\nabla_\parallel^2 \Phi + \frac{\partial^2 \Phi}{\partial y^2} = 0, \quad -h(\boldsymbol{r}) \le y \le \alpha\eta\,(t, \boldsymbol{r}),$$

$$\frac{\partial \Phi}{\partial t} + \frac{\alpha}{2}\left(\left(\nabla_\parallel \Phi\right)^2 + \left(\frac{\partial \Phi}{\partial y}\right)^2\right) + \eta = 0, \quad y = \alpha\eta\,(t, \boldsymbol{r}), \tag{3.55}$$

$$\frac{\partial \eta}{\partial t} + \alpha\left(\nabla_\parallel \Phi \cdot \nabla_\parallel \eta\right) - \frac{\partial \Phi}{\partial y} = 0, \quad y = \alpha\eta\,(t, \boldsymbol{r}),$$

$$\frac{\partial \Phi}{\partial y} + \left(\nabla_\parallel \Phi \cdot \nabla_\parallel h\right) = 0, \quad y = -h\,(\boldsymbol{r}).$$

Here, η is the free surface deviation, $\boldsymbol{r} \equiv (x, z)$ are the horizontal coordinates, $\alpha \equiv a/\lambda_0$ is the parameter of nonlinearity, a is a characteristic wave amplitude, λ_0 is a characteristic wavelength and $\nabla_\parallel \equiv (\partial/\partial \boldsymbol{r})$.

The velocity potential can be written in the following form:

$$\Phi\,(t, \boldsymbol{r}) = \alpha v_0\,(\boldsymbol{r})\,t + \varphi\,(t, \boldsymbol{r}). \tag{3.56}$$

This expression can be obtained by an appropriate choice of the average $\overline{\eta} = 0$. Note that the second condition on the free surface in the system of equations (3.55) implies that the derivative $\partial \Phi/\partial t$ is not equal to zero and the expansion for the function Φ should include a term containing t.

Let us go over from the variables t, $\boldsymbol{r}$ to the new variables, namely, phase S and coordinates $\boldsymbol{\xi}$:

$$\theta(t, \boldsymbol{\xi}) = \gamma S(t, \boldsymbol{r}), \quad \boldsymbol{\xi} = \gamma \boldsymbol{r}, \quad \gamma \ll 1, \tag{3.57}$$

which is equivalent to the assumption that the fluid depth variations are small. Introduce the notation

$$\boldsymbol{k} = \nabla_{\|} S, \quad \omega = -\frac{\partial S}{\partial t}. \tag{3.58}$$

Taking into account assumptions (3.56), (3.57) and the relations

$$\nabla_{\|} = \boldsymbol{k}\frac{\partial}{\partial S} + \gamma \nabla_{\xi}, \quad \nabla_{\xi} = \frac{\partial}{\partial \boldsymbol{\xi}},$$

$$\nabla_{\|}^2 = k^2\frac{\partial^2}{\partial S^2} + \gamma\left((\nabla_{\xi}\cdot\boldsymbol{k}) + (\boldsymbol{k}\cdot\nabla_{\xi})\right)\frac{\partial}{\partial S} + \gamma^2\nabla_{\xi}^2, \quad \frac{\partial}{\partial t} = -\omega\frac{\partial}{\partial S},$$

the system of equations (3.55) takes the following form:

$$k^2\frac{\partial^2\varphi}{\partial S^2} + \frac{\partial^2\varphi}{\partial y^2} + \gamma\frac{\partial}{\partial S}\Big((\nabla_{\xi}\cdot\boldsymbol{k})\,\varphi + (\boldsymbol{k}\cdot\nabla_{\xi}\,\varphi)\Big) + O\left(\gamma^2\right) = 0, \quad -h \le y \le \alpha\eta, \tag{3.59}$$

$$\alpha v_0 - \omega\frac{\partial\varphi}{\partial S} + \frac{\alpha}{2}\left(\left(\frac{\partial\varphi}{\partial S}\right)^2 k^2 + 2\gamma\left(\alpha t\,\nabla_{\xi}v_0 + \nabla_{\xi}\varphi\right)\boldsymbol{k}\frac{\partial\varphi}{\partial S} + O\left(\gamma^2\right) + \left(\frac{\partial\varphi}{\partial y}\right)^2\right) + \eta = 0, \quad y = \alpha\eta, \tag{3.60}$$

$$-\omega\frac{\partial\eta}{\partial S} + \alpha\left(\frac{\partial\varphi}{\partial S}\boldsymbol{k} + \gamma\nabla_{\xi}\varphi\right)\cdot\left(\frac{\partial\eta}{\partial S}\boldsymbol{k} + \gamma\nabla_{\xi}\eta\right) - \frac{\partial\varphi}{\partial y} = 0, \quad y = \alpha\eta, \tag{3.61}$$

$$\frac{\partial\varphi}{\partial y} + \gamma\frac{\partial\varphi}{\partial S}\left(\boldsymbol{k}\cdot\nabla_{\xi}h\right) + O\left(\gamma^2\right) = 0, \quad y = -h(\boldsymbol{\xi}). \tag{3.62}$$

Let us restrict ourselves to the case of Stokes waves, so that the solution to the system of equations (3.59)–(3.62) could be looked for in the form

$$\eta = \sum_{n=0}^{\infty}\sum_{m=1}^{\infty}\gamma^n\alpha^{m-1}\,\eta_{n,m}\cos(mS),$$

$$\varphi = \sum_{n=0}^{\infty}\sum_{m=1}^{\infty}\gamma^n\alpha^{m-1}\,v_{n,m}\cosh\big(mk\,(y+h)\big)\sin(mS), \tag{3.63}$$

$$(S,\,\omega,\,\boldsymbol{k}) = \sum_{n=0}^{\infty} \alpha^{2n}\left(S_{2n},\,\omega_{2n},\,\boldsymbol{k}_{2n}\right).$$

Substitute expansions (3.63) into the system of equations (3.59)–(3.62). Considering the case $\gamma \sim \alpha^2$ and equating the free terms to zero, we get the following relations from Eqs. (3.61) and (3.60):

$$\nu_{0,1} = \frac{\eta_{0,1}}{\omega_0\,\cosh(k_0 h)}, \tag{3.64}$$

$$\omega_0^2 = k_0\,\tanh(k_0 h). \tag{3.65}$$

Next by equating the terms proportional to α in Eq. (3.61) as well as the terms proportional to α and $\alpha\cos(2S_0)$ in Eq. (3.60) to zero, we get

$$\nu_0 = \frac{1}{4}\,\nu_{0,1}k_0\left(2\omega_0\,\eta_{0,1}\,\sinh(k_0 h) - k_0\,\nu_{0,1}\,\cosh(2k_0 h)\right), \tag{3.66}$$

$$\eta_{0,2} = \frac{1}{2}\,k_0\omega_0^2\nu_{0,1}\,\eta_{0,1}\,\sinh(k_0 h) + 2\omega_0\nu_{0,2}\,\cosh(k_0 h) - \frac{1}{4}\,k_0^2\nu_{0,1}^2. \tag{3.67}$$

Note that there are no terms of order α^0 and α^1 in Eqs. (3.59) and (3.62). Hence, we collect the coefficients at α^2. Then relation (3.59) yields

$$\nabla_{\|}\cdot\left(\boldsymbol{k}_0\nu_{0,1}\,\cosh\left(k_0(z+h)\right)\right) + \boldsymbol{k}_0\cdot\nabla_{\|}\left(\nu_{0,1}\,\cosh\left(k_0(y+h)\right)\right) = 0,$$
$$-h \le y \le \alpha\eta. \tag{3.68}$$

After the summation of the terms proportional to $\sin S_0$ and $\sin(3S_0)$, we get the following relations from Eq. (3.61):

$$\begin{aligned}
&\eta_{0,1}\omega_2 - h\nu_{0,1}\left(\boldsymbol{k}_0\cdot\boldsymbol{k}_2\right)\cosh(k_0 h) - \frac{1}{2}\nu_{0,1}\,\eta_{0,2}k_0^2\,\cosh(k_0 h)\\
&\quad - \frac{3}{8}\,\nu_{0,1}\eta_{0,1}^2k_0^3\,\sinh(k_0 h) - \nu_{0,2}k_0^2\,\eta_{0,1}\,\cosh(2k_0 h)\\
&- \nu_{0,1}\frac{\left(\boldsymbol{k}_0\cdot\boldsymbol{k}_2\right)}{k_0}\,\sinh(k_0 h) - k_0\nu_{1,1}\,\sinh(k_0 h) + \omega\eta_{1,1} = 0,
\end{aligned} \tag{3.69}$$

$$\begin{aligned}
&\omega_0\eta_{0,3} - \frac{1}{2}\,\nu_{0,1}\eta_{0,2}k_0^2\,\cosh(k_0 h) - \frac{1}{8}\,\nu_{0,1}\eta_{0,1}k_0^3\,\sinh(k_0 h)\\
&\quad - \nu_{0,2}k_0^2\,\eta_{0,1}\,\cosh(2k_0 h) - k_0\nu_{0,3}\,\sinh(3k_0 h) = 0.
\end{aligned} \tag{3.70}$$

By equating the coefficients at $\alpha^2 \cos S_0$ and $\alpha^2 \cos(3S_0)$ in relation (3.60), we have

$$
\begin{aligned}
&-\omega_2 \nu_{0,1} \cosh(k_0 h) - \omega_0 \nu_{0,1} h \frac{(\boldsymbol{k}_0 \cdot \boldsymbol{k}_2)}{k_0} \sinh(k_0 h) - \frac{1}{2} \omega_0 \nu_{0,1} \eta_{0,2} k_0 \sinh(k_0 h) \\
&- \frac{3}{8} \omega_0 \nu_{0,1} \eta_{0,1}^2 k_0^2 \cosh(k_0 h) - 2\omega_0 \nu_{0,2} k_0 \eta_{0,1} \sinh(k_0 h) - \nu_{1,1} \omega_0 \cosh(k_0 h) \\
&\quad + \frac{1}{2} k_0^3 \nu_{0,1}^2 \eta_{0,1} \sinh(2k_0 h) + k_0^2 \nu_{0,1} \nu_{0,2} \eta_{0,1} \cosh(k_0 h) \cosh(2k_0 h) \\
&\quad\quad + k_0^2 \nu_{0,1} \nu_{0,2} \sinh(k_0 h) \sinh(2k_0 h) + \eta_{1,1} = 0,
\end{aligned}
\tag{3.71}
$$

$$
\begin{aligned}
&- \frac{1}{2} \omega_0 \nu_{0,1} k_0 \eta_{0,2} \sinh(k_0 h) - \frac{1}{8} \omega_0 \nu_{0,1} \eta_{0,1}^2 k_0^2 \cosh(k_0 h) \\
&\quad - 2\nu_{0,2} \eta_{0,1} k_0 \omega_0 \sinh(2k_0 h) - 3\nu_{0,3} \omega_0 \cosh(3k_0 h) \\
&+ k_0^2 \nu_{0,2} \cosh(k_0 h) \cosh(2k_0 h) - k_0^2 \nu_{0,1} \nu_{0,2} \sinh(k_0 h) \sinh(2k_0 h) + \eta_{0,3} = 0.
\end{aligned}
\tag{3.72}
$$

Equations (3.64)–(3.72) can be transformed into the form convenient for numerical calculations. To this end, expressions (3.66) and (3.67) can be rewritten with taking into account relation (3.64):

$$
\nu_0 = -\frac{\eta_{0,1}^2 \omega_0}{4 \sinh^2(k_0 h)}, \tag{3.73}
$$

$$
\eta_{0,2} = \frac{k_0^2}{2\omega_0^2} \eta_{0,1}^2 \left(1 + \frac{3}{2 \sinh^2(k_0 h)} \right). \tag{3.74}
$$

Then for the coefficient $\nu_{0,2}$ we get the following expression:

$$
\nu_{0,2} = \frac{3\eta_{0,1}^2 \omega_0}{8 \sinh^4(k_0 h)}. \tag{3.75}
$$

Integrating equation (3.68) over y from $-h$ to $\alpha\eta$ yields

$$
\left(\nabla_{\parallel} \cdot \boldsymbol{c}_0 \right) \frac{\eta_{0,1}^2}{\omega_0} = 0, \tag{3.76}
$$

where the vector $\boldsymbol{c}_0$ is, in the point of fact, the wave group velocity in the linear approximation:

$$
\boldsymbol{c}_0 \equiv \frac{\partial \omega_0}{\partial \boldsymbol{k}_0} = \frac{1}{2\omega_0} \left(\tanh(k_0 h) + \frac{k_0 h}{\cosh^2(k_0 h)} \right) \frac{\boldsymbol{k}_0}{k_0} = \frac{\omega_0}{2k_0} \left(1 + \frac{2k_0 h}{\sinh(2k_0 h)} \right) \frac{\boldsymbol{k}_0}{k_0}.
$$

Eliminating the coefficient $\eta_{1,1}$ from relations (3.69) and (3.71) and taking into account (3.74) and (3.75), we have

$$\omega_2 = \left(\boldsymbol{c}_0 \cdot \boldsymbol{k}_2\right) + \eta_{0,1}^2 \Omega\left(k_0\right), \tag{3.77}$$

$$\Omega = \frac{\omega_0 k_0^2 \left(8 \sinh^4(k_0 h) + 8 \sinh(k_0 h) + 9\right)}{16 \sinh^4(k_0 h)}. \tag{3.78}$$

After some transformations, Eqs. (3.70) and (3.72) take the following form:

$$\nu_{0,3} = k_0 \eta_{0,1} \frac{Q}{2 \sinh^3(k_0 h)},$$

$$\eta_{0,3} = \frac{1}{8} \omega_0 k_0^2 \eta_{0,1} \cosh(k_0 h) + \frac{1}{2} \omega_0^3 \eta_{0,2} \eta_{0,1}$$

$$+ \frac{k_0^2}{\omega_0} \eta_{0,1} \nu_{0,2} \left(4 \sinh^2(k_0 h) - 1\right) + 3\omega_0 k_0 \eta_{0,1} Q \frac{\cosh(3k_0 h)}{2 \sinh^3(k_0 h)},$$

$$Q \equiv \frac{\eta_{0,2}}{2\omega_0 \cosh^2(k_0 h)} + 2\nu_{0,2} \left(1 - \sinh^2(k_0 h)\right).$$

Using relations (3.58), we get

$$\mathrm{d}S = (\boldsymbol{k} \cdot \mathrm{d}\boldsymbol{r}) - \omega \mathrm{d}t.$$

The system of canonical Hamiltonian equations corresponding to the above equation has the following form:

$$\frac{\mathrm{d}\boldsymbol{r}}{\mathrm{d}t} = \frac{\partial \omega}{\partial \boldsymbol{k}}, \quad \frac{\mathrm{d}\boldsymbol{k}}{\mathrm{d}t} = -\frac{\partial \omega}{\partial \boldsymbol{r}}. \tag{3.79}$$

Here, the frequency ω is determined by relations (3.63), (3.65), (3.77) and (3.78), so that

$$\omega(k) = \omega_0\left(k_0\right) + \alpha^2 \left(\boldsymbol{c}_0 \cdot \boldsymbol{k}_2\right) + \alpha^2 \Omega\, \eta_{0,1}^2. \tag{3.80}$$

Expanding $\omega_0(k)$ into power series over α^2, we have

$$\omega\left(k_0\right) = \omega_0(k) - \alpha^2 \left(\boldsymbol{c}_0 \cdot \boldsymbol{k}_2\right).$$

Substituting this expression into Eq. (3.80), we obtain the dispersion relation

$$\omega(k) = \omega_0(k) + \alpha^2 \eta_{0,1}^2 \Omega(k). \tag{3.81}$$

Relations (3.79) and (3.81) imply that the ray equations contain the parameter $\eta_{0,1}$ characterising the free surface elevation. Hence, the ray equations should be solved numerically together with the transport equation (3.76):

$$\begin{aligned} \frac{\mathrm{d}\boldsymbol{r}}{\mathrm{d}t} &= \frac{\partial \omega}{\partial \boldsymbol{k}} = \boldsymbol{c}_0 + \alpha^2 \frac{\partial}{\partial \boldsymbol{k}} \left(\Omega\, \eta_{0,1}^2 \right), \\ \frac{\mathrm{d}\boldsymbol{k}}{\mathrm{d}t} &= -\frac{\partial \omega_0}{\partial \boldsymbol{r}} - \alpha^2 \frac{\partial}{\partial \boldsymbol{r}} \left(\Omega\, \eta_{0,1}^2 \right), \end{aligned} \tag{3.82}$$

$$\left(\nabla_{\parallel} \cdot \boldsymbol{c}_0 \right) \frac{\eta_{0,1}^2}{\omega_0} = 0.$$

When the depth depends only on one coordinate, $h = h(z)$, an exact solution to the system of equations (3.82) can be obtained:

$$\eta_{0,1}^2 = \left(\frac{\eta_{0,1}^2 c_{0z}}{\omega_0} \right)_{z=0} \frac{\omega_0}{c_{0z}},$$

$$\omega = \omega_0 + \alpha^2 \left(\frac{\eta_{0,1}^2 c_{0z}}{\omega_0} \right)_{z=0} \frac{\omega_0 \Omega}{c_{0z}},$$

$$x = \int_{z_0}^{z} \frac{c_{gx}}{c_{gz}}\, \mathrm{d}z, \quad c_{gx} \equiv \frac{\partial \omega}{\partial k_x}, \quad c_{gz} \equiv \frac{\partial \omega}{\partial k_z},$$

where c_{gx} and c_{gz} are the group velocity components along the x and z axes.

Figures 3.13 and 3.14 show the ray paths and amplitudes of the waves with the initial incidence angle equal to $\pi/4$ and different initial wave steepnesses $\alpha = a/\lambda_0$. For $\alpha = 0.02$, the results are compared with the ray path obtained in Ref. [26] (dashed curve in Fig. 3.13). Figure 3.13 demonstrates that the proposed model describes the onset of the nonlinear wave capture by the coastal slope. The capture takes place sooner at larger initial wave steepnesses.

Figure 3.15 shows the results of the present model as compared to the approximation from Ref. [27] and the theoretical model proposed in Ref. [10]. The above-described model shows a closer agreement with the theoretical model from [10] in the initial part of the path, and it fits better the approximation from [27] in the end of the path. Figure 3.15 also demonstrates that the wavelength increases near the wave breaking point. This tendency seems to explain the fact that the length of the edge wave is larger than the length of the incident wave. Note that a more accurate description of edge waves in the framework of the outlined approach is not possible because of the presence of singular points. Nevertheless, the ray method is quite simple and is suitable for basic practical applications.

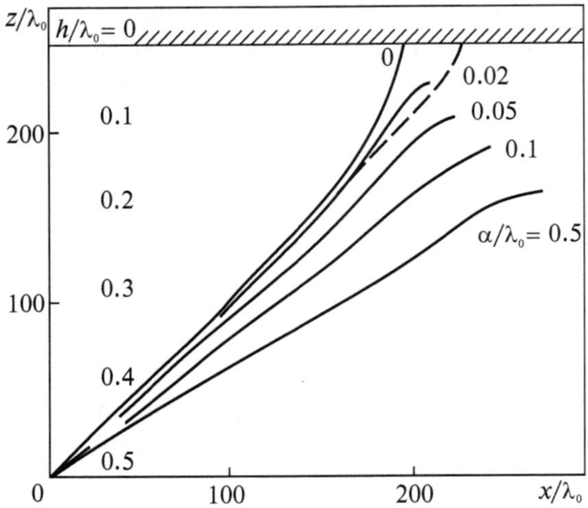

Fig. 3.13 Wave ray paths with the initial incidence angle equal to $\pi/4$ and various initial wave steepnesses a/λ_0

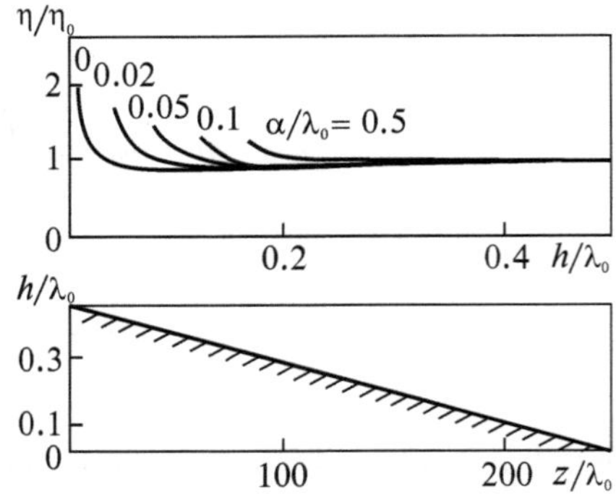

Fig. 3.14 Wave amplitudes for different initial wave steepnesses a/λ_0 at approaching the coastline

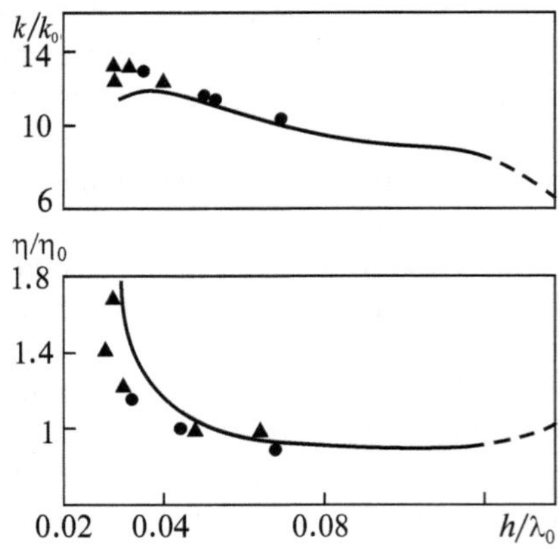

Fig. 3.15 Results of the present model (*solid curve*) as compared to the approximation from Ref. [27] (*triangles*) and the theoretical model proposed in Ref. [10] (*points*)

References

1. Andrews, D.G., McIntyre, M.E.: On wave-action and its relatives. J. Fluid Mech. **89**(4), 647–664 (1978)
2. Arthur, A.S.: Refraction of water wave by islands and shoals with circular bottom contours. Trans. Am. Geophys. Union. **27**(11), 533–545 (1946)
3. Banerjee, P.P., Korpel, A., Lonngren, K.E.: Self-refraction of nonlinear capillary-gravity waves. Phys. Fluids. **26**(9), 2393–2398 (1983)
4. Battjes, J.A.: Refraction of water waves. Proc. Am. Soc. Civ. Eng. **94**(WW4), 437–451 (1968)
5. Bellotti, U., Bornatici, M.: Conservation of wave action and radiative energy transfer in space- and time-varying media. Phys. Rev. E. **57**(5), 6088–6092 (1998)
6. Berkhoff, J., Booy, N., Radder, A.: Verification of numerical wave propagation models for simple harmonic linear water waves. Coastal Eng. **6**, 255–279 (1982)
7. Bigelow, H.B., Edmondson, W.T.: Wind Waves at Sea: Breakers and Surf. University Press of the Pacific (2002)
8. Boccotti, P.: Wave Mechanics for Ocean Engineering. Elsevier Oceanography Series. Elsevier Science, Amsterdam (2000)
9. Chu, V.H., Mei, C.C.: On slowly varying Stokes waves. J. Fluid Mech. **41**, 873–889 (1970)
10. Cokelet, E.D.: Steep gravity waves in water of arbitrary uniform depth. Philos. Trans. R. Soc. Lond. A. **286**, 183–230 (1977)
11. Dean, R.G., Dalrymple, R.A.: Water Wave Mechanics for Engineers and Scientists. Advanced Series on Ocean Engineering, vol. 2. World Scientific, Singapore (1991)
12. Guza, R.T., Davis, R.E.: Excitation of edge waves by waves incident on a beach. J. Geophys. Res. **79**, 1285–1291 (1974)
13. Hashimoto, M.: Geometrical optics of guided waves in waveguides. Prog. Electromagnet. Res. **13**, 115–147 (1996)
14. Headland, J.R., Chu, H.-L.: A numerical model for refraction of linear and Cnoidal waves. In: Proceedings of the 19th International Conference on Coastal Engineering (ICCE), pp. 1118–1131 (1984)
15. Ito, Y., Tanimoto, K.: A method of numerical analysis of wave propogation. In: Proceedings of the 13th International Conference on Coastal Engineering (ICCE), pp. 503–522 (1972)
16. Jonsson, I.G., Skovgaard, O.: Wave refraction across a shearing current. In: Proceedings of the 16th International Conference on Coastal Engineering (ICCE), pp. 722–741 (1978)
17. Jonsson, I.G., Christoffersen, J.B.: Current depth refraction of regular waves. In: Proceedings of the 19th International Conference on Coastal Engineering (ICCE), pp. 1103–1117 (1984)
18. Krylov, A.N.: Nyutonova teoriya astronomicheskoi refraktsii (Newton's theory of the astronomical refraction). In: A Collection of Works, vol. 6, pp. 151–225. Izdatelstvo AN SSSR, Moscow (1936)
19. Lappo, D.D., Ivanova, S.V., Kaplun, V.V., Mishchenko, S.S.: Metod rascheta nagruzki ot voln na obtekaemye pregrady (Method of calculating the wave load on the streamlined obstacles). Izvestiya VNIIG im. Vedeneeva **112**, 45–52 (1976)
20. LeBlond, P.H., Mysak, L.A.: Waves in the ocean. In: Elsevier Oceanography Series, vol. 20. Elsevier, Amsterdam (1978)
21. McCormick, M.: Ocean Wave Energy Conversion. Wiley, New York (1981)
22. Monin, A.S.: Theoretical Geophysical Fluid Dynamics. Kluwer, Dordrecht (1990)
23. Munk, W.H., Arthur, R.S.: Wave intensity along a refracted ray. Gravity waves. U.S. Nat. Bur. Stand. **521**, 95–108 (1952)
24. Munk, W.H., Traylor, M.A.: Refraction of ocean waves: process linking underwater topography to beach erosion. J. Geol. **15**, 1–26 (1947)
25. Nikishov, V.I., Selezov, I.T., Khomitskii, V.V.: Evolyutsiya i trasnformatsiya poverkhnostnykh gravitatsionnykh voln v beregovoi zone (Evolution and transformation of surface gravity waves in the coastal zone). Prykladna Hidromekhanika **12**(4), 63–70 (2010)
26. Oh, I.S., Grosch, C.E.: Numerical study of finite amplitude wave refraction. J. Waterway Port Coast. Ocean Eng. **111**(1), 78–95 (1985)

27. Peregrine, D.H., Ryrie, S.: Anomalous refraction and conjugate solutions of finite-amplitude water waves. J. Fluid Mech. **134**, 91–101 (1982)
28. Peak, S.D.: Wave refraction over complex nearshore bathymetry. Dissertation, Naval Postgraduate School, Monterey (2004)
29. Prasad, P.: A nonlinear ray theory. Wave Motion **20**, 21–31 (1994)
30. Provis, D.G.: Experimental studies of wave refraction. Waves on Water of Variable Depth. Lecture Notes in Physics, vol. 64, pp. 39–45. Springer, Berlin (1976)
31. Pyshkin, B.A.: Dinamika Beregov Vodokhranilishch (Dymanics of the Reservoir Banks). Naukova Dumka, Kiev (1973)
32. Riabouchinsky, D.: Sur l'analogie hydraulique des mouvements d'un fluide compressible. Mécanique des fluides. Comptes Rendus **195**(22), 998–999 (1932)
33. Rocliff, N.: Finite amplitude effects in free and forced edge waves. Math. Proc. Camb. Philos. Soc. **83**(3), 463–479 (1978)
34. Sager, G.: Refraktion von Meersvellen bei ebenem Abfall des Bodens. Gerlands Beitr. Geophys. **82**(3), 210–222 (1973)
35. Sager, G.: Refraktion von $\sqrt{gh}$ - Wellen bei hyperbolischen Abfall des Bodens. Gerlands Beitr. Geophys. **82**(4), 309–328 (1973)
36. Sager, G.: Refraktion von $\sqrt{gh}$ - Wellen bei Abfall des Bodens nach dem profil einer Parabel. Gerlands Beitr. Geophys. **82**(5), 361–378 (1973)
37. Selezov, I.T., Sorokina, V.V.: Issledovanie refraktsii poverkhnostnykh gravitatsionnykh voln pri lineinom izmenenii glubiny zhidkosti (Investigating the refraction of surface gravity waves at linear variations of the fluid depth). Gidromekhanika **34**, 3–8 (1976)
38. Selezov, I.T., Sorokina, V.V.: Numerical investigation of refraction of surface gravity waves by the ray method. Hydrotech. Constr. **15**(1), 46–50 (1981)
39. Selezov, I.T., Tkalich, P.V.: Rasprostranenie poverkhnostnykh voln v zhidkosti s nestatsionarnymi neodnorodnostyami (Propagation of surface waves in a fluid with nonstationary inhomogeneities). Doklkady AN USSR, Ser. A **3**, 28–31 (1985)
40. Selezov, I.T., Tkalich, P.V.: Nestatsionarnaya refraktsiya dlinnykh nelineinykh voln (Nonstationary refraction of long nonlinear waves). Doklkady AN USSR, Ser. A **2**, 30–33 (1986)
41. Selyuk, E.M.: Issledovaniya, raschety i prognozy vetrovogo volneniya na vodokhranilishchakh (Investigating, Calculating and Forcasting the Wind Disturbances on Reservoirs). Gidrometeoizdat, Leningrad (1961)
42. Shen, M.C.: Ray method for surface waves on fluid of variable depth. SIAM Rev. **17**(1), 38–56 (1975)
43. Shen, M.C., Meyer, R.E., Keller, J.B.: Spectra of water waves in channels and around islands. Phys. Fluids **11**, 2289–2304 (1968)
44. Shuleikin, V.V.: Refraktsiya voln na materikovoi otmeli (Wave refraction on a continental shoal). Izvestiya AN SSSR, Ser. Fizika i. Matematika **10**, 1355–1369 (1935)
45. Shuleikin, V.V.: Fizika morya (Physics of the Sea). Nauka, Moscow (1968)
46. Skovgaard, O., Bertelsen, J.A., Jonsson, I.G.: Computation of wave heights due to refraction and friction. J. Waterways Harbors Coast. Eng. Div. **101**(1), 15–32 (1975)
47. Skovgaard, O., Jonsson, I.G.: Current depth refraction using finite elements. In: Proceedings of the 15th International Conference on Coastal Engineering (ICCE), pp. 721–737 (1976)
48. Smirnov, G.N.: Okeanologiya (Oceanology). Vysshaya Shkola, Moscow (1974)
49. Smith, R.M.: Breaking wave criterion on a sloping beach. Dissertation, Naval Postgraduate School, Monterey (1976)
50. Sorokina, V.V.: Issledovanie refraktsii poverknostnykh gravitatsionnykh voln pri tablichnom zadanii relefa dna (Investigating the refraction of surface gravity waves in the case when the bottom relief is specified in tabular form). In: Matematicheskie Metody Issledovaniya Gidrodinamicheskikh Techenii (Mathematical Methods of Investigating the Hydrodynamical Flows), pp. 114–118. Naukova Dumka, Kiev (1978)
51. Stiassnie, M., Peregrine, D.H.: Shoaling of finite amplitude surface waves on water of slowly-varying depth. J. Fluid Mech. **97**, 783–805 (1980)

52. Stoker, J.J.: Water Waves: The Mathematical Theory with Applications. Wiley, New York (1992)
53. Svendsen, Ib A.: Introduction to nearshore hydrodynamics. In: Advanced Series on Ocean Engineering, vol. 24. World Scientific, Singapore (2006)
54. Tailleux, R.: A WKB analysis of the surface signature and vertical structure of long extratropical baroclinic Rossby waves over topography. Ocean Model. **6**, 191–219 (2004)
55. Tang, L.W., Lin, C.F.: Practical method for evaluating directional spectra after shoaling and refraction. In: Proceedings of the 20th International Conference on Coastal Engineering (ICCE), pp. 780–793 (1986)
56. Tsaitts, E.S.: Prirodnye issledovaniya refraktsii voln (Naturalistic investigation of the wave refraction). In: Dinamika Volnovykh i Tsirkulyatsionnykh Potokov (Dynamics of Wave and Circulation Flows), pp. 84–88. Naukova Dumka, Kiev (1966)
57. White, B.S., Fornberg, B.: On the chance of freak waves at sea. J. Fluid Mech. **355**, 113–138 (1998)
58. Whitham, G.B.: Linear and Nonlinear Waves. Wiley (1974)
59. Whitham, G.B.: Nonlinear effect in edge-waves. J. Fluid. Mech. **74**(2), 353–368 (1976)
60. Zheleznyak, M.I., Selezov, I.T., Sorokina, V.V., Tkalich, P.V.: Chislennoe modelirovanie refraktsii poverkhnostnykh voln pri nalichii kaustik (Numerical modeling of the refraction of surface waves in the presence of caustics). In: Problems of Designing and Exploiting the Engineering Structures and Port Installations, pp. 27–30. Mortekhinformreklama, Moscow (1986)

Chapter 4
Analytical and Numerical Solutions to the Wave Diffraction Problems

4.1 On the Formulation and Solution of Wave-Diffraction Problems

Diffraction of waves in a broad sense implies the deviation of wave motions from the laws of geometrical optics. This phenomenon is produced by the inhomogeneities in the medium that are either localised or distributed in space. This process occurs in various environments, the problem is described in many papers [4, 15, 21, 28, 29, 41–43, 45, 46, 49–55, 57, 59, 60, 62, 81].

Figure 4.1 shows the propagation of a plane wave (surface of discontinuity) along the positive direction of the *Ox* axis and its incidence on the inhomogeneity with variable properties in the domain $x > 0$ (e.g. an inhomogeneous bottom surface). In this case, the positions of the front surfaces are changed (Fig. 4.1a). A similar process occurs with the incidence of plane waves on an inhomogeneous cylindrical obstacle. Wave penetration through an inhomogeneous obstacle is accompanied by the deviation of rays (Fig. 4.1b). The solid lines correspond to the positions of wavefront surfaces at successive equal time intervals, the dashed lines (orthogonal to the wavefront surfaces) match against the positions of rays. When the wave reaches an inhomogeneity, both fronts and rays are significantly distorted, i.e. there occurs diffraction.

Three exact methods that provide solutions of the wave-diffraction problems are known: the method of sources based on the Huygens–Fresnel principle, when the diffracted field is represented by a superposition of secondary spherical waves emitted by each element; the Fourier method, when the function associated with the field is represented by a superposition of Fourier expansion components; the separation of variables method that can be applied to a certain class of partial differential equations. Applications of rigorous methods of mathematical physics to the problems of wave propagation, diffraction and scattering are restricted and concern classical problems only. Therefore, approximate analytical and numerical methods have been developed lately.

I.T. Selezov et al., *Wave Propagation and Diffraction*,
Foundations of Engineering Mechanics, DOI 10.1007/978-981-10-4923-1_4

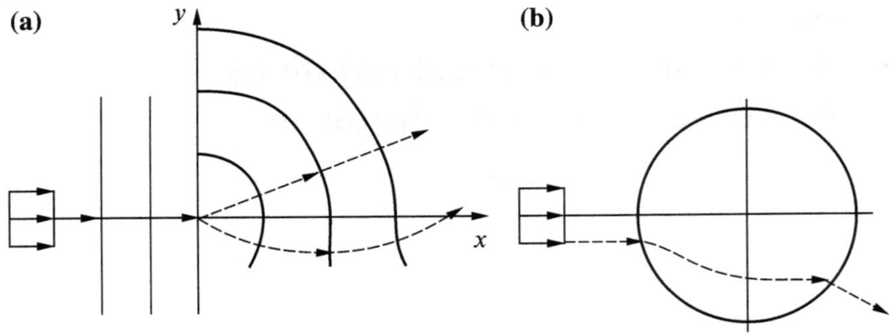

Fig. 4.1 A sketch of **a** refraction fields and **b** wave diffraction

The classical problem of the nonstationary wave diffraction theory is formulated as follows. Suppose that a closed smooth subdomain Ω^i with the boundary $\partial\Omega^i$ is contained in the domain Ω^e oriented in the orthogonal Cartesian coordinate system $x,\ y,\ z$ and a plane wave is propagating along the Ox axis. The diffraction problem reduces to the Cauchi problem or to the boundary-value problem for a partial differential equation (or a set of equations) whose coefficients possess first-kind discontinuities on the surface $\partial\Omega^i$ that is presented in a curvilinear coordinate system. In many cases, the fields associated with linear acoustic, electromagnetic or elastic oscillations can be described in terms of some scalar functions, e.g. the velocity potential in acoustics, electromagnetic potentials, scalar and vector potentials in the elastic theory.

Suppose that the field in the external domain Ω^e is described by the wave equation for the function $\varphi(x, y, z, t)$, as given by

$$\left(\nabla^2 - \frac{1}{c^2}\frac{\partial^2}{\partial t^2}\right)\varphi = 0. \tag{4.1}$$

The function φ is a sum of the incident field φ_i and the scattered field φ_s, i.e. $\varphi = \varphi_i + \varphi_s$, where φ_i is a given function associated with a plane wave propagating along the Ox axis, $\varphi_i(x, 0, 0, t) = F(x + ct)$.

Suppose that at time instant $t = 0$ the leading wavefront touches the surface $\partial\Omega^i$ at the point r_0. This implies that both the unknown function φ and its derivative over t satisfy the given initial conditions

$$\varphi|_{t=0} = \left.\frac{\partial\varphi}{\partial t}\right|_{t=0} = 0,\ \varphi_i = H(r_0 - x)F(x + ct). \tag{4.2}$$

The simplest version of the boundary conditions is given by the Dirichlet, Neumann or Leontovich conditions [8], i.e.

$$\varphi|_{\partial\Omega^i} = 0, \quad \left.\frac{\partial\varphi}{\partial \boldsymbol{n}}\right|_{\partial\Omega^i} = 0, \left(\frac{\partial\varphi}{\partial \boldsymbol{n}} + \alpha\frac{\partial\varphi}{\partial t}\right)_{\partial\Omega^i} = 0, \tag{4.3}$$

where $\boldsymbol{n}$ is the normal vector to the surface. Moreover, the solution should satisfy the decay condition at infinity. Substituting $\varphi = \varphi_i + \varphi_s$ in (4.1)–(4.3) yields a homogeneous equation for φ_s and inhomogeneous boundary conditions for the given function φ_i that satisfies the wave equation.

In the problems of wave diffraction, the incident and scattered fields are described in different coordinate systems. Thus, one of these should be transformed to another coordinate system.

The properties of the domain Ω^i, which in the diffraction theory can be associated with an inhomogeneity, an obstacle, a scatterer or a diffracting body, correspond to the three probable cases [43]: a perfect reflector or a nontransparent body, such that the field does not penetrate into the domain; a transparent object whose properties are permanent and differ from those of the environment, i.e. do not depend on the spatial coordinates within the domain Ω^i; an arbitrary inhomogeneity with space-coordinate-dependent properties within the domain Ω^i. In the latter two cases, we have to solve both exterior and interior problems. The first case includes a perfectly rigid body in an acoustic field that is a perfect conductor in the electromagnetic field, and a vertical cylinder that protrudes over a free surface in the field of surface gravity waves. The second and third cases are associated, e.g. with underwater circular protrusions of either constant or variable depths in the field of surface gravity waves.

With the function φ written in terms of a travelling monochromatic wave, i.e.

$$\varphi_i = \varphi_0 e^{i(kx+\omega t)}, \quad k = \omega/c, \tag{4.4}$$

we obtain a stationary diffraction problem. As follows from (4.4), the function to be found can be reduced to

$$\varphi(x, y, z, t) = \widetilde{\varphi}(x, y, z)e^{i\omega t}. \tag{4.5}$$

Within the context of (4.1) and (4.5), the stationary diffraction problem is formulated as a boundary-value problem for the Helmholtz equation, i.e.

$$(\nabla^2 + k^2)\widetilde{\varphi} = 0. \tag{4.6}$$

The solution is given, as before, by the sum $\widetilde{\varphi} = \widetilde{\varphi}_i + \widetilde{\varphi}_s$, where $\widetilde{\varphi}_i$ is a known given function that satisfies Eq. (4.6).

The boundary conditions for $\partial\Omega^i$ are similar. Moreover, in order to provide the uniqueness of the solution, we have to introduce the radiation and limitedness conditions at infinity, i.e. Sommerfeld's conditions. In the two- and three-dimensional cases these are given by

$$\lim_{r\to\infty} r\left(\frac{\partial\widetilde{\varphi}_s}{\partial r}+\mathrm{i}k\widetilde{\varphi}_s\right)=0,\quad \lim_{r\to\infty}\sqrt{r}\left(\frac{\partial\widetilde{\varphi}_s}{\partial r}+\mathrm{i}k\widetilde{\varphi}_s\right)=0,\quad \lim_{r\to\infty}\widetilde{\varphi}_s=0. \tag{4.7}$$

Plus in (4.7) corresponds to the representation exp[i(kr + ωt)], minus is associated with exp[i(kr − ωt)].

The problem of wave diffraction by a single obstacle is rather complicated even in the infinite domain, the difficulties build up immensely in multiconnected domains. For a semi-infinite domain, the solution can be constructed by the method of images, whose realisation requires only the knowledge of the solution in the infinite domain. This approach was applied to the stationary diffraction problems [55]. It should be noted that the method of images was employed mainly for the study of point charges in electrodynamics [21]. Many solutions are also known for the problems of wave diffraction and scattering by obstacles in infinite domains [15].

4.2 Wave Diffraction by a Partially Submerged Elliptical Cylinder

The great majority of papers dealing with the diffraction of surface gravity waves by vertical cylinders [2] consider mainly the scatterers with circular cross-sections [13, 57]. The configurations of real objects are, however, more complicated and hence the analysis of relevant diffracted wave fields is much more involved. The diffraction of surface gravity waves by elliptic cylinders partially submerged in a fluid was studied in Refs. [8, 70, 78].

Now we consider stationary wave diffraction by a partially submerged fixed cylinder with elliptical cross-section and nonflat base in terms of the linear theory of shallow water (Fig. 4.2). If the base face is slightly convex, the approximate solution can be found in terms of the expansion in power series of a small parameter.

Plane harmonic waves $\exp[\mathrm{i}(\boldsymbol{k}\cdot\boldsymbol{r}-\omega t)]$ are incident on a cylinder at an angle θ to the Ox axis, the factor $\exp(-\mathrm{i}\omega t)$ is omitted in what follows. The velocity potential of the incident wave is given by

$$\varphi_i = -\frac{\mathrm{i}ga_i}{\omega}\exp[\mathrm{i}k(x\cos\theta + y\sin\theta)], \tag{4.8}$$

where k is the wave number, a_i is the wave amplitude and ω is the circular frequency. The equation for the cylinder base face is given by

$$\Sigma_0(x,y) = H_0 - \left[h_0 \mp h \pm h\left(\frac{x^2}{a^2}+\frac{y^2}{b^2}\right)\right] = 0,\quad (x,y)\in S_0, \tag{4.9}$$

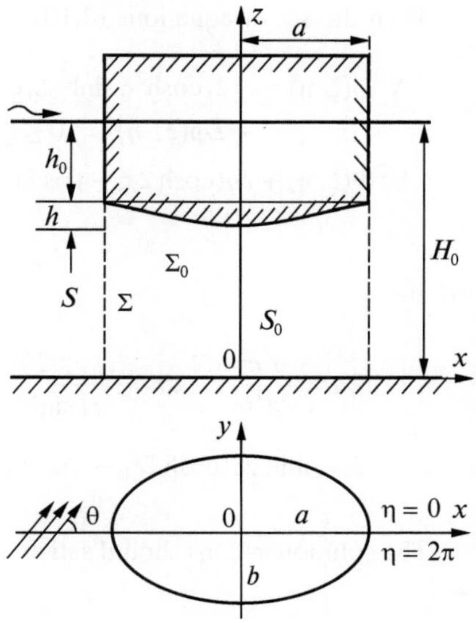

Fig. 4.2 Configuration of the problem

where $\sum_0 (x,y)$ is the surface function, the upper and lower signs correspond to the concave and convex base faces, respectively. In what follows we deal with the case of a convex base.

The velocity potentials in the domains S and S_0 are described, in view of (4.9), by the equations [57]

$$\begin{cases} 2h\left(\frac{x}{a^2}\frac{\partial\varphi}{\partial x}+\frac{y}{b^2}\frac{\partial\varphi}{\partial y}\right)+\left[H+h\left(\frac{x^2}{a^2}+\frac{y^2}{b^2}\right)\right]\nabla^2\varphi=0, & (x,y)\in S_0, \\ \nabla^2\Phi+\frac{\omega^2}{gH_0}\Phi=0, & (x,y)\in S, \end{cases} \tag{4.10}$$

where $H = H_0 - h_0 - h$. Moreover, the functions φ and Φ as well as their normal derivatives should satisfy the continuity condition on the interface Σ of the domains S and S_0, i.e.

$$\varphi|_\Sigma = \Phi|_\Sigma, \quad \left.\frac{\partial\varphi}{\partial n}\right|_\Sigma = \left.\frac{\partial\Phi}{\partial n}\right|_\Sigma, \tag{4.11}$$

while the function Φ should also satisfy the Sommerfeld radiation conditions.

We introduce an elliptic coordinate system, i.e.

$$x = c\cosh\xi\,\cos\eta, \quad y = c\,\sinh\xi\,\sin\eta, \quad c = \sqrt{a^2 - b^2}.$$

Then the set of equations (4.10) reduces to the one given by [34, 60, 74]

$$
\begin{aligned}
\nabla^2\varphi(\xi,\eta) &= \varepsilon\left[2\left(\cosh^2\xi\,\sinh^2\xi_0\cos^2\eta+\cosh^2\xi_0\sinh^2\xi\sin^2\eta\right)\nabla^2\varphi(\xi,\eta)\right.\\
&\quad\left.+L\varphi(\xi,\ \eta)\right],\quad 0\le\xi\le\xi_0,\quad 0\le\eta\le 2\pi,\\
\nabla^2\Phi(\xi,\eta) &+ 2q(\cosh 2\,\xi-\cos 2\eta)\Phi(\xi,\eta)=0,\quad \xi_0\le\xi<\infty,\quad 0\le\eta\le 2\pi,
\end{aligned}
\tag{4.12}
$$

where

$$
2q=\frac{\omega^2c^2}{2gH_0};\quad \varepsilon=-\frac{2h}{H\sinh^2 2\xi_0};\quad \xi_0=\ln\frac{a+b}{c};
$$

$$
L=\sinh 2\xi(\cosh 2\xi_0-\cos 2\eta)\frac{\partial}{\partial\xi}-\sin 2\eta(\cosh 2\xi_0-\cosh 2\xi)\frac{\partial}{\partial\eta}.
$$

The solution $\varphi(\xi,\eta)$ should satisfy the continuity condition at the segment $\xi=0$, i.e.

$$
\varphi^+|_{\xi=0}=\varphi^-|_{\xi=0},\quad \left.\frac{\partial\varphi^+}{\partial\xi}\right|_{\xi=0}=\left.\frac{\partial\varphi^-}{\partial\xi}\right|_{\xi=0},
\tag{4.13}
$$

where plus and minus correspond to the potential values at different sides of the interfocal segment.

In what follows we assume the deviation of the cylinder base surface from a plane to be small, $2|h|/(H_0-h_0-h)\sinh^2 2\xi_0=\varepsilon\ll 1$, and the first solution of set (4.10) to be given by a power series of the small parameter ε, i.e.

$$
\varphi=\varphi_0+\varepsilon\varphi_1+\varepsilon^2\varphi_2+\ldots
$$

The potential Φ is associated with the sum of the plane incident wave (4.8) and the wave reflected from the cylinder. It satisfies the Sommerfeld radiation conditions and thus can be written as

$$
\begin{aligned}
\Phi(\xi,\eta) &= -\frac{2ga_i\mathrm{i}}{\omega}\left\{\sum_{n=0}^{\infty}(-1)^n\left[\mathrm{Ce}_{2n}(\xi,q)-\frac{\mathrm{Ce}_{2n}'(\xi_0,q)}{\mathrm{Me}_{2n}'(\xi_0,q)}\mathrm{Me}_{2n}(\xi_0,q)\right]\right.\\
&\quad\times\mathrm{Ce}_{2n}(\eta,q)\mathrm{Ce}_{2n}(\theta,q)+(-1)^n\mathrm{i}\\
&\quad\times\left[\mathrm{Ce}_{2n+1}(\xi,q)-\frac{\mathrm{Ce}_{2n+1}'(\xi_0,q)}{\mathrm{Me}_{2n+1}'(\xi_0,q)}\mathrm{Me}_{2n+1}(\xi,q)\right]\\
&\quad\times\mathrm{Ce}_{2n+1}(\eta,q)\mathrm{Ce}_{2n+1}(\theta,q)+\sum_{n=0}^{\infty}\left(c_n^{(0)}+\varepsilon c_n^{(1)}+\varepsilon^2c_n^{(2)}+\cdots\right)\\
&\quad\left.\times\mathrm{Me}_n(\xi,q)\mathrm{Ce}_n(\eta,q)+\text{odd functions}\right\}.
\end{aligned}
\tag{4.14}
$$

Periodic solutions of Eq. (4.12) are determined by the periodic Mathieu functions $\mathrm{Me}_m(\xi, q)$, the even ones are given by $\mathrm{Ce}_m(\xi, \eta)$, the odd ones are denoted by $\mathrm{Se}_m(\xi, \eta)$. These form a complete orthogonal system of functions [74].

Suppose that the potential in the domain S_0 is given by

$$\varphi(\xi,\eta) = -\frac{2ga_i\mathrm{i}}{\omega}\left\{\sum_{n=0}^{\infty}\left[d_n^{(0)} + \varepsilon d_n^{(1)} + \varepsilon^2 d_n^{(2)} + \cdots\right]\cosh n\xi\,\cos n\eta + \text{odd functions}\right\}. \tag{4.15}$$

In the zero approximation, with the base of the elliptical cylinder being flat, the coefficients $c_n^{(0)}$ and $d_n^{(0)}$ reproduce the values obtained in Ref. [8]. We substitute (4.14) and (4.15) in the boundary conditions (4.11), multiply them by $\mathrm{Ce}_{2m}(\eta, q)$ and $\mathrm{Ce}_{2m+1}(\eta, q)$, integrate over η from 0 to 2π, and employ the orthogonality conditions for the Mathieu functions (Me_{2m} and Me_{2m+1} are the first-kind Mathieu–Hankel functions). Thus, we obtain expressions for the even and odd coefficients to be given by, respectively,

$$c_{2m}^{(0)} = \frac{H_0 - h_0}{H_0}\frac{1}{\mathrm{Me}'_{2m}(\xi_0, q)}\sum_{r=0}^{\infty} v_r 2r\sinh 2r\xi_0 A_{2r}^{(2m)}(q) d_{2r}^{(0)},$$

$$c_{2m+1}^{(0)} = \frac{H_0 - h_0}{H_0}\frac{1}{\mathrm{Me}'_{2m+1}(\xi_0, q)}\sum_{r=0}^{\infty}(2r+1)\sinh(2r+1)\xi_0 A_{2r+1}^{(2m+1)}(q) d_{2r+1}^{(0)},$$

$$\frac{2\mathrm{i}(-1)^m\mathrm{Ce}_{2m}(\theta, q)}{\pi\mathrm{Me}'_{2m}(\xi_0, q)} = \sum_{r=0}^{\infty} v_r A_{2r}^{(2m)}(q)\Big[\cosh 2r\xi_0 - \frac{H_0 - h_0}{H_0}\frac{\mathrm{Me}_{2m}(\xi_0, q)}{\mathrm{Me}'_{2m}(\xi_0, q)} 2r\sinh 2r\xi_0\Big] d_{2r}^{(0)},$$

$$\frac{2i(-1)^m\mathrm{i}\mathrm{Ce}_{2m+1}(\theta, q)}{\pi\mathrm{Me}'_{2m+1}(\xi_0, q)} = \sum_{r=0}^{\infty} A_{2r+1}^{(2m+1)}(q)\Big[\cosh(2r+1)\xi_0 - \frac{H_0 - h_0}{H_0}\frac{\mathrm{Me}_{2m+1}(\xi_0, q)}{\mathrm{Me}'_{2m+1}(\xi_0, q)}(2r+1)\sinh(2r+1)\xi_0\Big] d_{2r+1}^{(0)},$$

where $v_0 = 2$, $v_m = 1$, $m \geq 1$. The operators contained in (4.12) do not violate the parity of the functions; hence, the solutions for the even and odd functions can be found separately. Below we shall consider only the solutions for the even functions.

We write the first approximation for the solution in the domain $(\xi, \eta) \in S_0$ in terms of the sum $\varphi_1 = \varphi_{10} + \varphi_{1H}$, where

$$\varphi_{10} = -\frac{2ga_i\mathrm{i}}{\omega}\sum_{n=0}^{\infty} d_n^{(1)} \cosh n\xi \cos n\eta. \tag{4.16}$$

The inhomogeneous solution φ_{1H} is assumed to be a series similar to (4.16), it can be found from the equation

$$\nabla^2\varphi_1 = \left[\left(\cosh^2\xi_0 - \cosh^2\xi\right)(1+\cos 2\eta) + 2\cosh^2\xi_0\sinh^2\xi\right]\nabla^2\varphi_0 + L\varphi_0.$$

Having equated the coefficients before similar functions, we find that in the first approximation

$$\begin{aligned}\varphi_1(\xi,\eta) = &-\frac{2ga_i\mathrm{i}}{\omega}\left\{\sum_{n=0}^{\infty} d_n^{(1)}\cosh n\xi \cos n\eta + \frac{1}{8}\sum_{n=0}^{\infty} d_n^{(0)}\{n\cosh 2\xi_0\right.\\ &\times\left[\left(\frac{\cosh(2+n)\xi}{1+n} - \frac{\cosh(2-n)\xi}{1-n}\right)\cos n\eta\right.\\ &\left.+\left(\frac{\cos(2+n)\eta}{1+n} - \frac{\cos(2-n)\eta}{1-n}\right)\cosh n\xi\right]\\ &\left.-\frac{1}{2}[\cosh(2+n)\xi \cos(2-n)\eta + \cosh(2-n)\xi \cos(2+n)\eta]\right\}\\ &\left.+\frac{d_1^{(0)}(\cosh 2\xi_0 - 1)}{16}(\cosh 3\xi \cos\eta + \cosh\xi \cos 3\eta)\right\}.\end{aligned} \tag{4.17}$$

The solution $\varphi_1(\xi,\eta)$ satisfies the continuity condition (4.13). The unknown constants $c_n^{(1)}$ and $d_n^{(1)}$ are determined by the boundary relations (4.11). Thus, we obtain the final expression given by

$$\begin{aligned} c_{2m}^{(1)} &= \frac{H_0 - h_0}{H_0}\frac{1}{\mathrm{Me}'_{2m}(\xi_0, q)}\left[\sum_{r=0}^{\infty} 2r\sinh 2r\xi_0 A_{2r}^{(2m)}(q)d_{2r}^{(1)} + D_{2m}(\xi_0, q)\right],\\ c_{2m+1}^{(1)} &= \frac{H_0 - h_0}{H_0}\frac{1}{\mathrm{Me}'_{2m+1}(\xi_0, q)}\\ &\times\left[\sum_{r=0}^{\infty}(2r+1)\sinh(2r+1)\xi_0 A_{2r+1}^{(2m+1)}(q)d_{2r+1}^{(1)} + D_{2m+1}(\xi_0, q)\right].\end{aligned} \tag{4.18}$$

Here

$$D_{2m}(\xi_0,q)=\frac{\cosh 2\xi_0}{2}\sum_{r=1}^{\infty}d_{2r}^{(0)}\left\{rA_{2r}^{(2m)}(q)\left[\frac{1+r}{1+2r}\sinh(2+2r)\xi_0\right.\right.$$
$$\left.+\frac{r-1}{2r-1}\sinh(2r-2)\xi_0\right]+r^2\sinh 2r\xi_0\left[\frac{A_{2r+2}^{(2m)}(q)}{1+2r}+\frac{\mu_r A_{2r-2}^{(2)}(q)}{2r-1}\right]$$
$$-\frac{1}{4\cosh 2\xi_0}\left[(1+r)\sinh(2r+2)\xi_0\cdot\mu_r A_{2r-2}^{(2m)}(q)\right.$$
$$\left.\left.+(r-1)\sinh(2r-2)\xi_0 A_{2r+2}^{(2m)}(q)\right]\right\};$$

$$D_{2m+1}(\xi_0,q)=\frac{\cosh 2\xi_0}{16}\sum_{k=0}^{\infty}d_{2k+3}^{(0)}\left\{(2k+3)A_{2k+3}^{(2m+1)}(q)\left[\frac{2k+5}{k+2}\sinh(2k+5)\xi_0\right.\right.$$
$$\left.+\frac{2k+1}{k+1}\sinh(2k+1)\xi_0\right]$$
$$+(2k+3)^2\sinh(2k+3)\xi_0\left[\frac{A_{2k+5}^{(2m+1)}(q)}{k+2}+\frac{A_{2k+1}^{(2m+1)}(q)}{k+1}\right]$$
$$-\frac{1}{\cosh 2\xi_0}[(2k+5)\sinh(2k+5)\xi_0 A_{2k+1}^{(2m+1)}(q)$$
$$\left.+(2k+1)\sinh(2k+1)\xi_0 A_{2k+5}^{(2m+1)}(q)]\right\}$$
$$+\frac{d_1^{(0)}}{16}(\cosh 2\xi_0-1)\left[3\cosh 3\xi_0 A_1^{(2m+1)}(q)+\sinh\xi_0 A_3^{(2m+1)}(q)\right],$$
$$\mu_1=2,\ \mu_m=1,\ m\geq 2.$$

The coefficients $d_r^{(1)}$ are determined by the relations

$$\frac{H_0-h_0}{H_0}\frac{\mathrm{Me}_m(\xi_0,q)}{\mathrm{Me}_m'(\xi_0,q)}D_m(\xi_0,q)-G_m(\xi_0,q)=\sum_{r=0}^{\infty}v_rA_r^{(m)}(q)$$
$$\times\left[\cosh r\xi_0-\frac{H_0-h_0}{H_0}\frac{\mathrm{Me}_m(\xi_0,q)}{\mathrm{Me}_m'(\xi_0,q)}r\sinh r\xi_0\right]d_r^{(1)},\quad m=0,1,2,\ldots,\tag{4.19}$$

where

$$G_{2m}(\xi_0, q) = \frac{\cosh 2\xi_0}{8} \sum_{k=1}^{\infty} d_{2k}^{(0)} \Bigg\{ 2k \left[\frac{\cosh(2+2k)\xi_0}{1+2k} + \frac{\cosh(2k-2)\xi_0}{2k-1} \right] A_{2k}^{(2m)}(q)$$
$$+ 2k \cosh 2k\xi_0 \left[\frac{A_{2k+2}^{(2m)}(q)}{2k+1} + \frac{\mu_k A_{2k-2}^{(2m)}(q)}{2k-1} \right] - \frac{1}{2\cosh 2\xi_0}$$
$$\times \left[\mu_k \cosh(2+2k)\xi_0 A_{2k-2}^{(2m)}(q) + \cosh(2k-2)\xi_0 A_{2k+2}^{(2m)}(q) \right] \Bigg\};$$

$$G_{2m+1}(\xi_0, q) = \frac{\cosh 2\xi_0}{16} \sum_{r=1}^{\infty} d_{2r+1}^{(0)} \Bigg[(2r+1) \left(\frac{\cosh(2r+3)\xi_0}{1+r} + \frac{\cosh(2r-1)\xi_0}{r} \right)$$
$$\times A_{2r+1}^{(2m+1)}(q) + (2r+1)\cosh(2r+1)\xi_0 \left[\frac{A_{2r+3}^{(2m+1)}(q)}{r+1} + \frac{A_{2r-1}^{(2m+1)}(q)}{r} \right]$$
$$- \frac{1}{\cosh 2\xi_0} \left[\cosh(2r+3)\xi_0 A_{2r-1}^{(2m+1)}(q) + \cosh(2r-1)\xi_0 A_{2r+3}^{(2m+1)}(q) \right] \Bigg]$$
$$+ \frac{d_1^{(0)}(\cosh 2\xi_0 - 1)}{16} \left[\cosh 3\xi_0 A_1^{(2m+1)}(q) + \cosh \xi_0 A_3^{(2m+1)}(q) \right].$$

The unknown coefficients $c_n^{(1)}$ and $d_n^{(1)}$ are determined by the solution of the set (4.18), (4.19). This concludes the construction of the solutions in the first approximation, i.e. the potential φ in the domain under the cylinder S_0 and the potential Φ outside the cylinder, in the domain S. The higher order solutions can be constructed in the similar manner.

Now let us find wave forces and moments. In terms of the potentials $\varphi(\xi, \eta)$ and $\Phi(\xi, \eta)$, the hydrodynamic pressure can be written as

$$p_1(\xi, \eta) = -\rho \frac{\partial \varphi}{\partial t}, \; p_2(\xi, \eta) = -\rho \frac{\partial \Phi}{\partial t}.$$

Suppose the angle of incidence is equal to zero, $\theta = 0°$. In view of the symmetry of the wave process with respect to the Ox axis, we have

$$F_y = M_z = M_x = 0.$$

The other components of the total force are given by

$$F_x = -h_0 \int_0^{2\pi} p_2(\xi_0, \eta) \cos(n, x) T \, d\eta - \int_0^{2\pi} \int_0^{\xi_0} p_1(\xi, \eta) \cos(n, x) T^2 d\xi \, d\eta, \quad (4.20)$$

$$F_z = \int_0^{\xi_0}\int_0^{2\pi} p_1(\xi,\,\eta)\cos(n,\,x)T^2 d\xi\, d\eta, \tag{4.21}$$

$$\begin{aligned} M_y = &- \int_{H_0-h_0}^{H_0} (z - H_0 - h_0)dz \int_0^{2\pi} p_0(\xi_0,\eta)\cos(n,\,x)T\, d\eta \\ &- \int_0^{\xi_0}\int_0^{2\pi} p_1(\xi,\eta)xT^2\cos(n,z)d\eta d\xi, \\ &+ \int_0^{\xi_0}\int_0^{2\pi} p_1(\xi,\eta)\cos(n,x)T^2(H_0 - h_0 - z_1)d\xi\, d\eta, \end{aligned} \tag{4.22}$$

where

$$\cos(n,x) = c/T\sinh\xi_0\cos\eta;\;\; T = c\sqrt{(\cosh 2\xi_0 - \cos 2\eta)/2};$$

$T\,d\eta$ is the linear element along the ellipse; $T^2 d\eta d\xi$ is an element of the elliptical surface; n and n_1 are unit normals to the curved and base surfaces of the cylinder;

$$z_1 = H_0 - h_0 - h\left(1 - \frac{x^2}{a^2} - \frac{y^2}{b^2}\right).$$

Having substituted the expressions for $\varphi(\xi,\,\eta)$ and $\Phi(\xi,\,\eta)$ in (4.20)–(4.22), we have

$$\begin{aligned} F_x = &-2\rho g\, a_i\, \pi h_0\, b\Bigg\{\Bigg[-\frac{2}{\pi}\sum_{n=0}^{\infty}\frac{(-1)^n \mathrm{Ce}_{2n+1}(\theta,q)A_1^{(2n+1)}}{\mathrm{Me}'_{2n+1}(\xi_0,q)} \\ &+ \sum_{n=0}^{\infty}\left(c_{2n+1}^{(0)} + \varepsilon\, c_{2n+1}^{(1)}\right)\mathrm{Me}_{2n+1}(\xi_0,q)A_1^{(2n+1)}\Bigg] \\ &+ \frac{c^2 h}{4a^2 b}\Bigg(\left(d_1^{(0)} - d_3^{(0)}\right)\left(\frac{\sinh 4\xi_0}{2} + \sinh 2\,\xi_0\right) \\ &+ d_1^{(0)}(\cosh 2\xi_0 - 1)\left(\frac{\sinh 6\xi_0}{3} - \sinh 2\xi_0\right) \\ &+ d_3^{(0)}\left[3\cosh 2\xi_0\left(\frac{\sinh 6\xi_0}{6} - \frac{\sinh 4\xi_0}{4} - \sinh 2\xi_0\right) - \frac{\sinh 8\xi_0}{4} - \sinh 2\xi_0\right] \\ &+ d_5^{(0)}\left[\frac{\sinh 8\xi_0}{4} + \frac{\sinh 6\xi_0}{3} - \frac{5\cosh 2\xi_0}{2}\left(\frac{\sinh 6\xi_0}{3} + \frac{\sinh 4\xi_0}{2}\right)\right]\Bigg)\Bigg\}, \end{aligned} \tag{4.23}$$

$$F_z = \frac{\rho g a_i \pi c^2}{2}\Big\{\sinh 2\xi_0\left(2d_0^{(0)} - 2d_2^{(0)}\right) + \varepsilon\Big[\sinh 2\xi_0\left(2d_0^{(1)} - 2d_2^{(1)}\right) + \frac{d_2^{(0)}}{48}(3\sinh 6\xi_0 + \sinh 2\xi_0) - \frac{d_4^{(0)}}{48}(3\sinh 6\xi_0 + 4\sinh 2\xi_0)\Big]\Big\} \tag{4.24}$$

$$\begin{aligned} M_y = & -\pi\rho g a_i c^3 h_0\Big\{d_1^{(0)}\cosh\xi_0 + \varepsilon\Big(d_1^{(1)}\cosh\xi_0 \\ & + \frac{1}{16}\left[\frac{d_3^{(0)}}{2}(\cosh 5\xi_0 + 3\cosh\xi_0) + d_1^{(0)}(\cosh 2\xi_0 - 1)\cosh 3\xi_0\right]\Big)\Big\} \\ & - c\int\limits_0^{\xi_0}\int\limits_0^{2\pi} p_1(\xi,\eta)\cos(n_1, z)\, xT^2\cosh\xi\cos\eta\, d\xi\, d\eta \\ & + h\int\limits_0^{\xi_0}\int\limits_0^{2\pi} p_1(\xi,\eta)\cos(n_1, x)T^2\left(1 - \frac{\cosh^2\xi\cos^2\eta}{\cosh^2\xi_0} - \frac{\sinh^2\xi\sin^2\eta}{\sinh^2\xi_0}\right)d\xi\, d\eta. \end{aligned} \tag{4.25}$$

Figures 4.3, 4.4 and 4.5 show the calculated results that follow from formulas (4.23)–(4.25). The maximum values of the forces and moment per period are given in the dimensionless form, i.e.

$$F_x^* = \frac{F_x}{2\pi\,\rho g a_i\, b\, h_0},\ F_z^* = \frac{F_z}{\pi\,\rho g a_i\, a\, b},\ M_y^* = \frac{M_y}{\pi\,\rho g a_i\, a^3}.$$

Figure 4.3 demonstrates the values of the horizontal force F_x^* applied to the curved surface of the cylinder. The convex base increases the horizontal load on the cylinder.

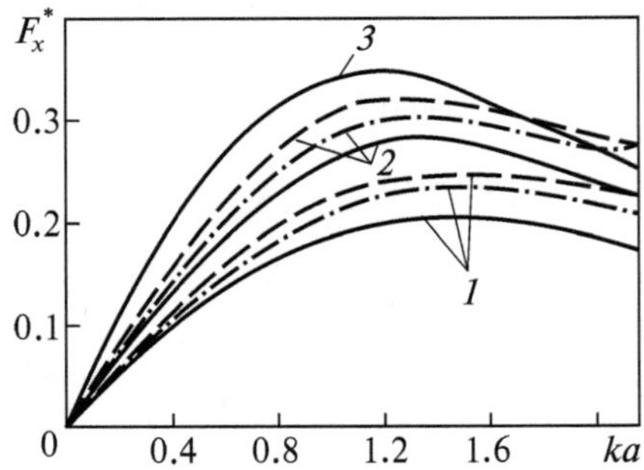

Fig. 4.3 The horizontal force as a function of the wave number for the cylinder submersion depths (*1*) $h_0/H_0 = 0.3$, (*2*) 0.4, (*3*) 0.5 and ellipticity $b/a = 0.5$ in the cases when $h = 0$ (*solid curve*), $h = 0.05$ (*dot-dash curve*), and $h = 0.1$ (*dashed curve*)

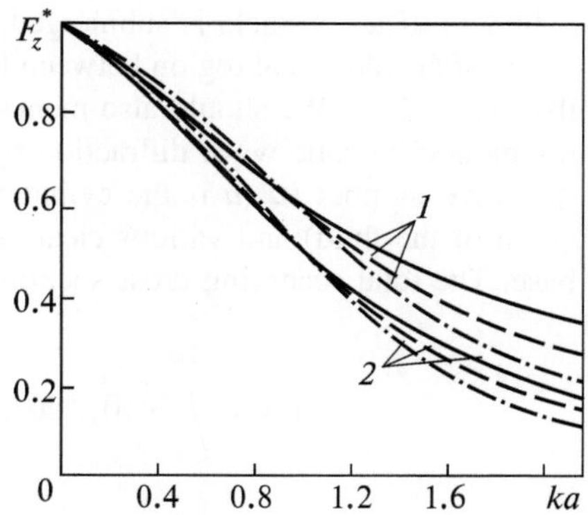

Fig. 4.4 The vertical force F_z as a function of the wave number in the case of a convex base surface ($h/h_0 = 0.3$) for (*1*) $h/H_0 = 0.1$, (*2*) 0.4 and ellipticities $b/a = 0.9$ (*solid curve*), 0.7 (*dashed curve*), 0.5 (*dot-dash curve*)

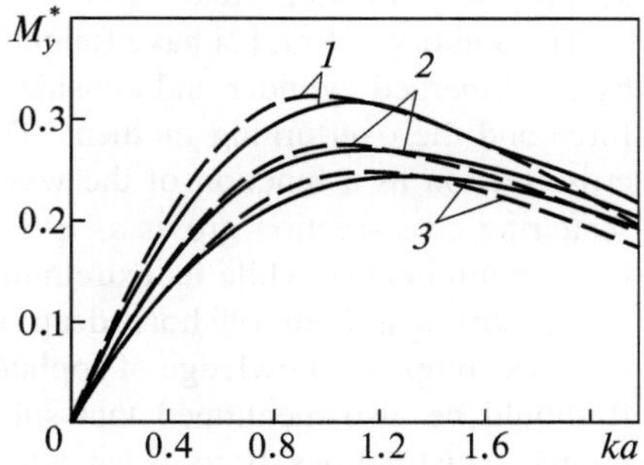

Fig. 4.5 The overturning moment as a function of the wave number for a flat-base cylinder with the ellipticity $b/a = 0.9$ (*dashed curves*) and for a circular cylinder with $b/a = 1$ (*solid curves*), the submersion depth h_0/H_0 being (*1*) 0.5, (*2*) 0.4, (*3*) 0.3

Figure 4.4 shows the influence of ellipticity on the vertical force F_z^* for the cylinders with (1) $h/H_0 = 0.1$ and (2) $h/H_0 = 0.4$. The solid, dashed and dot-dash lines correspond to the semi-axes ratio $b/a = 0.9$, 0.7 and 0.5. For the wave numbers $ka \geq 1.3$, the ellipticity notably favours the decrease of the load F_z^*.

Figure 4.5 illustrates the comparison of the results for the overturning moment M_y^* along the *Oy* axis in the case of small-ellipticity cylinder ($b/a = 0.9$) with a flat base (dashed curve) and the known solution for a circular cylinder (solid curve) [57]. The difference between the moments for the elliptical and circular cylinders is seen to be not too large, i.e. the calculations of the moment for a slightly elliptical cylinder can be carried out in terms of the formulas for the circular cylinder with an accuracy to within 5%.

4.3 Wave Diffraction by a Submerged Circular Cylinder

Diffraction interaction of surface gravity waves with various structures and constructions was studied by many authors [2, 5, 12, 22, 57, 63]. The relevant wave-diffraction problems have been associated mainly with protruding vertical cylinders. If an obstacle is submerged, the problem is much more complicated by virtue of an additional region between the lower face of the submerged obstacle and the free surface. We should also mention Ref. [35] dealing with the dependence of the monochromatic wave diffraction by a partially submerged circular cylinder on the wave number *ka* (*a* is the cylinder radius) for various values of *d*/*a* (*d* is the depth of the fluid) and various clearings *h*/*d* between the bottom and the cylinder base. The total scattering cross-section is given by [36]

$$Q = a \int_0^{2\pi} |A(\theta)|^2 d\theta, \; A(\theta) = \sum_{m=0}^{\infty} \varepsilon_m A_m \cos m\theta,$$

where $A(\theta)$ is the scattering amplitude. In the far region, i.e. at $ka \to 0$, the asymptotic estimation yields $Q \sim 4a$.

The authors of Ref. [2] have found the solutions of the wave-diffraction problem by a submerged cylinder and considered the horizontal force, the vertical buoyant force and the overturning moment. They have also calculated the total scattering cross-section as a function of the wave number *ka* and have shown that the total scattering cross-section grows as *a*/*H* increases (*a* is the cylinder radius, *H* is the submersion depth), while its extremum is shifted towards greater values of *ka*.

Constructing both offshore deep water and onshore shallow water systems requires complete knowledge of applied loads, especially the extremum wave loads. It should be also mentioned that submerged breakwaters that can be treated as mobile constructions become an attractive alternative of the shore shields with admissible moderate energy penetration. In such cases complex wave fields are induced and hence solutions of the diffraction–theory problems are required [12]. Reference [22] deals with numerical and experimental studies of wave-absorption parameters of a submerged caisson breakwater in view of the optimum efficiency of the wave absorber. The absorber is shown to decrease the low-frequency reflected waves with rather high efficiency.

Lately, much attention is given to the effects that accompany the wave diffraction. Reference [63] deals with experimental studies of wave slamming under the diffraction by a vertical cylinder. In some cases the authors consider a vertical caisson breakwater with circular cross-section under the impact of surface gravity waves and do not treat the diffraction problem. The resultant lateral force is assumed to be known and to be applied to the breakwater axis.

Here, the potential formulation in terms of the shallow water wave model is employed to analyse the influence of the depth of submersion of a circular cylinder with circular base on the wave loads. It should be noted that in the case of shallow water wave model (longwave approximation) the problem of surface gravity wave diffraction is similar to the problem of acoustic wave diffraction by a transparent obstacle [62]. We assume that plane monochromatic waves $\exp[i(kx + \omega t)]$ are incident on an obstacle from infinity and are diffracted by this obstacle. The total field is equal to the sum of the incident (φ_i) and scattered (φ_s) fields, i.e. $\varphi = \varphi_i + \varphi_s$.

The diffraction problem reduces to the calculation of the velocity potentials $\varphi_k(r, \theta, t)$, $k = 1, 2, 3$ in the three domains Ω_1, Ω_2 and Ω_3 (Fig. 4.6). The potentials are the solutions of the wave equation

$$\frac{\partial^2 \varphi_k}{\partial r^2} + \frac{1}{r}\frac{\partial \varphi_k}{\partial r} + \frac{1}{r^2}\frac{\partial^2 \varphi_k}{\partial \theta^2} - \frac{1}{\left(c_k^{\mathrm{sh}}\right)^2}\frac{\partial^2 \varphi_k}{\partial t^2} = 0, \quad c_k^{\mathrm{sh}} = \sqrt{gH_k}, \tag{4.26}$$

that satisfy the matching conditions at the domain boundaries, i.e.

$$\begin{aligned} \varphi_3|_{r=b} &= \varphi_2|_{r=b}, \quad H_3 \frac{\partial \varphi_3}{\partial r}\bigg|_{r=b} = H_2 \frac{\partial \varphi_2}{\partial r}\bigg|_{r=b}, \\ \varphi_2|_{r=a} &= \varphi_1|_{r=a}, \quad H_2 \frac{\partial \varphi_2}{\partial r}\bigg|_{r=a} = H_1 \frac{\partial \varphi_1}{\partial r}\bigg|_{r=a}. \end{aligned} \tag{4.27}$$

Moreover, the solution φ_1 should satisfy the regularity condition for $r \to 0$, i.e.

$$|\varphi_1| < \infty \text{ at } r = 0. \tag{4.28}$$

The perturbation is simulated by plane monochromatic waves that propagate along the negative direction (see Fig. 4.6), i.e.

$$\varphi_i(x, t) = Ae^{\mathrm{i}(kx + \omega t)}. \tag{4.29}$$

We expand (4.29) in a series of cylindrical functions, substitute the expansion in (4.26)–(4.28), and thus obtain the relevant problem for the Helmholtz equation. Then we impose the Sommerfeld radiation conditions on the function φ_{3s}, i.e.

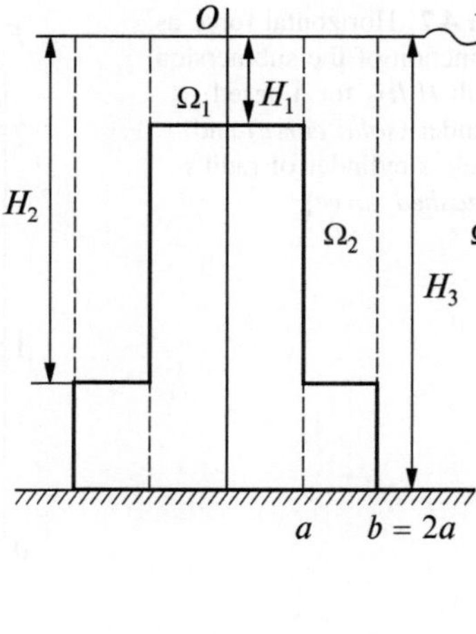

Fig. 4.6 Configuration of the problem

$$\lim_{r\to\infty}\sqrt{r}\left(\frac{\partial\varphi_{3s}}{\partial r}+\mathrm{i}k_3\varphi_{3s}\right)=0,\qquad \lim_{r\to\infty}\varphi_{3s}=0. \tag{4.30}$$

The solutions that satisfy the conditions (4.28), (4.30) can be written as

$$\varphi_1=\sum_{m=0}^{\infty} d_m J_m(k_1 r)\cos m\theta, \tag{4.31}$$

$$\varphi_2=\sum_{m=0}^{\infty}\left[b_m J_m(k_2 r)+c_m N_m(k_2 r)\right]\cos m\theta, \tag{4.32}$$

$$\varphi_3=\sum_{m=0}^{\infty}\left[\varepsilon_m \mathrm{i}^m A J_m(k_3 r)+a_m H_m^{(2)}(k_3 r)\right]\cos m\theta, \tag{4.33}$$

where ε_m is the Neumann factor,

$$\varepsilon_m=\begin{cases}1 & \text{at}\quad m=0,\\ 2 & \text{at}\quad m\geq 0,\end{cases}$$

A is the incident wave amplitude, $J_m(k_1 r)$ is the Bessel function, $N_m(k_2 r)$ is the Neumann function; $H_m^{(2)}(k_3 r)$ is the Hankel function of the second kind. The arbitrary constants a_m, b_m, c_m, and d_m should be found from the matching conditions (4.27) with (4.31)–(4.33) substituted for φ_k.

Figure 4.7 shows the calculated results for $H_2 = 0.76\,H_3$. Both the decrease of the submersion depth and the increase of the wave number are accompanied by the approach to the resonance range, though for different reasons. In the first case, the reason is that the model is approximate; in the second case, the result is associated

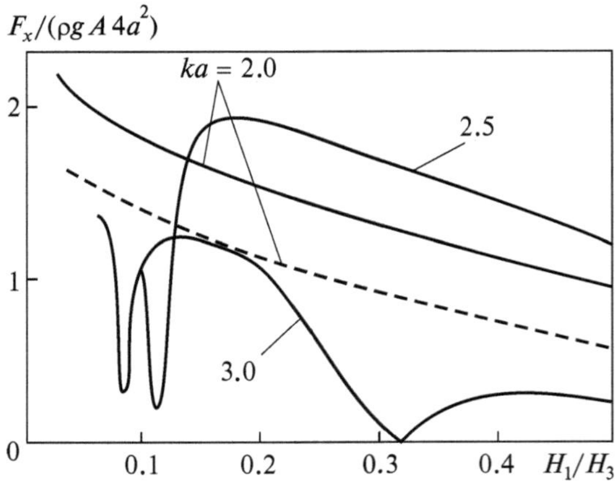

Fig. 4.7 Horizontal force as a function of the submersion depth H_1/H_3 for a based cylinder (*solid curve*) and baseless cylinder of radius a (*dashed curve*)

with the method employed, since the convergence of the series is slower for greater wave numbers.

To summarise the above reasoning, we have to note that the analysis of the solution to the problem of surface gravity wave diffraction by a submersed vertical cylinder provides understanding of the effect of the obstacle submersion degree on the horizontal wave force and the overturning moment [58]. Both the decrease of the clearing H_1/H_3 and the increase of the wave number ka are accompanied by the approach to the resonance range, where the reliability of the results gets worse. Numerous calculation results provide evidence for strong increase of the space-time oscillations for narrower clearings between the obstacle base and the free surface. Calculations of the horizontal wave force and the overturning moment show that the wave number should not be greater than $2\sqrt{\pi}$. We note that analysing the total scattering cross-section in terms of the finite-depth-fluid model is of crucial importance for the studies of the wave-damping properties of the scatterer.

4.4 Scattering of Magnetoacoustic Cylindrical Waves by a Cylinder

The scattering of magnetoacoustic waves in an electrically conducting compressible medium was considered for the first time in Ref. [53]. A two-dimensional problem is considered. A cylindrical wave is emitted by a lumped axial source O_1 located at distance b from the axis O_2 of the object under consideration and is scattered by an absolutely rigid immobile perfectly conducting circular cylinder (Fig. 4.8). The outer medium is described by the equations of magnetohydrodynamics in the linear acoustic approximation. The medium is assumed to be in an undisturbed rest state, to have weak electric conductivity and to be exposed to a strong magnetic field.

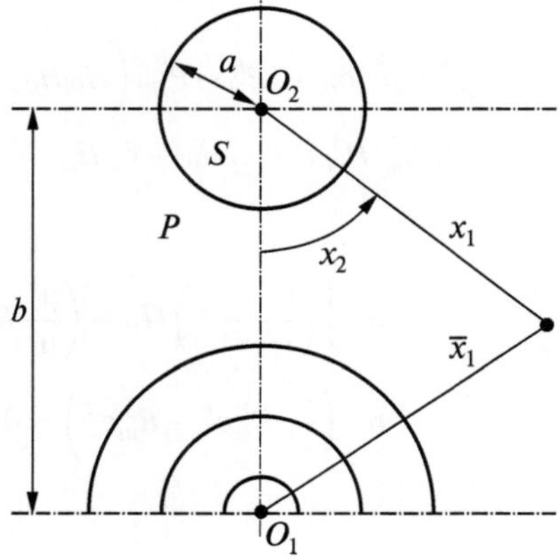

Fig. 4.8 Configuration of the problem

The magnetohydrodynamic equations in the approximation $R_m \ll 1$ and $P_H \gg 1$ are given by [40, 56]

$$\rho_0 \frac{\partial \boldsymbol{v}}{\partial t} = -c_0^2 \nabla \rho + P_H R_m (\boldsymbol{v} \cdot \boldsymbol{H}_0) \cdot \boldsymbol{H}_0$$

$$\frac{\partial \rho}{\partial t} + \rho_0 \nabla \cdot \boldsymbol{v} = 0, \tag{4.34}$$

$$\nabla \cdot \boldsymbol{h} = \boldsymbol{v} \cdot \boldsymbol{H}_0, \quad \nabla \cdot \boldsymbol{e} = -\frac{\partial \boldsymbol{h}}{\partial t},$$

$$\nabla \cdot \boldsymbol{h} = 0, \quad \nabla \cdot \boldsymbol{e} = 0,$$

Here

$$P_H = \frac{\mu H_0^2}{\rho_0 c_0^2}$$

is the magnetic pressure, $R_m = \mu \sigma c_0 a$ is the magnetic Reynolds number, μ is the magnetic susceptibility, σ is the electric conductivity, a is some typical scattering distance (radius), c_0 is the speed of disturbance propagation corresponding to the isoentropic state, $\boldsymbol{H}_0$ is the given vector of the strength of the external magnetic field, ρ_0 is the density in the undisturbed state, $\rho(\boldsymbol{x})$ is the disturbed density, $\boldsymbol{x} = (x_1, x_2, x_3)$ is the vector of spatial coordinates in the orthogonal (cylindrical) system associated with the scatterer, ∇ is the Hamilton operator, $(\,\cdot\,)$ is the scalar product, t is time, $\boldsymbol{v}$, $\boldsymbol{h}$ and $\boldsymbol{e}$ are the vectors of disturbed velocity, disturbed magnetic field strength and disturbed electric field strength.

The relevant matching conditions on the interface R between the external medium P and the internal medium S express the continuity of velocities, total stress tensors and electromagnetic fields. They are given in general form by

$$\boldsymbol{n} \cdot (\boldsymbol{v}^P - \boldsymbol{v}^S) = 0,$$

$$\begin{aligned}\Big[&-(p^P - p^S)\delta_{ik} + (P_H^P - P_H^S)\left(H_{0i}H_{0k} - \frac{1}{2}\delta_{ik}H_0^2\right) + P_H^P R_m^P (H_{0i}h_k^P + h_i^S H_{0k} - \delta_{ik}\boldsymbol{H}_0 \cdot \boldsymbol{h}^S) \\ &- P_H^S R_m^S (H_{0i}h_k^S + h_i^S H_{0k} - \delta_{ik}\boldsymbol{H}_0 \cdot \boldsymbol{h}^S)\Big] n_i = 0 \; (i, k = 1, 2, 3),\end{aligned} \tag{4.35}$$

$$\boldsymbol{n} \cdot \left[\left(\frac{\mu^P}{\mu^S} - 1\right)\boldsymbol{H}_0 + \left(\frac{\mu^P}{\mu^S} R_m^P \boldsymbol{h}^P - R_m^S \boldsymbol{h}^S\right)\right] = 0, \quad \boldsymbol{n} \cdot R_m^p \boldsymbol{h}^p = \boldsymbol{j},$$

$$\boldsymbol{n} \cdot \left(\frac{\varepsilon^P}{\varepsilon^S} R_m^P \boldsymbol{e}^P - R_m^S \boldsymbol{e}^S\right) = 0, \quad \boldsymbol{n} \cdot R_m^P \boldsymbol{e}^P = 0,$$

where $\boldsymbol{n}$ is the normal vector to the surface, δ_{ik} is the Kronecker delta, $p(\boldsymbol{x})$ is the disturbed pressure, ε is the electric susceptibility and $\boldsymbol{j}$ is the current vector. For the problem under consideration, the following relations hold

$$H_{01} = H_{02} = 0, \quad H_{03} \neq 0, \quad \frac{\partial}{\partial x_3}() \equiv 0. \tag{4.36}$$

Moreover, we assume that conditions of no electric charge and equal magnetic and electric susceptibilities of the two media are satisfied, i.e. $\mu^p = \mu^s$, $\varepsilon^p = \varepsilon^s$.

In the cylindrical coordinate system (x_1, x_2, x_3) associated with the scatterer, Eqs. (4.34) reduce to

$$\begin{gathered} \rho_0 \frac{\partial v_1}{\partial t} = -c_0^2 \frac{\partial \rho}{\partial x_1} - P_H R_m v_1, \quad \rho_0 \frac{\partial v_2}{\partial t} = -c_0^2 \frac{\partial \rho}{\partial x_2} - P_H R_m v_2, \\ \frac{\partial \rho}{\partial t} + \rho_0 \frac{1}{x_1} \left[\frac{\partial}{\partial x_1}(v_1 x_1) + \frac{\partial}{\partial x_2} v_2 \right] = 0, \end{gathered} \tag{4.37}$$

$$\begin{gathered} \frac{1}{x_1} \left[\frac{\partial}{\partial x_1}(h_1 x_1) + \frac{\partial h_2}{\partial x_2} \right] = 0, \quad \frac{1}{x_1} \frac{\partial h_3}{\partial x_2} = v_2, \quad \frac{\partial h_3}{\partial x_2} = v_1, \\ \frac{1}{x_1} \left[\frac{\partial}{\partial x_1}(e_1 x_1) + \frac{\partial e_2}{\partial x_2} \right] = 0. \end{gathered} \tag{4.38}$$

The matching conditions (4.35) correspond to the undisturbed surface $x_1 = 1$ and, by virtue of perfect conductivity of the cylinder, reduce to the form given by

$$v_1^P = 0, \quad R_m^P h_3^P = j_3, \quad e_2^P = 0, \tag{4.39}$$

while the other conditions are satisfied identically. Moreover, for $x_1 \to \infty$ the functions of interest should satisfy the radiation and boundedness conditions in the external domain.

We note that Eqs. (4.37) contain terms associated with the magnetic field effect, form a closed system, and can be considered individually. As follows from the boundary conditions (4.39), the coupling between the "electromagnetic" and "acoustic" values vanishes in the approximation of interest. Hence we first find ρ and $\boldsymbol{v}$ from Eqs. (4.37) and (4.39), and then obtain $\boldsymbol{h}$ and $\boldsymbol{e}$ from (4.38) taking into account (4.39).

In order to study the type of the emitted wave, we have to solve the radiation problem in the coordinate system $(\bar{x}_1, \bar{x}_2, \bar{x}_3)$ associated with the radiator (Fig. 4.8). Suppose that the functions of interest are given by

$$f(\bar{x}_1, \bar{x}_2, \bar{x}_3, t) = f^*(\bar{x}_1, \bar{x}_2, \bar{x}_3) e^{-i\omega t}. \tag{4.40}$$

The asterisks and the factor $e^{-i\omega t}$ are omitted in what follows.

Within the context of (4.40) and the axial symmetry condition

$$\frac{\partial}{\partial \bar{x}_2}() \equiv 0,$$

Eqs. (4.37) and (4.38) yield two subsystems, i.e.

$$\rho_0 \mathrm{i}\,\omega\, v_1 = c_0^2 \frac{\partial \rho}{\partial \bar{x}_1} + P_H R_m v_1, \quad \mathrm{i}\,\omega\,\rho = \rho_0 \frac{1}{\bar{x}_1}\frac{\partial}{\partial \bar{x}_1}(v_1 \bar{x}_1) \tag{4.41}$$

and

$$\frac{\partial h_3}{\partial \bar{x}_1} = v_1, \quad \frac{1}{\bar{x}_1}\frac{\partial}{\partial \bar{x}_1}(e_2 \bar{x}_1) = \mathrm{i}\,\omega\, h_3, \quad \frac{1}{\bar{x}_1}\left[\frac{\partial}{\partial \bar{x}_1}(e_2 \bar{x}_1)\right] = 0. \tag{4.42}$$

Equations (4.41) can be reduced to the Bessel equation

$$\frac{\partial^2 \rho}{d\bar{x}_1^2} + \frac{1}{\bar{x}_1}\frac{d\rho}{d\bar{x}_1} - p^2 \rho = 0, \tag{4.43}$$

where

$$p^2 = \frac{\omega^2}{c_0^2} + \mathrm{i}\frac{\omega}{\rho_0 c_0^2} P_H R_m.$$

The solution of Eq. (4.43) is given by

$$\rho = H_0^{(1)}(p\bar{x}_1). \tag{4.44}$$

Here $H_0^{(1)}(p\bar{x}_1)$ is the Hankel function of the first kind, ω and p satisfy the conditions

$$\omega > 0, \quad \mathrm{Re}\, p > 0, \quad \mathrm{Im}\, p > 0, \quad \mathrm{Im}\, p^2 > 0.$$

Equations (4.41) and (4.42) yield expressions for v_1, h_3, e_2, i.e.

$$\begin{gathered} v_1 = -\frac{c_0^2 p}{\rho_0 \mathrm{i}\,\omega - P_H R_m} H_1^{(1)}(p\bar{x}_1),\ h_3 = \frac{c_0^2}{\rho_0 \mathrm{i}\,\omega - P_H R_m} H_1^{(1)}(p\bar{x}_1), \\ e_2 = \frac{\mathrm{i}\,\omega\, c_0^2}{p(\rho_0 \mathrm{i}\,\omega - P_H R_m)} H_1^{(1)}(p\bar{x}_1), \end{gathered} \tag{4.45}$$

All the other components are equal to zero, i.e.

$$v_2 = 0,\ h_1 = 0,\ h_2 = 0,\ e_1 = 0, e_3 = 0.$$

In the external domain, the wave scattered by a cylinder is described by Eqs. (4.37); the latter can be transformed to the form

$$x_1^2 \frac{\partial^2 \rho}{\partial x_1^2} + \frac{\partial^2 \rho}{\partial x_2^2} + \frac{\partial \rho}{\partial x_1} + x_1^2 p^2 \rho = 0. \tag{4.46}$$

The solution of Eq. (4.46) with condition (4.44) is given by

$$\rho = \sum_{\nu=0}^{\infty} C_\nu H_\nu^{(1)}(px_1)\cos \nu x_2. \tag{4.47}$$

The series of the form (4.47) absolutely and uniformly converge pointwise outside the cylinder and on its surface (with p and x_2 being fixed) [17].

We employ the Neumann addition theorem [27], and thus write an expression for the diffracted field in the coordinate system associated with the scatterer for $x_1 < b$, i.e.

$$\rho^d = \sum_{\nu=0}^{\infty} [\varepsilon_\nu J_\nu(px_1) H_\nu^{(1)}(pb) + C_\nu H_\nu^{(1)}(px_1)] \cos \nu x_2. \tag{4.48}$$

Equations (4.38), (4.45) and (4.48) yield v_1, v_2 and h_3 to be given by

$$v_1^d = -\frac{c_0^2}{R_m P_H - \rho_0 \mathrm{i}\,\omega} \sum_{\nu=0}^{\infty} [\varepsilon_\nu J_\nu'(px_1) H_\nu^{(1)}(pb) + C_\nu H_\nu^{(1)\prime}(px_1)] \cos \nu x_2, \tag{4.49}$$

$$v_2^d = \frac{c_0^2}{R_m P_H - \rho_0 \mathrm{i}\omega} \sum_{\nu=0}^{\infty} [\varepsilon_\nu J_\nu(px_1) H_\nu^{(1)}(pb) + C_\nu H_\nu^{(1)}(px_1)] \frac{\nu}{x_1} \sin \nu x_2, \tag{4.50}$$

$$h_3^d = -\frac{c_0^2}{R_m P_H - \rho_0 \mathrm{i}\omega} \sum_{\nu=0}^{\infty} [\varepsilon_\nu J_\nu(px_1) H_\nu^{(1)}(pb) + C_\nu H_\nu^{(1)}(px_1)] \cos \nu x_2. \tag{4.51}$$

It follows from (4.38) that in the class of perturbations under consideration $e_3 \equiv 0$, and hence $h_1 \equiv 0$, $h_2 \equiv 0$.

The coefficient C_ν is determined by condition (4.39) and expression (4.49). Thus, we have

$$C_\nu = -\varepsilon_\nu \frac{J_{\nu-1}(p) - J_{\nu+1}(p)}{H_{\nu-1}^{(1)}(p) - H_{\nu+1}^{(1)}(p)} H_\nu^{(1)}(pb). \tag{4.52}$$

The components e_1 and e_2 of the electric field intensity are described by the equations

$$\frac{1}{x_1}\left[\frac{\partial}{\partial x_1}(e_2 x_1)-\frac{\partial e_1}{\partial x_2}\right]=\mathrm{i}\,\omega\, h_3,\ \frac{1}{x_1}\left[\frac{\partial}{\partial x_1}(e_1 x_1)-\frac{\partial e_2}{\partial x_2}\right]=0. \tag{4.53}$$

Thus, we obtain an inhomogeneous equation for the scattered field from (4.53) and (4.51), i.e.

$$\frac{\partial^2 e_1}{\partial x_1^2}+\frac{3}{x_1}\frac{\partial e_1}{\partial x_1}+\frac{1}{x_1^2}e_1+\frac{1}{x_1^2}\frac{\partial^2 e_1}{\partial x_2^2}=-\frac{c_0^2\mathrm{i}\,\omega}{R_m P_H-\rho_0\mathrm{i}\,\omega}\sum_{\nu=0}^{\infty}\nu C_\nu H_\nu^{(1)}(px_1)\sin\nu x_2. \tag{4.54}$$

The homogeneous equation corresponding to (4.54) is solved by the method of separation of variables. Thus, we obtain an Euler equation [19] whose solution should be chosen within the context of the conditions at infinity. The particular solution to Eq. (4.54) should be found by varying the arbitrary constants.

The complete solution of Eq. (4.54) is given by

$$\begin{aligned} e_1 = {} & \sum_{k=0}^{\infty}\bar{C}_k x_1^{-1-k}\sin kx_2+\frac{1}{2}\frac{c_0^2\mathrm{i}\omega}{R_m P_H-\rho_0\mathrm{i}\omega} \\ & \times\sum_{\nu=0}^{\infty}C_\nu\left[x_1^{-1-\nu}\int_1^{x_1}\frac{H_\nu^{(1)}(p\xi)}{\xi^{-1-\nu}}\mathrm{d}\xi-x_1^{-1+\nu}\int_1^{x_1}\frac{H_\nu^{(1)}(p\xi)}{\xi^{-1+\nu}}\mathrm{d}\xi\right]\sin\nu x_2. \end{aligned} \tag{4.55}$$

Substituting this expression in the second Eq. (4.53) yields

$$\begin{aligned} e_2 = {} & -\sum_{k=0}^{\infty}\bar{C}_k x_1^{-1-k}\cos kx_2-\frac{1}{2}\frac{c_0^2\mathrm{i}\omega}{R_m P_H-\rho_0\mathrm{i}\omega} \\ & \times\sum_{\nu=0}^{\infty}C_\nu\left[x_1^{-1-\nu}\int_1^{x_1}\frac{H_\nu^{(1)}(p\xi)}{\xi^{-1-\nu}}\mathrm{d}\xi+x_1^{-1+\nu}\int_1^{x_1}\frac{H_\nu^{(1)}(p\xi)}{\xi^{-1+\nu}}\mathrm{d}\xi\right]\cos\nu x_2. \end{aligned} \tag{4.56}$$

In view of the boundary condition (4.39), the constant $\bar{C}_k$ in expressions (4.54) and (4.55) is equal to zero.

Now we construct a solution for the distant scattered field ρ in the longwave approximation. In this case, asymptotic expressions for large px_1 are substituted for the Hankel functions in solution (4.47); C_0 and C_1 are dominant among the coefficients C_ν determined by formula (4.52). Moreover, we consider only small values of $P_H R_m$. Then all the functions contained in the solution can be regarded as an asymptotic expansion in $P_H R_m$. Thus, we obtain the solution in terms of several analytic expressions and can analyse them.

Some rather cumbersome transformations yield the expressions for ρ given by

$$\mathrm{Re}\rho = -2\exp\left[-\frac{1}{2}\frac{P_H R_m}{\rho_0 c_0}(b+x_1)\right]\left[1+\frac{1}{16}\left(\frac{P_H R_m}{\omega\rho_0}\right)\right]^{-2} \times \left(\cos kb - \frac{P_H R_m}{\omega\rho_0}\sin kb\right)(1+2\cos x_2), \tag{4.57}$$

$$\mathrm{Im}\rho = -2\exp\left[-\frac{1}{2}\frac{P_H R_m}{\rho_0 c_0}(b+x_1)\right]\left[1+\frac{1}{16}\left(\frac{P_H R_m}{\omega\rho_0}\right)\right]^{-2} \times \left(\sin kb + \frac{P_H R_m}{\omega\rho_0}\cos kb\right)(1+2\cos x_2), \tag{4.58}$$

where

$$k = \frac{\omega}{c_0}.$$

The factor $(1/kx_1)^{1/2}\exp(ikx_1)$ is omitted in the right-hand parts of (4.57) and (4.58); the lefthand parts are normalised to the factor $4(kb)^{1/2}k^{-2}$. The field remotedness criterion and maximum relative error for given calculation accuracy ς_e are given, respectively, by

$$px_1 \geq \frac{4\nu^2-1}{8\varsigma_e}, \quad \frac{1}{2}p^2\ln\frac{p}{2} \leq \varsigma_e.$$

With no magnetic field, when $R_m P_H = 0$, expressions (4.57) and (4.58) yield the known solution of the acoustic problem [64].

It is easy to show within the context of (4.57) and (4.58) that, for fixed b and x_1, the scattered field amplitude decreases exponentially as either undisturbed magnetic field intensity H_0 or electric conductivity σ increases. Moreover, variations of the magnetic field intensity or electric conductivity influence the positions of zeros and extremums of the functions Reρ and Imρ on the arc coordinate x_2. The latter observation is important for the determination of the beam pattern of the scattered field.

4.5 Wave Diffraction by a System of Cylinders

A system of arbitrarily arranged circular cylinders is studied. Interaction of surface gravity waves with large cross-section obstacles is considered for the case when the diffraction effects are dominant and the influence of the viscosity can be disregarded. In the case of a system of obstacles, the wave field is much more involved since the diffracted waves are reflected from each obstacle. In particular, these effects were studied in Refs. [7, 9, 18, 24, 25, 30–33, 39, 67–69, 72, 73, 81]. The theory of diffraction of electromagnetic, acoustic and elastic waves in multiconnected domains was considered in Refs. [6, 10, 17, 65]. The authors of Ref. [79] showed that the porosity can considerably decrease the hydrodynamical loads on the cylinders, Ref. [23] deals with the problem of steep Stokes waves, Ref. [80] gives the solution for truncated vertical cylinders.

Here we apply the linear theory of small-amplitude waves to find the solution to the problem of diffraction of surface gravity waves by a system of arbitrarily arranged vertical circular cylinders in a fluid of finite depth [1]. Analytic expressions are derived for the wave forces and overturning moments applied to each cylinder of the system. Extremum properties of the wave loads are considered as functions of the wave number. Maximum loads are shown to appear at a non-frontal obstacle. As an example, a system of obstacles consisting of two three-cylinder rows is considered. For some certain ratios of the cylinder diameters, intercylinder distances, the angle of wave incidence and the wavelength, the maximum values of the hydrodynamic forces applied to the cylinders calculated with regard for the interaction of diffracted fields can be much greater than those disregarding the latter interaction.

Suppose that plane regular waves are incident on a system of arbitrarily arranged vertical circular cylinders of radius a_s along the unit vector $\boldsymbol{n}$ at angle α to the positive direction of the Ox axis, the cylinders being submerged in a fluid of finite depth H (Fig. 4.9). The fluid is supposed to be ideal and incompressible, the motion is assumed to be irrotational. Each scatterer is described in an individual polar

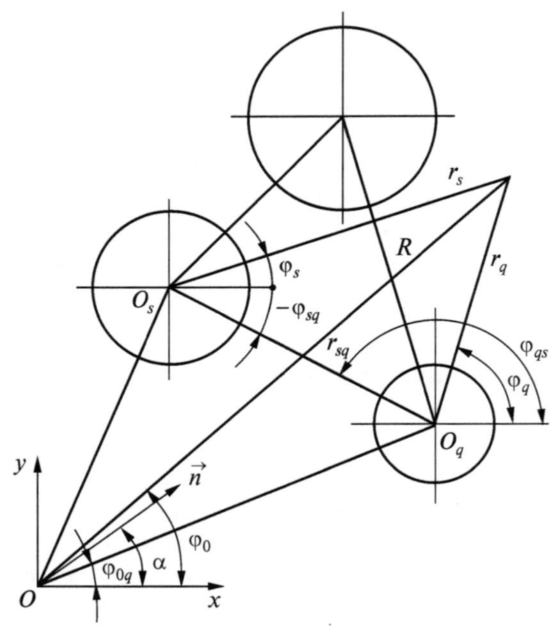

Fig. 4.9 Configuration of the problem

coordinate system (r_s, θ_s) whose centre coincides with the centre of the obstacle. The coordinate system *Oxy* is also introduced, as shown in Fig. 4.9.

The velocity potential $\Phi_s(x_s, y_s, z_s, t)$ is determined by the Laplace equation, the latter in the polar system is given by

$$\frac{1}{r_s}\frac{\partial}{\partial r_s}\left(r_s\frac{\partial \Phi_s}{\partial r_s}\right) + \frac{1}{r_s^2}\frac{\partial^2 \Phi_s}{\partial \theta_s^2} + \frac{\partial^2 \Phi_s}{\partial z_s^2} = 0. \tag{4.59}$$

The function Φ_s is assumed to satisfy the conditions given below, i.e.

$$\text{at the bottom:} \quad \left(\frac{\partial \Phi_s}{\partial z_s}\right)_{z_s = -H} = 0, \tag{4.60}$$

$$\text{on the cylinder surface:} \quad \left(\frac{\partial \Phi_s}{\partial r_s}\right)_{r_s = a_s} = 0, \tag{4.61}$$

$$\text{on the free surface of the fluid:} \quad \frac{\partial^2 \Phi_s}{\partial t^2} + g\frac{\partial \Phi_s}{\partial z_s} = 0, \quad z_s = 0, \tag{4.62}$$

as well as radiation and limitedness conditions at infinity, i.e.

$$\lim_{r_s \to \infty} \sqrt{r_s}\left(\frac{\partial \Phi_s}{\partial r_s} - ik\Phi_s\right) = 0. \tag{4.63}$$

The velocity potential for the diffracted field can be written as

$$\Phi = \mathrm{i}g\frac{A}{\omega}\frac{\cosh k(z+H)}{\cosh kH}\left[e^{\mathrm{i}k(\boldsymbol{n}\cdot\boldsymbol{R})} + \Phi_s\right]e^{-\mathrm{i}\omega t}, \tag{4.64}$$

where Φ_s is the potential of the scattered field, $\boldsymbol{R}$ is the radius vector of the point under consideration.

The wave number $k = 2\pi/L$, the oscillation frequency $\omega = 2\pi/T$ and fluid depth H satisfy the relation

$$\omega^2 = kg \tanh kH.$$

Hence, in the local coordinates of the s-th cylinder we have

$$\Phi_s = \mathrm{i}g\frac{A}{\omega}\frac{\cosh k(z_s+H)}{\cosh kH}\left[e^{\mathrm{i}k(\boldsymbol{n}\cdot\boldsymbol{r}_{0s}) + \mathrm{i}k(\boldsymbol{n}\cdot\boldsymbol{r}_s)} + \Phi_s^j\right]e^{-\mathrm{i}\omega t}. \tag{4.65}$$

We substitute (4.65) in the Laplace equation (4.59) and thus find that the scattered field potential Φ_s^j should satisfy the Helmholtz equation

$$\frac{1}{r_s}\frac{\partial}{\partial r_s}\left(r_s\frac{\partial \Phi_s^j}{\partial r_s}\right)+\frac{1}{r_s^2}\frac{\partial^2 \Phi_s^j}{\partial \theta_s^2}+k^2\Phi_s^j=0. \tag{4.66}$$

The factor $\exp(-\mathrm{i}\,\omega\, t)$ is omitted here.

We present the velocity potential of the incident wave in the s-th coordinate system as an expansion in terms of cylindrical functions, i.e.

$$e^{\mathrm{i}k(\boldsymbol{n}\cdot\boldsymbol{R})}=e^{ikr_{0s}\cos(\theta_{0s}-\alpha)}\sum_{n=0}^{\infty}\varepsilon_n \mathrm{i}^n J_n(kr_s)\cos n\,(\theta_s-\alpha), \tag{4.67}$$

where $\varepsilon_0=1;\varepsilon_n=2,n\geq 1,2,\ldots$.

The velocity potential of the scattered field with the Sommerfeld radiation and decay condition taken into account is given by

$$\Phi_s^j=\sum_{n=0}^{\infty}\begin{pmatrix}C_n^{(s)}\\ B_n^{(s)}\end{pmatrix}H_n^{(1)}(kr_s)\begin{pmatrix}\cos n\theta_s\\ \sin n\theta_s\end{pmatrix}+\left(\sum_{q=1}^{s-1}+\sum_{q=s+1}^{N}\right)\sum_{n=0}^{\infty}\begin{pmatrix}C_n^{(q)}\\ B_n^{(q)}\end{pmatrix}H_n^{(1)}\begin{pmatrix}\cos n\theta_q\\ \sin n\theta_q\end{pmatrix}. \tag{4.68}$$

In order to match the local coordinate systems to the axis of the s–th cylinder, we employ the addition theorems for the cylindrical functions, i.e.

$$\begin{aligned}H_m^{(1)}\begin{pmatrix}\cos m\theta_s\\ \sin m\theta_s\end{pmatrix}=\sum_{n=0}^{\infty}J_n(kr_q)\Bigg\{\Bigg[&H_{m-n}^{(1)}(kr_{sq})\begin{pmatrix}\cos(m-n)\theta_{sq}\\ \sin(m-n)\theta_{sq}\end{pmatrix}\\ &+(-1)^n\delta_n H_{m+n}^{(1)}(kr_{sq})\begin{pmatrix}\cos(m+n)\theta_{sq}\\ \sin(m+n)\theta_{sq}\end{pmatrix}\Bigg]\cos n\theta_{sq}\\ &\pm\Bigg[H_{m-n}^{(1)}(kr_{sq})\begin{pmatrix}\sin(m-n)\theta_{sq}\\ \cos(m-n)\theta_{sq}\end{pmatrix}\\ &-(-1)^n\delta_n H_{m+n}^{(1)}(kr_{sq})\times\begin{pmatrix}\sin(m+n)\theta_{sq}\\ \cos(m+n)\theta_{sq}\end{pmatrix}\Bigg]\sin n\theta_q\Bigg\}.\end{aligned} \tag{4.69}$$

Then the velocity potential Φ_s can be written as

$$\Phi_s = \mathrm{i}g\frac{A}{\omega}\frac{\cosh k(z_s+H)}{\cosh kH}\left\{e^{\mathrm{i}kr_{0s}\cos(\theta_{0s}-\alpha)}\sum_{n=0}^{\infty}\varepsilon_n \mathrm{i}^n J_n(kr_s)\cos n(\theta_s-\alpha)\right.$$
$$+\sum_{n=0}^{\infty}\begin{pmatrix}C_n^{(s)}\\ B_n^{(s)}\end{pmatrix}H_n^{(1)}(kr_s)\begin{pmatrix}\cos n\theta_s\\ \sin n\theta_s\end{pmatrix}+\left(\sum_{q=1}^{s-1}+\sum_{q=s+1}^{N}\right)\sum_{n=0}^{\infty}\begin{pmatrix}C_n^{(q)}\\ B_n^{(q)}\end{pmatrix}\sum_{m=0}^{\infty}J_m(kr_s)$$
$$\times\left[\left[H_{n-m}^{(1)}(kr_{sq})\begin{pmatrix}\cos(n-m)\theta_{sq}\\ \sin(n-m)\theta_{sq}\end{pmatrix}+(-1)^m\delta_m H_{n+m}^{(1)}(kr_{sq})\begin{pmatrix}\cos(n+m)\theta_{sq}\\ \sin(n+m)\theta_{sq}\end{pmatrix}\right]\cos m\theta_s\right.$$
$$\left.\left.+\left[H_{n-m}^{(1)}(kr_{sq})\begin{pmatrix}\sin(n-m)\theta_{sq}\\ \cos(n-m)\theta_{sq}\end{pmatrix}-(-1)^m\delta_m H_{n+m}^{(1)}(kr_{sq})\begin{pmatrix}\sin(n+m)\theta_{sq}\\ \cos(n+m)\theta_{sq}\end{pmatrix}\right]\sin m\theta_s\right]\right\}. \tag{4.70}$$

To find the unknown coefficients C_n and B_n, we substitute (4.70) in the boundary condition (4.61), and thus obtain two systems of linear algebraic equations given by

$$\begin{aligned}C_n^{(s)}+\left(\sum_{q=1}^{s-1}+\sum_{q=s+1}^{N}\right)\sum_{m=0}^{\infty}\left[\gamma_{mn}^{(s,\,q)}C_m^{(q)}+\delta_{mn}^{(s,\,q)}B_m^{(q)}\right]=K_n^{(s)},\\ B_n^{(s)}+\left(\sum_{q=1}^{s-1}+\sum_{q=s+1}^{N}\right)\sum_{m=0}^{\infty}\left[\alpha_{mn}^{(s,\,q)}B_m^{(q)}+\beta_{mn}^{(s,\,q)}C_m^{(q)}\right]=F_n^{(s)},\end{aligned} \tag{4.71}$$

where

$$\gamma_{mn}^{(s,\,q)}=\frac{J_n'(ka_s)}{H_n^{(1)\prime}(ka_s)}\left[H_{m-n}^{(1)}(kr_{sq})\cos(m-n)\theta_{sq}+(-1)^n\delta_n H_{m-n}^{(1)}(kr_{sq})\cos(m+n)\theta_{sq}\right];$$
$$\gamma_{mn}^{(s,\,q)}=\frac{J_n'(ka_s)}{H_n^{(1)\prime}(ka_s)}\left[H_{m-n}^{(1)}(kr_{sq})\sin(m-n)\theta_{sq}+(-1)^n\delta_n H_{m+n}^{(1)}(kr_{sq})\sin(m+n)\theta_{sq}\right];$$
$$\alpha_{mn}^{(s,\,q)}=\frac{J_n'(ka_s)}{H_n^{(1)\prime}(ka_s)}\left[H_{m-n}^{(1)}(kr_{sq})\cos(m-n)\theta_{sq}-(-1)^n\delta_n H_{m+n}^{(1)}(kr_{sq})\cos(m+n)\theta_{sq}\right];$$
$$\beta_{mn}^{(s,\,q)}=\frac{J_n'(ka_s)}{H_n^{(1)\prime}(ka_s)}\left[H_{m-n}^{(1)}(kr_{sq})\sin(m-n)\theta_{sq}-(-1)^n\delta_n H_{m+n}^{(1)}(kr_{sq})\sin(m+n)\theta_{sq}\right];$$

$$K_n^{(s)}=-\frac{e^{\mathrm{i}kr_{0s}\cos(\theta_{0s}-\alpha)}\varepsilon_n \mathrm{i}^n J_n'(ka_s)\cos n\alpha}{H_n^{(1)\prime}(ka_s)};$$
$$F_n^{(s)}=-\frac{e^{\mathrm{i}kr_{0s}\cos(\theta_{0s}-\alpha)}\varepsilon_n \mathrm{i}^n J_n'(ka_s)\sin n\alpha}{H_n^{(1)\prime}(ka_s)}$$

System (4.71) can be regularised by means of the substitution [17]. We replace the unknowns C_n and B_n by interrelated new variables, i.e.

$$\bar{C}_n^{(s)} = C_n^{(s)} H_n^{(1)\prime}(ka_s),\ \bar{B}_n^{(s)} = B_n^{(s)} H_n^{(1)\prime}(ka_s).$$

In this way, the system of linear algebraic equations (4.71) becomes convergent. The convergence of such systems is proved in Ref. [17]; the solution can be found by means of the reduction method or in terms of successive approximations.

The hydrodynamic pressure of the total wave field on the cylinder is governed by the equation

$$p_s = \rho \frac{\partial \Phi_s}{\partial t} \quad \text{for} \quad r_s = a_s, \tag{4.72}$$

or, in the detailed form,

$$\begin{aligned} p_s = \rho g A \frac{\cosh k(z_s + H)}{\cosh kH} \Bigg\{ & e^{ikr_{0s}\cos(\theta_{0s} - \alpha)} \sum_{n=0}^{\infty} \varepsilon_n \mathrm{i}^n J_n(kr_s) \cos n(\theta_s - \alpha) \\ & + \sum_{n=0}^{\infty} \begin{pmatrix} C_n^{(s)} \\ B_n^{(s)} \end{pmatrix} H_n^{(1)}(kr_s) \begin{pmatrix} \cos n\theta_s \\ \sin n\theta_s \end{pmatrix} + \left(\sum_{q=1}^{s-1} + \sum_{q=s+1}^{N} \right) \\ & \times \sum_{n=0}^{\infty} \begin{pmatrix} C_n^{(q)} \\ B_n^{(q)} \end{pmatrix} \sum_{m=0}^{\infty} J_m(kr_s) \left[H_{n-m}^{(1)}(kr_{sq}) \begin{pmatrix} \cos(n-m)\theta_{sq} \\ \sin(n-m)\theta_{sq} \end{pmatrix} \right. \\ & \left. + (-1)^m \delta_m H_{m-n}^{(1)}(kr_{sq}) \begin{pmatrix} \cos(n+m)\theta_{sq} \\ \sin(n+m)\theta_{sq} \end{pmatrix} \right] \cos m\theta_s \\ & \mp \left[H_{n-m}^{(1)}(kr_{sq}) \begin{pmatrix} \sin(n-m)\theta_{sq} \\ \cos(n-m)\theta_{sq} \end{pmatrix} - (-1)^m \delta_m H_{n+m}^{(1)}(kr_{sq}) \begin{pmatrix} \sin(n+m)\theta_{sq} \\ \cos(n+m)\theta_{sq} \end{pmatrix} \right] \sin m\theta_s \Bigg\} e^{-\mathrm{i}\omega t}. \end{aligned} \tag{4.73}$$

Formula (4.73) provides a possibility to express the hydrodynamic force and the overturning moment along the *Ox* and *Oy* axes in terms of pressure. We have

$$\begin{aligned} F_x &= \int_0^d \int_0^{2\pi} p_s a_s \cos \theta_s \mathrm{d}\theta_s \mathrm{d}z, \\ F_y &= \int_0^d \int_0^{2\pi} p_s a_s \sin \theta_s \mathrm{d}\theta_s \mathrm{d}z, \\ M_x &= \int_0^d \int_0^{2\pi} p_s a_s z \sin \theta_s \mathrm{d}\theta_s \mathrm{d}z, \\ M_y &= \int_0^d \int_0^{2\pi} p_s a_s z \cos \theta_s \mathrm{d}\theta_s \mathrm{d}z. \end{aligned} \tag{4.74}$$

Thus, we have derived analytic expressions for the wave forces and overturning moments applied to a system of vertical circular columns that are arbitrarily arranged in space.

For example, we consider a system of two three-cylinder rows of circular vertical cylinders (Fig. 4.10). The system consists of six cylinders; hence in order to find approximate values of the diffraction coefficients, we have to solve a system of $12(N+1)$ equations with $12(N+1)$ unknowns that follows from the expressions (4.71) truncated to the M-th order, i.e. with $n, m = 0, 1, \ldots, M; N = 0, 1, 2 \ldots, M$. It was shown in Ref. [17] that the truncation order can be found from the relations $m,\ n \geq M = [2ka]$, where $[2ka]$ implies the integer part of $2ka$. In our calculation of the diffraction coefficients, system (4.71) was truncated to the order $m,\ n \geq M \approx [2ka+2]$. For greater truncation orders, the absolute error of the calculation of wave forces and overturning moments does not exceed 1%.

Figure 4.10 shows the distributions of the absolute values of wave forces F_x applied to each cylinder along the Ox axis with the interaction of diffracted fields (solid curves) taken into account for various angles α of wave incidence on the cylinder system versus wave number ka. Here and below the interrow distance is $b/a = 2$, the intercylinder distance in a row is $l/a = 4$, the fluid depth is $d/a = 5$. It can be seen that, within the considered range of wavenumber variation, the absolute values of F_x attain their maxima at $ka \approx 0.52$ for the first, second, fifth and sixth cylinders. Moreover, the values of F_x increase as the wave incidence angle decreases. The dashed curves demonstrate the dependence of the horizontal load F_x for a single cylinder regardless of the interaction of diffracted fields. Comparing these curves shows that, in the points of local maxima, the values of F_x calculated

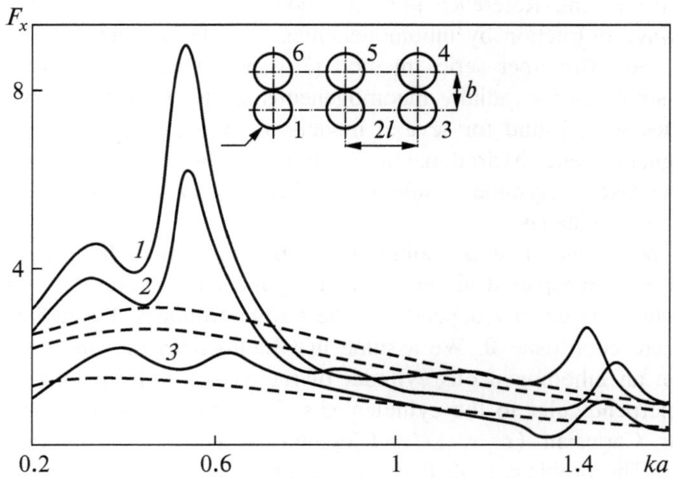

Fig. 4.10 The modulus of the lateral force F_x as a function of the wave number ka (*solid curves*). *Dashed curves* correspond to the case of a single cylinder. The wave incidence angle is (*1*) $\alpha = 0$, (*2*) $\pi/6$, (*3*) $\pi/3$

with taking into account the interaction of diffracted fields are much greater than the similar values of F_x calculated regardless of the interaction. For example, the value of F_x is three times greater than the similar value of F_x for a single cinder at $\alpha = 0$ and $ka \approx 0.52$.

Calculations of the lateral forces F_y show that interaction of diffracted fields causes some redistribution of the projections of forces applied to the cylinder system. Namely, the force F_x is greater and the force F_y is smaller than the corresponding values of F_x and F_y plotted regardless of the interaction.

4.6 Wave Scattering by an Asymmetrically Inhomogeneous Cylinder

Nowadays, only few exact solutions of the problems of wave scattering by local transparent inhomogeneities are known, all these concern radially symmetric inhomogeneities with the quadratic law of variation of scatterer properties [47]. The main complications are caused by the need to find the solutions of partial differential equations with variable coefficients. An elaborate review of relevant studies is given in [47, 48]. Here we mention only some papers. In Ref. [61], two classes of new exact analytic solutions were obtained in the closed form. References [11, 37, 38] deal with the electromagnetic wave diffraction by a radially inhomogeneous plasma cylinder. The authors of [20, 66] consider the problem for a scalar wave field. The majority of the studies seem to concern plasma inhomogeneities in view of the required diagnostics of the electron density radial distribution in the course of plasma confinement. Reference [47] also considers the main approaches to the study of wave diffraction by inhomogeneities, e.g. the asymptotic Born approximation [3, 36], Bremmer series method [76], power series method [47]. Exact analytical solutions for radially inhomogeneous scatterers in terms of convergent power series were found for several models of inhomogeneous media [62], i.e. electromagnetic field, hydrodynamic medium, shallow water, membrane, bent plate. In the case of asymmetric inhomogeneities, constructing new exact solutions is far more complicated.

Here a new class of exact solutions of the problem of hydrodynamic wave diffraction by a transparent inhomogeneous cylinder-type scatterer is obtained for the case when the density depends on the two coordinates, the radial coordinate r and the arc coordinate θ. We assume that monochromatic pressure waves are incident on an inhomogeneous cylinder of radius $r = r_0$ from infinity along the direction perpendicular to the cylinder axis. The cylinder is associated with the rectangular Cartesian (x, y, z) and cylindrical (r, θ, z) coordinate systems (Fig. 4.11). The problem is to find the scattered pressure field p_s in the exterior domain Ω^e filled with a homogeneous medium with density ρ_0 and the pressure field p in the interior domain Ω filled with an inhomogeneous medium with density $\rho(r, \theta)$.

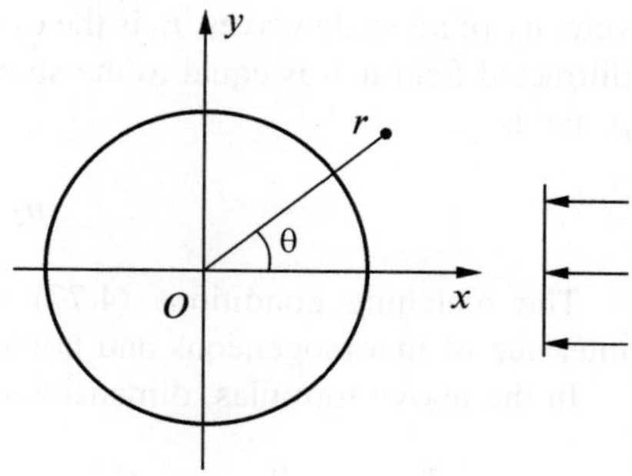

Fig. 4.11 Configuration of the problem

The mathematical formulation of the problem includes the Helmholtz equation in the exterior domain, i.e.

$$\frac{\partial^2 p_2}{\partial r^2}+\frac{1}{r}\frac{\partial p_2}{\partial r}+\frac{1}{r^2}\frac{\partial^2 p_2}{\partial \theta^2}+k_2^2 p_2=0 \quad \text{in} \quad \Omega^e \tag{4.75}$$

the Helmholtz equation with variable coefficients in the interior domain, i.e.

$$\begin{aligned}&\frac{\partial^2 p}{\partial r^2}+\frac{1}{r}\frac{\partial p}{\partial r}-\frac{1}{\rho(r,\theta)}\frac{\partial \rho(r,\theta)}{\partial r}\frac{\partial p}{\partial r}+\frac{1}{r^2}\frac{\partial^2 p}{\partial \theta^2}\\&-\frac{1}{\rho(r,\theta)r^2}\frac{\partial \rho(r,\theta)}{\partial \theta}\frac{\partial p}{\partial \theta}+k_0^2(r)p=0 \quad \text{in} \quad \Omega,\end{aligned} \tag{4.76}$$

the matching conditions given by

$$p|_{r=1}=p_2|_{r=1}, \quad \frac{1}{\rho(r,\theta)}\frac{\partial p}{\partial r}|_{r=1}=\frac{\partial p_2}{\partial r}|_{r=1}, \tag{4.77}$$

the regularity conditions for p,

$$p=0 \quad \text{at} \quad r=0, \tag{4.78}$$

the Sommerfeld radiation conditions for p_s,

$$\lim_{r\to\infty} r\left(\frac{\partial p_s}{\partial r}-\mathrm{i}k_2 p_s\right)=0, \quad \lim_{r\to\infty} p_s=0. \tag{4.79}$$

and the expression for the incident wave

$$p_i=p_{0i}\,e^{\mathrm{i}(k_2 r\cos\theta+\omega t)}. \tag{4.80}$$

Here t is time, p_{0i} is the amplitude value of the pressure in the incident wave, k_2 and $k_0(r)$ are wave numbers, $\omega = k_2 c_0$ is the circular frequency, c_0 is the propagation velocity of acoustic waves, r_0 is the cylinder radius, p_2 is the pressure in the exterior diffracted field that is equal to the sum of pressures of the incident p_i and scattered p_s fields,

$$p_2 = p_i + p_s.$$

The matching conditions (4.77) imply that pressures and velocities on the interface of inhomogeneous and homogeneous media are equal.

In the above formulas, dimensionless values with asterisks were introduced,

$$r^* = \frac{r}{r_0},\ \rho^* = \frac{\rho}{\rho_0},\ p^* = \frac{p}{p_0},\ k_2^* = k_2 r_0 = \frac{\omega}{c_0} r_0,\ k_0^*(r) = k_0(r) r_0 = \frac{\omega}{c_0(r)} r_0,$$

and the asterisks are omitted in what follows.

Below the sound velocity $c_0(r)$, the wave number $k_0(r)$ in (4.76) are taken to be constant. This assumption is known to be valid for such problems with high accuracy [62].

We separate the variables in Eq. (4.75) and thus obtain a system given by

$$\begin{aligned} &\frac{d^2 S}{dr^2} + \frac{1}{r}\frac{dS}{dr} + \left(k_2^2 - \frac{m^2}{r^2}\right) = 0, \\ &\frac{d^2 Y}{d\theta^2} + m^2 Y = 0. \end{aligned} \tag{4.81}$$

where $p_2 = S(r)\,Y(\theta)$. The separation of variables in Eq. (4.76) yields a system given by

$$\begin{aligned} &\frac{d^2 R}{d r^2} + \frac{1}{r}\frac{d R}{d r} - \frac{1}{\rho_r(r)}\frac{d\rho_r(r)}{d r}\frac{d R}{d r} + \left(k_0^2(r) - \frac{m^2}{r^2}\right) R = 0, \\ &\frac{d^2 T}{d\theta^2} - \frac{1}{\rho_\theta(\theta)}\frac{d\rho_\theta(\theta)}{d\theta}\frac{d T}{d\theta} + m^2 T = 0, \end{aligned} \tag{4.82}$$

where $p = R(r)\,T(\theta)$. The total field is periodic in θ, hence all the values of the parameter m in (4.81) are integers.

The solution of the first equation of (4.81) can be written in terms of cylindrical functions, i.e.

$$S_m = C_m H_m^{(1)}(k_2 r) + D_m H_m^{(2)}(k_2 r), \tag{4.83}$$

where $H_m^{(1)}(k_2 r)$ and $H_m^{(2)}(k_2 r)$ are Hankel functions of the first and second kinds. According to the radiation principle, the field of interest should be characterised by waves propagating to infinity along radial directions. Within the context of the asymptotic representations [27] for the Hankel functions with large arguments, $|z| \gg 1$, the term $D_m H_m^{(2)}(k_2 r)$ in the solution (4.83) for $r \to \infty$ corresponds to a wave propagating from the source, thus the radiation conditions are satisfied for $C_m = 0$.

The solution of the second equation of the system (4.81) can be written as

$$Y_m = A_m \cos m\theta + B_m \sin m\theta. \quad (4.84)$$

The field of the incident wave is even with respect to $\theta = 0$, so we put $B_m = 0$. Then the general solution of Eq. (4.75) is given by an infinite series

$$p_s = \sum_{m=0}^{\infty} a_m H_m^{(2)}(k_2 r) \cos m\theta. \quad (4.85)$$

with unknown coefficients a_m.

Exact solutions to the equations of system (4.82) associated with the field in the interior domain cannot be found for arbitrary functions $\rho_r(r)$, $\rho_\theta(\theta)$, $c_0^2(r)$. In what follows, we consider the case when the density variations are given by

$$\rho(r,\theta) = e^{a r_0 (1-r)} \sin^2\theta. \quad (4.86)$$

The system of equations (4.82) with ansatz (4.86) taken into account reduces to

$$\begin{aligned} &\frac{\mathrm{d}^2 R}{\mathrm{d} r^2} + \left(\frac{1}{r} + a r_0\right)\frac{\mathrm{d}R}{\mathrm{d}r} + \left(k_0^2(r) - \frac{m^2}{r^2}\right)R = 0, \\ &\frac{\mathrm{d}^2 T}{\mathrm{d}\theta^2} - 2\cot\theta\frac{\mathrm{d}T}{\mathrm{d}\theta} + m^2 T = 0. \end{aligned} \quad (4.87)$$

The solution of the first equation of (4.82) is given by [19]

$$\begin{aligned} R_m(r) = r^m \exp\left(-\left(a r_0 + \sqrt{(a r_0)^2 - 4k^2(r)}\right) r/2\right)\left(\sqrt{(a r_0)^2 - 4k^2(r)}\right)^{1/2+m}, \\ \times {}_1F_1\left(1/2 + m + \frac{a r_0}{2\sqrt{(a r_0)^2 - 4k^2(r)}},\ 2m+1,\ \sqrt{(a r_0)^2 - 4k^2(r)}\,r\right) \end{aligned} \quad (4.88)$$

where ${}_1F_1(a,b,c)$ is the degenerate hypergeometric function.

Substituting $T(\theta) = \eta(\xi)$, and $\xi = \theta + \pi/2$ in the second equation of (4.87), we obtain

$$\eta'' + 2\eta' \tan\xi + m^2\eta = 0. \quad (4.89)$$

This equation has a solution given by

$$\eta = \cos^3 \xi \left(\frac{1}{\cos \xi} \frac{d}{d\xi} \left[\frac{1}{\cos \xi} \frac{d}{d\xi} \nu \right] \right), \tag{4.90}$$

where ν runs through all the solutions of the equation $\nu'' + m^2 + 1 = 0$, i.e.

$$\nu = C_1 \cos\left(\xi\sqrt{m^2+1}\right) + C_2 \sin\left(\xi\sqrt{m^2+1}\right), \quad m^2+1>0.$$

Therefore,

$$\begin{aligned}\eta(\xi) = {} & \sin\xi\left(- C_1\sqrt{m^2+1}\sin\left(\xi\sqrt{m^2+1}\right) + C_2\sqrt{m^2+1}\cos\left(\xi\sqrt{m^2+1}\right)\right) \\ & + \cos\xi\left(- C_1\left(m^2+1\right)\cos\left(\xi\sqrt{m^2+1}\right) - C_2\left(m^2+1\right)\sin\left(\xi\sqrt{m^2+1}\right)\right),\end{aligned} \tag{4.91}$$

$$\begin{aligned}T(\theta) = {} & \cos\theta\left(- C_1\sqrt{m^2+1}\sin\left((\theta+\pi/2)\sqrt{m^2+1}\right) + C_2\sqrt{m^2+1}\right. \\ & \times \cos\left((\theta+\pi/2)\sqrt{m^2+1}\right)\Big) + \sin\theta\Big(- C_1\left(m^2+1\right) \\ & \times \cos\left((\theta+\pi/2)\sqrt{m^2+1}\right) - C_2\left(m^2+1\right)\sin\left((\theta+\pi/2)\sqrt{m^2+1}\right).\end{aligned} \tag{4.92}$$

We expand (4.92) in Fourier series of the system of trigonometric functions $1/2$, $\cos x$, $\sin x$, $\cos 2x$, … and make use of the evenness of the incident wave to put the coefficient before $\sin m\theta$ equal to zero. Thus, we have

$$T_m = b_m \cos m\theta,$$

where b_m are unknown coefficients. Thus, the general solution of the system (4.87) can be written in the form

$$p = \sum b_m R_m(r) \cos m\theta. \tag{4.93}$$

The pressure in the plane incident wave can be presented as an expansion in terms of cylindrical functions, i.e.

$$p_i = \sum_{m=0}^{\infty} \varepsilon_m \mathrm{i}^m J_m(k_2 r) \cos m\theta, \tag{4.94}$$

where

$$\varepsilon_m = \begin{cases} 1, & m=0, \\ 2, & m\geq 1 \end{cases}$$

is the Neumann factor.

We substitute solutions (4.85), (4.93) and expression (4.94) in the matching condition (4.77) to find the coefficients to be given by

$$a_0 = -\frac{J_0(k_2) + J_0'(k_2)}{H_0^{(2)}(k_2) + H_0^{(2)\prime}(k_2)}, \tag{4.95}$$

$$a_m = \frac{\varepsilon_m \mathrm{i}^m [k_2 \rho_0 J_m'(k_2) R_m(1) - J_m(k_2) R_m'(1)]}{[H_m^{(2)}(k_2) R_m'(m) - k_2 \rho_0 H_m^{(2)\prime}(k_2) R_m(1)]}, \quad m \geq 1, \tag{4.96}$$

$$b_m = \frac{\varepsilon_m \mathrm{i}^m k_2 \rho_0 [H_m^{(2)\prime}(k_2) J_m(k_2) - H_m^{(2)}(k_2) J_m'(k_2)]}{[H_m^{(2)}(k_2) R_m'(m) - k_2 \rho_0 H_m^{(2)\prime}(k_2) R_m(1)]}, \quad m \geq 1. \tag{4.97}$$

Solutions (4.88) and (4.93) and expressions (4.94)–(4.97) for the coefficients were used to calculate the total scattered energy that is usually described by the total scattering cross-section given by

$$Q = \int_0^{2\pi} |f(r, \theta)|^2 \mathrm{d}\theta,$$

where $f(r, \theta)$ is the angular distribution factor for the scattered radiation along the direction of θ.

In order to estimate the influence of the asymmetric inhomogeneity on the wave scattering, we compare the total scattering cross-sections for symmetric and asymmetric density distributions calculated in terms of (4.86) with the coefficient $ar_0 = \ln 5$. Figure 4.12a, b shows the calculated values of Q versus the wave number $k_2 r_0$ plotted in logarithmic scale along the ordinate. The results are given for the wavenumber ranges $k_2 r_0 \in [0, 2]$ and $k_2 r_0 \in [0, 10]$ whose upper values correspond to the wavelengths $\lambda/2r_0 \cong 1.5$ and $\lambda/2r_0 \cong 0.3$, respectively. The plots suggest the following conclusions: (1) both symmetric and asymmetric density distributions show local resonance effects; (2) in the wavenumber range of interest, increase of the scattered energy is observed for the asymmetric density distribution and (3) the influence of the arc coordinate on the density distribution is manifested only for low frequencies.

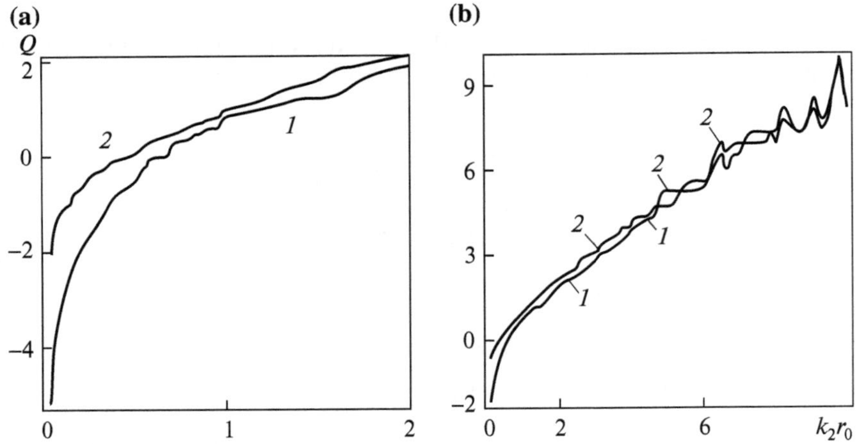

Fig. 4.12 Influence of the scatterer assymetry on the total scattering cross-section Q: *1* symmetric inhomogeneity, *2* asymmetric inhomogeneity

4.7 Numerical-Analytical Method of an Auxiliary Boundary for Studying the Wave Diffraction by a Vertical Obstacle

We consider a stationary wave diffraction problem with the solution of the form $\exp(-\mathrm{i}\,\omega\,t)$ (t is the time coordinate, ω is the frequency). The relevant problem reduces to the exterior problem for the second-order elliptic equation in an open regular domain $\Omega^e \subset R^3$. The domain Ω^e is exterior with respect to the domain Ω^i bounded by an arbitrary smooth surface $\partial\Omega^i$ formed by the rotation of some curve about a straight axis Ox_3. The domain Ω^i is closed in the three-dimensional case and open cylindrical in the two-dimensional case.

The problem is to find a scalar function $\varphi = \varphi(x_1, x_2, x_3, t)$ that satisfies the equation

$$\nabla^2\varphi = 0 \ \text{ in } \ \Omega^e \tag{4.98}$$

and the boundary condition

$$\frac{\partial\varphi}{\partial n} = 0 \text{ or } \varphi = 0 \ \text{ at } \ \partial\Omega^i. \tag{4.99}$$

The function φ satisfies the radiation condition $|\nabla\varphi| \to 0$ at $|r| \to \infty$. The incident field $\varphi^i = f$ is also given.

According to the above-mentioned method, the domain Ω^i is enveloped by an arbitrary spherical or cylindrical surface $\partial\Omega^s$. Let Ω^{se} be an external domain outside this auxiliary surface. Then the problem is solved numerically by means of the spline-collocation method in the domain $\Omega^s = \Omega^e \backslash \Omega^{se}$; outside $\partial\Omega^s$ an analytic solution can be found. The numerical and analytic solutions are matched on the interface $\partial\Omega^s$. This approach can be regarded as *IS*-mapping of the initial exact continuum model *I* to a set of approximate (simplified) models *S* from which a model should be chosen in a way to be realised by means of the spline-collocation method.

Now we consider the problem of wave diffraction by a vertical body of revolution in a fluid of finite depth (Fig. 4.13) [44]. The surface $\partial\Omega^i$ is generated by the function

$$z = g(r), \quad r \in [a, b]. \tag{4.100}$$

The problem is considered in the cylindrical coordinate system (r, θ, z). The θ-dependent variable in Eq. (4.98) is separated, then the solutions that satisfy the first condition in (4.99) can be written as

$$\varphi(r, \theta, z) = \sum_{m=0}^{\infty} \varphi_m(r, z) \cos m\theta, \quad (r, z) \in \Omega^s, \tag{4.101}$$

$$\varphi_s(r, \theta, z) = A \frac{\cosh k(z+1)}{\cosh k} \sum_{m=0}^{\infty} a_m H_m^{(2)}(kr) \cos m\theta, \quad (r, z) \in \Omega^{se}, \tag{4.102}$$

$$f(r, \theta, z) = A \frac{\cosh k(z+1)}{\cosh k} \sum_{m=0}^{\infty} \varepsilon_m \, \mathrm{i}^m I_m(kr) \cos m\theta, \tag{4.103}$$

where k is the wave number, ε_m is the Neumann factor, a_m are unknown coefficients, $H_m^{(2)}$ and I_m are Hankel and Bessel functions, respectively. The fluid depth is normalised so that in the dimensionless variables it is equal to unity.

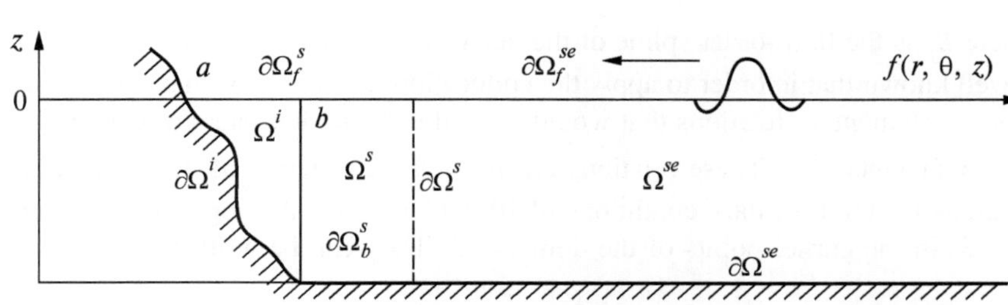

Fig. 4.13 Configuration of the problem of surface wave diffraction in the cylindrical coordinate system (r, θ, z). The *dashed line* designates the auxiliary surface $\partial\Omega^s$

The unknown functions $\varphi_m(r,z)$ should be given by the solutions of the equation [62]

$$\frac{1}{r}\frac{\partial}{\partial r}\left[r\frac{\partial\varphi_m(r,z)}{\partial r}\right]+\frac{\partial^2\varphi_m(r,z)}{\partial z^2}-\frac{m^2}{r^2}\varphi_m(r,\ z)=0\ \text{ in }\ \Omega^s. \tag{4.104}$$

The boundary conditions are given by

$$\frac{\partial\varphi_m(r,z)}{\partial \boldsymbol{n}}=0\ \text{ at }\ \partial\Omega^i\cup\partial\Omega^s\cup\partial\Omega^{se}, \tag{4.105}$$

$$\frac{\partial\varphi_m(r,z)}{\partial z}-k\,\tanh k\,\varphi_m(r,z)=0,\quad \partial\Omega_f^s\cup\partial\Omega_f^{se}, \tag{4.106}$$

$$\begin{aligned}&\frac{\partial\varphi_m(r,z)}{\partial r}H_m^{(2)}(kr)-\varphi_m(r,z)H_m^{(2)'}(kr)=A\frac{\cosh k(z+1)}{\cosh k}k\varepsilon_m\,\mathrm{i}^m\\&\times\left[I_m'(kr)H_m^{(2)}(kr)-I_m(kr)H_m^{(2)'}(kr)\right]\ \text{at}\ \partial\Omega^s.\end{aligned} \tag{4.107}$$

Within the framework of the above assumptions, the problem (4.104)–(4.107) is well-posed because the domain Ω^s does not contain singular points of Eq. (4.104) and the boundary conditions (4.105)–(4.107) completely determine the uniqueness of the solution in the class of twice-differentiable functions.

We divide the domain Ω^s by a rectangular grid with $(N+1)$ and $(M+1)$ cells along the coordinate lines Or and Oz, respectively. Let the solution $\varphi_m(r,z)$ be approximated by cubic splines of the form [75, 82]

$$\varphi_m(r,z)=\sum_{i=-1}^{N+1}\sum_{j=-1}^{M+1}b_{ij}^{(m)}B_i^3(r)B_j^3(z), \tag{4.108}$$

where B_j^3 is the third-order spline of the defect 1, $b_{ij}^{(m)}$ is an unknown coefficient. It is well known that in order to apply the collocation technique, we have to introduce four supplementary relations that would close the algebraic system of equations for the coefficients $b_{ij}^{(m)}$. These relations are formulated in terms of the compatibility relations for the boundary conditions (4.105)–(4.107) for the approximate solution (4.108) in the corner points of the domain Ω^s. They can be written as

$$\frac{\partial^2\varphi_m(r,z)}{\partial r\,\partial z}n_r+k\,\tanh k\frac{\partial\varphi_m(r,z)}{\partial z}n_z=0,\quad \partial\Omega^i\cap\partial\Omega_f^s, \tag{4.109}$$

$$\frac{\partial^2\varphi_m(r,z)}{\partial r\partial z}=0,\quad \partial\Omega^i\cap\partial\Omega_b^s\ \text{ and }\ \partial\Omega^s\cap\partial\Omega_b^s, \tag{4.110}$$

$$\frac{\partial^2 \varphi_m(r,z)}{\partial r\, \partial z} - k \tanh k \frac{\partial \varphi_m(r,z)}{\partial r} = 0, \quad \partial\Omega^s \cap \partial\Omega_f^s, \tag{4.111}$$

where

$$n_r = \frac{g'(r)}{\left(1 + (g'(r))^2\right)^{1/2}}, \quad n_z = \frac{1}{\left(1 + (g'(r))^2\right)^{1/2}}.$$

We substitute (4.108) into (4.104)–(4.107) and (4.109)–(4.111), and thus obtain a system of algebraic equations given by

$$A\boldsymbol{b}_m = \boldsymbol{Y}, \tag{4.112}$$

where $\boldsymbol{b}_m$ is the vector (matrix) of unknown coefficients in (4.108), $\boldsymbol{b}_m = \left\{ b_{ij}^m \right\}$; A is a known square matrix of the size $(N + 3) \times (M + 3) \times (N + 3)\ (M + 3)$; $\boldsymbol{Y}$ is the vector of right-hand parts.

The vectors of the wave force $\boldsymbol{F}$ and moment $\boldsymbol{M}$ are given by the expressions

$$\boldsymbol{F} = \int_0^{2\pi} \int_{\partial\Omega^i} r p \boldsymbol{n}\, \mathrm{d}r\, \mathrm{d}z\, \mathrm{d}\theta, \quad \boldsymbol{M} = \int_0^{2\pi} \int_{\partial\Omega^i} r p \boldsymbol{n} R\, \mathrm{d}r\, \mathrm{d}z\, \mathrm{d}\theta, \tag{4.113}$$

where $p = -\mathrm{i}\,\omega\, \rho\, \varphi(r, \theta, z)$.

The method was tested for two problems whose exact solutions are known. These are the analytic solution of the problem of wave diffraction in water by a vertical circular cylinder [14] and the numerical solution for the diffraction by a circular cone [16]. Figure 4.14 gives the results for a new problem for a cone over a circular cylinder. The calculation accuracy for the scattered field considerably depends on the wave number $ka = \pi(2a/\lambda)$, where λ is the wavelength of the incident field, $2a$ is the maximum size of the scatterer (diameter in the problem under consideration). The accuracy is the highest in the longwave approximation $ka \ll 1$. As ka increases, we approach the resonance range $ka \sim 1$, and the accuracy of the series solutions decreases. For large wave numbers, $ka \gg 1$, the series solution does not hold and we have to find asymptotic solutions. Within the context of the above calculations, the method provides satisfactory results within the range from $2a/\lambda \ll 1$ to the resonance range $2a/\lambda > 1$.

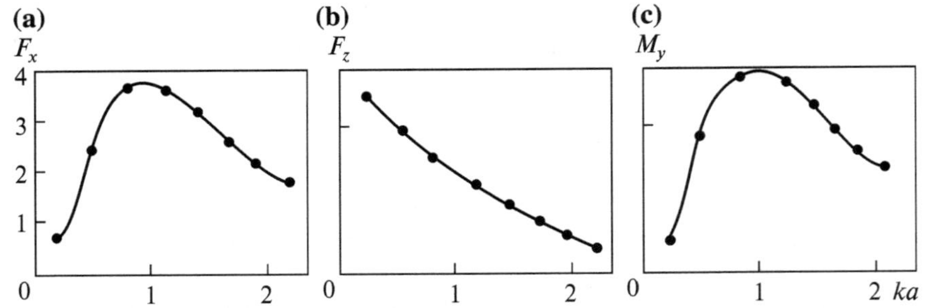

Fig. 4.14 **a** Lateral force F_x, **b** vertical force F_z, and **c** overturning moment M_y versus wave number for a cone over circular cylinder

4.8 Numerical Study of the Acoustic Wave Diffraction by a Body of Revolution

4.8.1 Symmetric Inhomogeneities

Now we consider the applications of the spline-function and finite-element methods to the problem of wave diffraction by symmetric inhomogeneities. The approach was tested by comparing the results with the known exact solutions; the relevant algorithms proved to be rather efficient. The authors of Ref. [62] deal with the stationary problems of wave diffraction in acoustic, hydrodynamic and elastic media for the cases of axial and central symmetry. Separation of variables is shown to reduce the analysis to the solution of total differential equations of the form

$$L^n[y(x)] = r(x) \quad (n = 2, 4;\ x \in [a, b]),$$
$$U_j(a, b) = c_j \quad (j = 1, \ldots, n), \tag{4.114}$$

where L^n is the linear operator with variable coefficients of the order n, U_j are linear forms formed by the values of the function $y(x)$ to be calculated and its derivatives up to the order $(n - 1)$ at points $x = a$ and $x = b$ whose expressions are determined by the boundary conditions, $r(x)$ and c_j are known functions and constants.

We consider problem (4.114) by means of the spline-collocation method and approximate the solutions to be found in the basis of finite B-splines. Suppose that a partition $\Delta\colon a = x_0 < x_1 < \cdots < x_N = b$ is given on $[a,\ b]$. The consideration is restricted to the case when collocation nodes coincide with the grid points [75, 82]. We write the solution in terms of a series expansion in the basis of normalised B-splines of the order $(n + 1)$, i.e.

$$y(x) = \sum_{i=-n/2}^{N+n/2} d_i B_i^{(n+1)}(x), \tag{4.115}$$

and supplement the grid with the points

$$x_{-(n+1)} < x_{-n} < \cdots < x_{-1} < x_0, \quad x_{N+n+1} > \cdots > x_{N+1} > x_N,$$

Substituting (4.115) in (4.114) yields

$$\sum_{j=0}^{n} d_{i+j-n/2} L^n \left[B_{i+j-n/2}^{(n+1)}(x_i) \right] = r(x_i) \quad (i=0, 1, \ldots, N), \tag{4.116}$$

$$\sum_{j=0}^{n} \left[B_{j-n/2}^{(n+1)}(a) \bar{K}_{j,l}^{(1)} + B_{j+N-n/2}^{(n+1)}(b) \bar{K}_{j,l}^{(2)} \right] \boldsymbol{d} = c_l \quad (l=1, 2, \ldots, n), \tag{4.117}$$

where $\boldsymbol{d} = (d_{-n/2}, d_{-n/2+1}, \ldots, d_{N+n/2-1}, d_{N+n/2})^T$; $\bar{K}_{j,l}^{(1)}$, $\bar{K}_{j,l}^{(2)}$ are known vector rows of the dimension $(N+n+1)$ that correspond to the type of the boundary conditions. It should be noted that (4.116) and (4.117) make a system of linear algebraic equations whose solutions provide the unknown coefficients in (4.115).

Now we apply the above algorithm to consider the plane wave diffraction by an inhomogeneous cylinder. For a pressure wave p_i that is incident normally to the axis of an infinite inhomogeneous cylinder with the variable parameter $\rho(r)$ for $r \in [0, 1]$, we have

$$p_i = e^{ik_0 r \cos\theta} = \sum_{m=0}^{\infty} \varepsilon_m \mathrm{i}^m J_m(k_0 r) \cos m\theta, \tag{4.118}$$

$$\varepsilon_m = \begin{cases} 1, & m=0, \\ 2, & m \neq 0. \end{cases}$$

The time factor $\exp(\mathrm{i}\omega t)$ is omitted here and in what follows. It can be shown [62] that the scattered field $p_s(r > 1)$ that satisfies the Sommerfeld radiation condition and the field $p_0(0 \leq r \leq 1)$ in the interior domain are given, respectively, by

$$\begin{aligned} p_s(r, \theta) &= \sum_{m=0}^{\infty} a_m H_m^{(2)}(k_0 r) \cos m\theta, \\ p_0(r, \theta) &= \sum_{m=0}^{\infty} R_m(r) \cos m\theta, \end{aligned} \tag{4.119}$$

where a_m are unknown coefficients, $R_m(r)$ is a function that satisfies the condition

$$R_m'' + \left(\frac{1}{r} - \frac{\rho'}{\rho}\right)R_m' + (k_0^2\rho - m^2)R_m = 0 \tag{4.120}$$

and the boundary conditions

$$\begin{aligned} &R_m(1) = \varepsilon_m \mathrm{i}^m J_m(k_0) + a_m H_m^{(2)}(k_0), \\ &\frac{1}{\rho(1)} R_m'(1) = k_0[\varepsilon_m \mathrm{i}^m J_m'(k_0) + a_m H_m^{(2)'}(k_0)]. \end{aligned} \tag{4.121}$$

We note that the solution of Eq. (4.120) should be regular for $r = 0$. Hence, in order to employ the numerical method (4.115)–(4.117), we have to consider the two intervals, $[0, \varepsilon)$ and $(\varepsilon, 1]$, where ε is a small parameter. Moreover, for $r \in [0, \varepsilon)$ we put $\rho = \rho(\varepsilon)$. Then the solution in the vicinity of $r = 0$, with the regularity condition taken into account, is given by

$$p_\varepsilon(r, \theta) = \sum_{m=0}^{\infty} b_m J_m(k_0\sqrt{\rho(\varepsilon)}r) \cos m\theta,$$

and the boundary conditions

$$\begin{aligned} &R_m(\varepsilon) = b_m J_m(k_0\sqrt{\rho(\varepsilon)}\varepsilon), \\ &R_m'(b) = b_m k_0\sqrt{\rho(\varepsilon)} J_m'(k_0\sqrt{\rho(\varepsilon)}\ \varepsilon) \end{aligned} \tag{4.122}$$

are satisfied on the boundary of the inner sublayer.

For the case under consideration, we write the solution of Eq. (4.119) for $r \in (\varepsilon, 1]$ in form (4.115), i.e.

$$R_m(r) = \sum_{i=-1}^{N+1} d_i B_i^{(3)}(r). \tag{4.123}$$

Then it is not difficult to show that relations (4.121) determine Eqs. (4.117) in the spline-collocation method while (4.122) are supplementary equations that can be used to find the unknown constants a_m and b_m in (4.121) and (4.122). Equations (4.116), (4.121), and (4.122) with (4.120) and (4.123) taken into account, make a system of linear algebraic equations for the unknown constants entering the solutions to be found.

The above algorithm for $\varepsilon = 0.01$ on a uniform grid with the step $h = (1-\varepsilon)/(N-1)$, where $N = 30$, was applied to calculate the values of the scattered field on the surface of an inhomogeneous cylinder $(r = 1)$. Two patterns of the inhomogeneity parameter variation were considered:

$$\rho(r) = \rho_c + (1 - \rho_c)\tan\frac{\pi r}{4}$$
$$\rho(r) = \rho_c + (1 - \rho_c)r^2.$$

Comparison of the numerical results obtained in [28] with the exact solutions derived in [62] shows that spline approximations provide an efficient tool for the treatment of diffraction problems.

4.8.2 Asymmetric Inhomogeneities

Now we consider the application of the method of surface finite elements to calculating the wave diffraction by the bodies of revolution of noncanonical shapes. Nowadays, the most complete knowledge is available of the processes of wave interaction with the bodies of canonical shapes. The reason is that in these cases the solutions can be found in terms of special functions, the corresponding techniques being worked out rather elaborately. However, the realisation of the solutions is in some cases nontrivial even in terms of special functions. Calculating the diffracted field parameters in the case of wave interaction with the bodies of noncanonical shapes requires efficient approximate computational methods.

Here, we give the results of numerical studies of wave diffraction by the bodies of noncanonical shapes with the use of the Kirchhoff integral formula [26, 71, 77]. We employ the velocity potential for a homogeneous medium to consider the incidence of a plane wave of unit intensity and arbitrary orientation,

$$\varphi_i = e^{ik(\boldsymbol{e}_\alpha \cdot \boldsymbol{r} + ct)}, \quad c = 1,$$

on a body with positive Gaussian curvature at any point of the surface. Here, $\boldsymbol{e}_\alpha$ is the unit vector that determines the direction of wave propagation.

We write the diffracted field potential in the form $\varphi_s = \psi e^{ikt}$. Then taking into account the Sommerfeld radiation condition and the direction of the normal (Fig. 4.15), we obtain from the Kirchhoff formula for ψ an equation given by

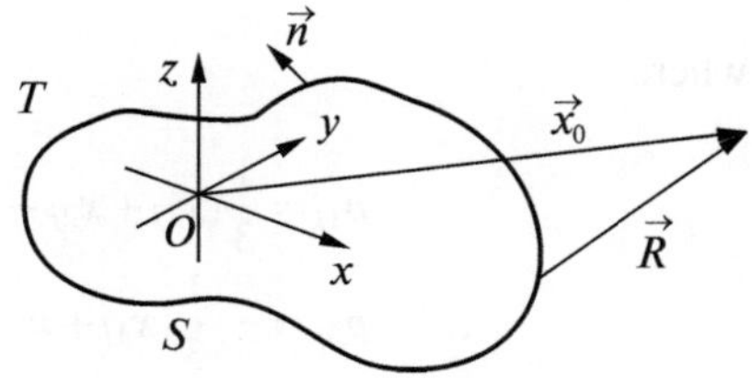

Fig. 4.15 Application of the Kirchhoff formula

$$\psi(\boldsymbol{x}_0) = \frac{\Omega}{4\pi}\int\limits_S \left[\psi\frac{\partial}{\partial \boldsymbol{n}}\left(\frac{e^{-\mathrm{i}kR}}{R}\right) - \frac{\partial\psi}{\partial \boldsymbol{n}}\frac{e^{-\mathrm{i}kR}}{R}\right]\mathrm{d}s,$$
$$R = \sqrt{(x-x_0)^2 + (y-y_0)^2 + (z-z_0)^2},\ \boldsymbol{x}_0 = (x_0, y_0, z_0), \tag{4.124}$$
$$\Omega = \begin{cases} 2, & \boldsymbol{x}_0 \in S, \\ 1, & \boldsymbol{x}_0 \in T. \end{cases}$$

On the surface of the body, the impenetrability condition should be satisfied, i.e.

$$\frac{\partial\psi}{\partial \boldsymbol{n}} = -\mathrm{i}k(\boldsymbol{e}_\alpha \cdot \boldsymbol{n})e^{\mathrm{i}k(\boldsymbol{e}_\alpha\cdot\boldsymbol{r})}. \tag{4.125}$$

Before describing the solving algorithm for the example of a body of revolution, we specify the method of dividing the surface into finite elements. We direct the Ox axis of the rectangular Cartesian coordinate system along the symmetry axis of the body. Then we divide the surface by N sections in the lateral direction and by M surface layers in the longitudinal direction. Each obtained curvilinear quadrangle is then divided by the diagonal into two curvilinear triangles. Thus, the body surface S is covered by a set of triangular curvilinear elements S_j, and Eq. (4.124) with (4.125) taken into account reduces to

$$\psi(\boldsymbol{x}_0) = \frac{\Omega}{4\pi}\sum_{j=1}^{2NM}\int\limits_{S_j}\frac{e^{-\mathrm{i}kR}}{R}\left[\mathrm{i}k(\boldsymbol{e}_\alpha\cdot\boldsymbol{n})e^{\mathrm{i}k(\boldsymbol{e}_\alpha\cdot\boldsymbol{x})} - \frac{(1+\mathrm{i}kR)}{R^2}(\boldsymbol{R}\cdot\boldsymbol{n})\psi\right]\mathrm{d}s. \tag{4.126}$$

We note that the choice of triangular elements as basic elements in the partition of the surface provides, along with the suitable way of shape description, also the construction of linear approximating polynomials with the use of nodal values of the approximated functions.

In order to construct the universal procedure of obtaining the influence matrices, we put the jth element in correspondence with an equilateral triangle of the auxiliary plane $\xi O\eta$ with the vertices numerated as shown in Fig. 4.16. Then the surface is approximated by the dependence given by

$$\boldsymbol{x} = \boldsymbol{p}_{1j} + \boldsymbol{p}_{2j}\xi + \boldsymbol{p}_{3j}\eta, \tag{4.127}$$

where

$$\boldsymbol{p}_{1j} = \frac{1}{3}(\boldsymbol{x}_{1j} + \boldsymbol{x}_{2j} + \boldsymbol{x}_{3j}),\quad \boldsymbol{p}_{2j} = \frac{1}{2}(\boldsymbol{x}_{1j} - \boldsymbol{x}_{2j}),$$
$$\boldsymbol{p}_{3j} = -\frac{1}{3}(\boldsymbol{x}_{1j} + \boldsymbol{x}_{2j} - 2\boldsymbol{x}_{3j}), \tag{4.128}$$

while the external normal to the surface and the Jacobian of the transformation to the local coordinate system are given by

$$\boldsymbol{n}^{(j)} = \frac{\boldsymbol{p}_{2j} \times \boldsymbol{p}_{3j}}{\left|\boldsymbol{p}_{2j} \times \boldsymbol{p}_{3j}\right|}, \quad I = \left|\boldsymbol{p}_{2j} \times \boldsymbol{p}_{3j}\right|. \tag{4.129}$$

Moreover, by analogy with the description of the surface for the node numeration shown in Fig. 4.16, the change of the potential within the jth element is determined, according to (4.127) and (4.128), by the formula

$$\psi = (1, \xi, \eta) \begin{pmatrix} \frac{1}{3} & \frac{1}{3} & \frac{1}{3} \\ \frac{1}{2} & -\frac{1}{2} & 0 \\ -\frac{1}{3} & -\frac{1}{3} & \frac{2}{3} \end{pmatrix} \begin{pmatrix} \psi_{1j} \\ \psi_{2j} \\ \psi_{3j} \end{pmatrix}. \tag{4.130}$$

In the manner similar to the analysis in Ref. [26], we present all the integrands except the potential by their values in the characteristic point of the element for which we put $\xi = \eta = 0$. Then having performed necessary transformations of (4.126) with (4.127)–(4.130) taken into account, we can show that the equation for the scattered field potential reduces to

$$\begin{aligned} &\psi(\boldsymbol{x}_0) + \frac{\Omega}{8\pi} \sum_{j=1}^{2NM} \frac{e^{-ikR_j(\boldsymbol{x}_0)}}{R_j^3(\boldsymbol{x}_0)} [1 + ikR_j(\boldsymbol{x}_0)](\boldsymbol{R}_j(\boldsymbol{x}_0) \cdot \widetilde{\boldsymbol{n}}^{(j)}) \sum_{m=1}^{3} \psi_{mj} \\ &= \frac{3ik\Omega}{8\pi} \sum_{j=1}^{2NM} \frac{(\boldsymbol{e}_\alpha \cdot \widetilde{\boldsymbol{n}}^{(j)})}{R_j(\boldsymbol{x}_0)} e^{ik[(\boldsymbol{e}_\alpha \cdot \boldsymbol{x}_j) - R_j(\boldsymbol{x}_0)]}, \end{aligned} \tag{4.131}$$

where

$$\boldsymbol{x}_j = \boldsymbol{p}_{1j}; \quad R_j(\boldsymbol{x}_0) = \boldsymbol{x}_j - \boldsymbol{x}_0; \quad \widetilde{\boldsymbol{n}}^{(j)} = \boldsymbol{p}_{2j} \times \boldsymbol{p}_{3j}.$$

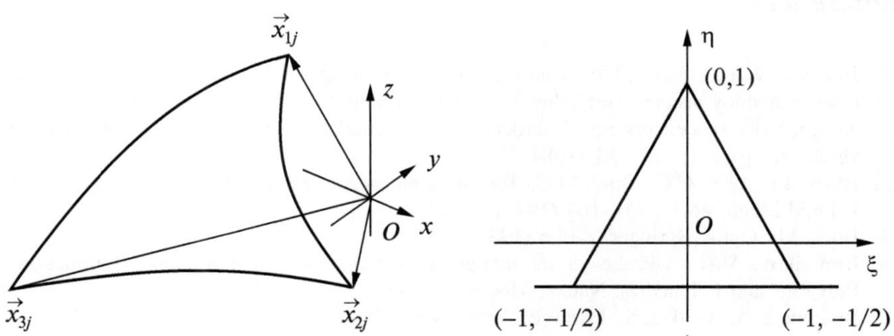

Fig. 4.16 An example of introducing an auxiliary plane

For the velocity field in an arbitrary point, we obtain an expression

$$\begin{aligned}\boldsymbol{v}(\boldsymbol{x}_0) = \frac{\Omega}{8\pi}\sum_{j=1}^{2NM}\frac{e^{-\mathrm{i}kR_j(\boldsymbol{x}_0)}}{R_j^3(\boldsymbol{x}_0)}\langle 3\mathrm{i}k(\boldsymbol{e}_\alpha\cdot\widetilde{\boldsymbol{n}}^{(j)})[1+\mathrm{i}kR_j^2(\boldsymbol{x}_0)]\boldsymbol{R}_j(\boldsymbol{x}_0)e^{\mathrm{i}\,k(\boldsymbol{e}_\alpha\cdot\boldsymbol{x}_j)} \\ -\left\{\frac{\boldsymbol{R}_j(\boldsymbol{x}_0)}{R_j(\boldsymbol{x}_0)}(\boldsymbol{R}_j(\boldsymbol{x}_0)\cdot\widetilde{\boldsymbol{n}}^{(j)})\left[\mathrm{i}k(1+\mathrm{i}kR_j(\boldsymbol{x}_0))+\frac{1}{R_j(\boldsymbol{x}_0)}(3+2\mathrm{i}kR_j(\boldsymbol{x}_0))\right]\right. \\ \left. -\ [1+\mathrm{i}kR_j(\boldsymbol{x}_0)]\widetilde{\boldsymbol{n}}^{(j)}\right\}\sum_{m=1}^{3}\psi_{mj}\rangle.\end{aligned} \tag{4.132}$$

To find the potential distribution over the body surface, we have to impose the condition that (4.131) must be satisfied in the nodes of the partition grid and thus obtain a system of linear algebraic equations of the form

$$A\boldsymbol{\psi} = \boldsymbol{c}, \tag{4.133}$$

where A, $\boldsymbol{\psi}$, $\boldsymbol{c}$ are the square matrix and column vectors of the dimension $(NM+2)$. We note that (4.131) and (4.132) contain the values of the velocity potential in the nodal points of the element.

The technique was tested by comparing the approximate results obtained for a rigid sphere to the given exact solutions (cf. Ref. [28]). The results demonstrate that the choice of the element shape under the triangulation of the surface can either improve or worsen the accuracy of the solution. The elements elongated along the direction of wave incidence decrease the accuracy, whereas the elements elongated along the wave front provide higher accuracy. Thus, the recommended optimum shape of the partition should be near-equilateral triangular elements. It should also be noted that the accuracy of the solution for given partition depends on the frequency of the incident wave. Thus, the number of divisions should be increased for greater frequencies.

References

1. Belyaev, V.A., Selezov, I.T., Yakovlev, V.V.: Difraktsiya regulyarnykh voln na proizvolno raspolozhennoy sisteme vertikalnykh krugovykh tsilindrov (Diffraction of regular waves by an arbitrarily spaced system of vertical circular cylinders). Izvestiya AN SSSR. Mekhanika zhidkosti i gaza **1**, 177–181 (1984)
2. Black, J.L., Mei, C.C., Bray, M.C.: Radiation and scattering of water waves by rigid bodies. J. Fluid Mech. **46**(1), 151–164 (1971)
3. Born, M.: Optik. Springer, Berlin (1933)
4. Borovikov, V.A.: Difraktsiya na mnogougolnikakh i mnogogrannikakh (Diffraction by Polygons and Polyhedra). Nauka, Moscow (1966)
5. Brebbia, C.A., Walker, S.: Dynamic Analysis of Offshore Structures. Butterworth-Heinemann Ltd., London (1979)

6. Cai, L.-W.: Evaluation of layered multiple-scattering method for antiplane shear wave scattering from gratings. J. Acoust. Soc. Am. **120**(1), 49–61 (2006)
7. Chakrabarti, S.K.: Wave forces on multiple vertical cylinders. J. Waterway Port Coast. Ocean Div. **104**(2), 147–161 (1978)
8. Chen, H.S., Mei, C.C.: Wave forces on a stationary platform of elliptical shape. J. Ship Res. **17**(2), 61–71 (1973)
9. D'Este, F., Codiglia, R., Contento, G.: On the side wall effects on the nonli-near wave loads on a TLP-like structure in a wave tank: analysis of the nonlinear near-trapping. In: Proceedings of the 12th International Offshore and Polar Engineering Conference, Kitakyushu, Japan, 26–31 May 2002, pp. 196–202
10. Doyle, T.E.: Iterative simulation of elastic wave scattering in arbitrary dispersions of spherical particles. J. Acoust. Soc. Am. **119**(5), 2599–2610 (2006)
11. Fejer, J.A.: Scattering of electromagnetic waves by a plasma cylinder. Phys. Fluids **7**(3), 439–445 (1964)
12. Ferrante, V., Vicinanza, D.: Spectral analysis of wave transformation behind submerged breakwater. In: Proceedings of the 16th International Offshore and Polar Engineering Conference, San Francisco, USA, 28 May–2 June 2006, vol. 3, pp. 727–733
13. Garrett, C.R.: Wave forces on a circular doc. J. Fluid Mech. **46**(1), 129–139 (1971)
14. Haskind, M.D.: Gidrodinamicheskaya teoriya kachki korablya (The Hydrodynamic Theory of Ship Rolling). Nauka, Moscow (1973)
15. Hönl, H., Maue, A.W., Westpfahl, K.: Theorie der Beugung. Springer, Berlin (1961)
16. Isaacson, M.: Fixed and floating axisymmetric structure in waves. J. Waterway Port Coast. Ocean Eng. **108**(2), 180–189 (1982)
17. Ivanov, E.A.: Difraktsiya elektromagnitnykh voln na dvukh telakh (Diffraction of Electromagnetic Waves by Two Bodies). Nauka i Tekhnika, Minsk (1968)
18. Kagemoto, H., Murai, M., Saito, M., Molin, B., Malenica, S.: Experimental and theoretical analysis of the wave decay along a long array of vertical cylinders. J. Fluid Mech. **456**, 113–135 (2002)
19. Kamke, E.: Handbook of Ordinary Differential Equations. Chelsea Publishing (1976)
20. Karlsson, A., Kristensson, G.: Wave splitting in the time domain for a radially symmetric geometry. Wave Motion **12**(3), 197–211 (1990)
21. Katsenelenbaum, B.Z.: High-Frequency Electrodynamics. Wiley-VCH, Weinheim (2006)
22. Kee, S.T.: Submerged horizontal and vertical membrane wave barrier. In: Proceedings of the 15th International Offshore and Polar Engineering Conference, Seoul, South Korea, 19–24 June 2005, vol. 3, pp. 587–594
23. Kim, J.W., Kyoung, J.H., Ertekin, R.C., Bai, K.J.: Finite-element computation of wave-structure interaction between steep Stokes waves and vertical cylinders. J. Waterway Port Coast. Ocean Eng. **5**, 337–347 (2006)
24. Kim, M.H.: Interaction of waves with N vertical circular cylinders. J. Waterway Port Coast. Ocean Eng. **119**(6), 671–689 (1993)
25. Kim, M.H., Natvig, B.J., Mercier, R.S., Gu, G., Wu, C.: PC-based computation for second-order wave loads on large volume multi-column structures. In: Proceedings of the 5th International Offshore and Polar Engineering Conference, Hague, Netherlands, 11–16 June 1995, pp. 561–570
26. Kornilov, I.E.: Primenenie metoda konechnykh elementov k raschetu obtekaniya mnogo-svyaznykh oblastei potentsialnym potokom szhimaemoi zhidkosti (Finite-element analysis of the circumfluence of multiply-connected domains by a potential flow of compressible fluid). In: Matematicheskie Metody Issledovaniya Gidrodinamicheskikh Techenii (Mathematical Methods of Investigating the Hydrodynamical Flows), pp. 38–41. Naukova Dumka, Kiev (1978)

27. Kratzer, A., Franz, W.: Transzendente Funktionen. Akademische Verlagsgesellschaft Geest & Portig, Leipzig (1960)
28. Kryvonos, Yu.G., Kornilov, I.E., Noga, Yu.V., Selezov, I.T.: Chislennoe issledovanie volnovykh polei ot proizvolnykh tel vrashcheniya i neodnorodnostei (Numerical investigation of the wave fields from arbitrary bodies of revolution and inhomogeneities). Preprint, Institute of Cybernetics AN USSR, Kiev (1983)
29. Kryvonos, Yu.G., Selezov, I.T.: Upravlenie diagrammoi napravlennosti pri rasseyanii tsilindricheskoi akusticheskoi volny na tsilindre v poluogranichennoi oblasti (Controlling the directional diagram during the scattering of a cylindrical acoustic wave by a cylinder in a semibounded domain). In: Kibernetika i vychislitelnaya tekhnika. Slozhnye sistemy upravleniya (Cybernetics and Computational Techniques. Complex Control Systems), vol. 8, pp. 102–109. Naukova Dumka, Kiev (1971)
30. Linton, C.M., Evans, D.V.: The interaction of waves with arrays of vertical circular cylinders. J. Fluid Mech. **215**, 549–569 (1990)
31. Malenica, S., Taylor, R.E., Huang, J.B.: Second-order water wave diffraction by an array of vertical cylinders. J. Fluid Mech. **390**, 349–373 (1999)
32. Massel, S.B.: Interaction of water waves with cylinder barrier. J. Waterways Harbors Coast Eng. Div. **102**(2), 165–187 (1976)
33. Mavrakos, S.A., Koumoutsakos, P.: Hydrodynamic interaction among vertical axisymmetric bodies restrained in waves. Appl. Ocean Res. **9**(3), 128–140 (1987)
34. McLachlan, N.W.: Theory and Application of Mathieu Functions. Dover, New York (1964)
35. Miles, J.W., Gilbert, F.: Scattering of gravity waves by a circular dock. J. Fluid Mech. **34**(4), 783–793 (1968)
36. Morse, P.M., Feshbach, H.: Methods of Theoretical Physics. Parts 1 & 2. McGraw-Hill, New York (1953)
37. Nozaka, M., Takaku, K.: Scattering of electromagnetic waves by a cylindrically inhomogeneous plasma. J. Phys. Soc. Jpn. **24**(1), 172–184 (1968)
38. Parkinson, R.G., Kharadly, M.M.Z.: Experimental study of scattering of electromagnetic waves by radially inhomogeneous cylinders. Can. J. Phys. **49**(23), 2989–2996 (1971)
39. Peter, M.A., Meylan, M.N.: Water-wave scattering by a semi-infinite periodic array of arbitrary bodies. J. Fluid Mech. **575**, 473–494 (2007)
40. Selezov, I.T.: Rasprostranenie magnitouprugikh voln napryazheniya ot tsilindricheskoi polosti v provodyashchei srede (Propagation of magnetoelastic stress waves from a cylindrical cavity in the conducting medium). Zhurnal prikladnoi mekhaniki i tekhnicheskoi fiziki **2**, 15–20 (1969)
41. Selezov, I.T.: Dyfraktsiia khvyl na tsylindri v pruzhnomu pivprostori (Wave diffraction by a cylinder in an elastic half-space). Dopovidi AN URSR. Ser. A **3**, 259–264 (1970)
42. Selezov, I.T.: Diffraction of waves by convex bodies in semibounded regions. Soviet Appl. Mech. **6**(3), 258–263 (1970)
43. Selezov, I.T.: Nekotorye zadachi difraktsii i rasseyaniya voln prostranstvennymi neodnorodnostyami v okeane i atmosphere (Some problems of wave diffraction and scattering by spatial inhomogeneities in the ocean and atmosphere). In: Teoreticheskie i eksperimentalnye issledovaniya poverkhnostnykh i vnutrennikh voln (Theoretical and Experimental Studies of Surface and Internal Waves), pp. 7–18. Institute of Marine Hydrophysics, Sevastopol (1980)
44. Selezov, I.T.: Numerical solution of the boundary value problems of the wave diffraction by arbitrary revolution surfaces. In: Numerical Methods. Colloquia Mathematica Societatis János Bolyai, vol. 50, pp. 473–480. North-Holland, Amsterdam (1987)
45. Selezov, I.T.: Modelirovanie volnovykh i difraktsionnykh protsessov v sploshnykh sredakh (Modeling the Wave and Diffraction Processes in Continuous Media). Naukova Dumka, Kiev (1989)
46. Selezov, I.T.: Rasprostranenie i transformatsiya poverkhnostnykh gravitatsionnykh voln v zhidkosti konechnoi glubiny (Propagation and transformation of surface gravity waves in a finite-depth fluid). In: Itogi nauki i tekhniki. Ser. Mekhanika zhidkosti i gaza (Science and Technology Review. Ser. Mechanics of Fluids and Gases), vol. 24, pp. 3–76. VINITI, Moscow (1990)

47. Selezov, I.T.: Diffraction of waves by radially inhomogeneous inclusions. Phys. Express **1**(2), 101–115 (1993)
48. Selezov, I.T.: Propagation and diffraction of waves in locally inhomogeneous and bound structures. Int. J. Fluid Mech. Research **23**(1–2), 81–95 (1996)
49. Selezov, I.T., Korsunskii, S.V.: Nestatsionarnye i nelineynye volny v elektroprovodyashchikh sredakh (Nonstationary and Nonlinear Waves in Electrically Conducting Media). Naukova Dumka, Kiev (1991)
50. Selezov, I.T., Kryvonos, Yu.G.: Rasseyanie akusticheskikh voln na tsilindre v poluogra-nichennoi oblasti (Scattering of acoustic waves by a cylinder in a semibounded domain). Gidromekhanika **15**, 99–106 (1969)
51. Selezov, I.T., Kryvonos, Yu.G.: Rozsiiannia akustychnoi khvyli na sferi v napivobmezhenii oblasti (Scattering of an acoustic wave by a sphere in a semibounded domain). Dopovidi AN URSR. Ser. A **7**, 632–636 (1969)
52. Selezov, I.T., Kryvonos, Yu.G.: Rozsiiannia elektromahnitnykh khvyl na tsylindri v napivobmezhenii oblasti (Scattering of electromagnetic waves by a cylinder in a semibounded domain). Dopovidi AN URSR. Ser. A. **4**, 367–372 (1970)
53. Selezov, I.T., Kryvonos, Yu.G.: Statsionarna zadacha rozsiiannia tsylindrychnoi mahnito-akustychnoi khvyli na idealnoprovidnomu tsylindri (A stationary problem of scattering of a cylindrical magnetoacoustic wave by an ideally conducting cylinder). Dopovidi AN URSR. Ser. A **2**, 169–173 (1971)
54. Selezov, I.T., Kryvonos, Yu.G., Yakovlev, V.V.: Rasseyanie voln lokalnymi neodnorodnostyami v sploshnykh sredakh (Wave Scattering by Local Inhomogeneities in Continuous Media). Naukova Dumka, Kiev (1985)
55. Selezov, I.T., Lazarenko, M.A.: Rozsiiannia ta dyfraktsiia pruzhnykh khvyl na sferi, rozmischenii u pivprostori (Scattering and diffraction of elastic waves by a sphere located in a half-space). Dopovidi AN URSR. Ser. A **2**, 179–182 (1966)
56. Selezov, I.T., Selezova, L.V.: The flow of an ionized gas around an oscillating boundary in the presence of a magnetic field. Magnetohydrodynamics **3**(1), 5–10 (1967)
57. Selezov, I.T., Sidorchuk, V.N., Yakovlev, V.V.: Transformatsiya voln v pribrezhnoi zone shelfa (Wave Transformation in the Coastal Shelf Zone). Naukova Dumka, Kiev (1983)
58. Selezov, I.T., Svechnikova, K.M.: Difraktsiya poverkhnostnykh gravitatsionnykh voln na zatoplennom tsilindre, raspolozhennom na krugovom osnovanii (Diffraction of surface gravity waves by a submerged cylinder located on a circular base). Dinamicheskie sistemy **24**, 33–37 (2008)
59. Selezov, I.T., Tkachenko, V.A., Yakovlev, V.V.: Difraktsiya voln tsunami na podvodnom beregovom sklone krugovogo ostrova (Diffraction of tsunami waves by an underwater coastal slope of a circular island). In: Generatsiya tsunami i vykhod voln na bereg (Tsunami Generation and Wave Run-Up on the Coast), pp. 41–46. Radio i svyaz, Moscow (1984)
60. Selezov, I.T., Vanina, E.G.: Difraktsiya gidrodinamicheskikh voln na polupogruzhennom vertikalnom ellipticheskom tsilindre s neploskim tortsom (Diffraction of hydrodynamic waves by a partially submerged elliptical cylinder with nonflat base). Dopovidi AN URSR. Ser. A **5**, 33–37 (1986)
61. Selezov, I.T., Yakovlev, V.V.: Problems of plane-wave diffraction by a variable-density cylinder. Soviet Physics—Acoustics **23**(5), 460–462 (1977)
62. Selezov, I.T., Yakovlev, V.V.: Difraktsiya voln na simmetrichnykh neodnorodnostyakh (Wave Diffraction by Symmetric Inhomogeneities). Naukova Dumka, Kiev (1978)
63. Sheikh, R., Swan, C.: Wave slamming on vertical surface-piercing cylinders: the role of nonlinear wave scattering. In: Proceedings of the 15th International Offshore and Polar Engineering Conference, Seoul, South Korea, 19–24 June 2005, vol. 3, pp. 652–659
64. Shenderov, E.L.: Difraktsiya tsilindricheskoi zvukovoi volny na tsilindre (Diffraction of a cylindrical acoustic wave by a cylinder). Akusticheckii zhurnal **7**(3), 370–374 (1961)
65. Shenderov, E.L.: Volnovye zadachi gidroakustiki (Wave Problems of Hydroacoustics). Sudostroyenie, Leningrad (1972)

66. Sleeman, B.D.: Acoustic scattering by inhomogeneous media. Q. J. Mech. Appl. Math. **33**(4), 373–383 (1980)
67. Spring, B.H., Monkmeyer, P.L.: Interaction of plane waves with vertical cylinders. In: Proceedings of the 14th International Conference on Coastal Engineering, Copenhagen, Denmark, 24–28 June 1974, vol. 3, pp. 1828–1847. American Society of Civil Engineers, New York (1975)
68. Sun, L., Li, Y., Teng, B., Zhang, N.: Wave action on structures of combined cylinders. China Ocean Eng. **19**(3), 375–384 (2005)
69. Swan, C., Sheikh, R.: The interaction between steep waves and a surface-piercing column. Philos. Trans. R. Soc. Lond. A **373**(2033), 20140114 (2015)
70. Takayama, T., Goda, Y.: Wave forces on a semi-submerged vertical cylinder of elliptical shape. Coast. Eng. Jpn. **16**, 147–164 (1973)
71. Tikhonov, A.N., Samarskii, A.A.: Equations of Mathematical Physics. Dover Publication, New York (2011)
72. Tsubogo, T.: Near-resonance of diffraction waves by an array of plates on shallow water in oblique waves. In: Proceedings of the 10th International Offshore and Polar Engineering Conference, Seattle, USA, 27 May–2 June 2000, vol. 3, pp. 154–164
73. Twersky, V.: Multiple scattering of radiation by an arbitrary configuration of parallel cylinders. J. Acoust. Soc. Am. **24**(1), 42–46 (1952)
74. Vanina, E.G.: Rasseyanie poverkhnostnykh gravitatsionnykh voln ellipticheskimi neodnorodnostyami (Scattering of surface gravity waves by elliptical inhomogeneities). Dopovidi AN URSR. Ser. A **4**, 29–32 (1984)
75. Vasilenko, V.A.: Splain-funktsii: teoriya, algoritmy, programmy (Spline Functions: Theory, Algorithms, Programs). Nauka, Novosibirsk (1983)
76. Vissers, F.P.L.M., Kamp, L.P.J., Sluijter, F.W.: Some generalizations of radial Bremmer series. Wave Motion **13**(1), 43–51 (1991)
77. Walker, S.: Boundary elements in fluid-structure interaction problems of rotational shells. Appl. Math. Model. **4**, 345–350 (1980)
78. Williams, A.N.: Wave diffraction by elliptical breakwaters in shallow water. Ocean Eng. **12** (1), 25–43 (1985)
79. Williams, A.N., Li, W.: Water wave interaction with an array of bottom-mounted surface-piercing porous cylinders. Ocean Eng. **27**, 841–866 (2000)
80. Yilmaz, O.: Hydrodynamic interactions of waves with group of truncated vertical cylinders. J. Waterway Port Coast. Ocean Eng. **124**, 272–279 (1998)
81. Young, J.W., Bertrand, J.C.: Multiple scattering by two cylinders. J. Acoust. Soc. Am. **58**(6), 1190–1195 (1975)
82. Zavyalov, Yu.S., Kvasov, B.I., Miroshnichenko, V.D.: Metody splain-funktsii (Methods of Spline Functions). Nauka, Moscow (1980)

Chapter 5
Wave Diffraction by Convex Bodies in Semi-infinite Domains

5.1 Formulation of the Wave-Diffraction Problems in Semi-infinite Domains

Solving the problems of wave diffraction by a single object is rather complicated even for an infinite domain [1, 3–5, 9, 11, 12], and the difficulties grow significantly for multiply connected domains. For a semi-infinite domain, the problem can be solved by means of the method of images whose realisation requires only the knowledge of the solution for an infinite domain. This approach is applied in electrodynamics to the solution of stationary problems concerning point charges [7, 8]. The studies of various fields show that in all the cases a flat boundary located in the vicinity of the scatterer produces strongly oscillating fields. Physically this implies the superposition of perturbed fields. This effect of a flat boundary is considered in Refs. [14–19] for some specified problems. For example, the dependence on the distance from the boundary to the scatterer is studied for the acoustic wave scattering by a sphere.

We consider the fields associated with linear acoustic, electromagnetic or elastic oscillations that can be described in terms of scalar functions, e.g. the velocity potential in acoustics, electromagnetic potentials, scalar and vector potentials in the elastic theory. In these cases, the diffraction problem reduces to the Cauchy problem or to the boundary-value problem for the wave equation (or system of equations) in some domain Q with the coefficients having first-kind discontinuities on some surface R.

Suppose that the surface R_1 of a perfectly rigid obstacle is smooth, closed and coincides with the constant coordinate surface. The obstacle is located in a semi-infinite domain Q_2 with the flat boundary R_2. The surface R_2 is given in the Cartesian coordinate system xyz by the equation $y = h$. The surface R_1 is given in a curvilinear orthogonal coordinate system with the origin (0, 0, 0) in the xyz system. A plane wave with the front parallel to R_2 is propagating from infinity along the y-axis (Fig. 5.1) [13].

I.T. Selezov et al., *Wave Propagation and Diffraction*,
Foundations of Engineering Mechanics, DOI 10.1007/978-981-10-4923-1_5

Fig. 5.1 Configuration of the problem

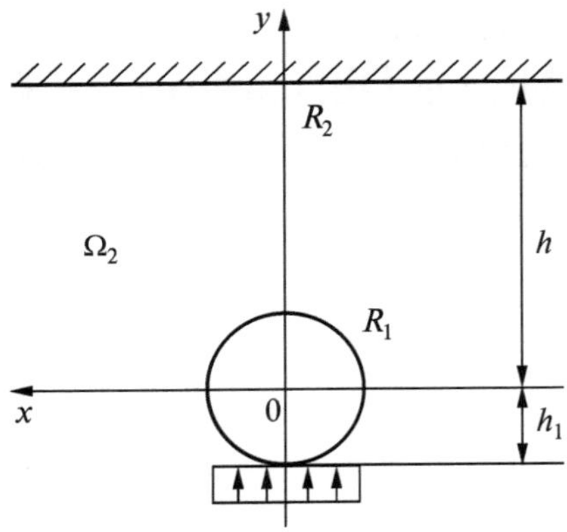

The nonstationary problem reduces to the equation

$$\nabla^2\varphi - \frac{1}{c^2}\frac{\partial^2\varphi}{\partial t^2} = 0, \quad (5.1)$$

where c is the wave speed. The function φ can be written as a sum of the two functions associated with the incident and scattered fields, i.e. $\varphi = \varphi_i + \varphi_s$. Let the leading wavefront be in contact with the obstacle surface at $t = 0$. Then for a function describing the scattered wave field we have

$$f(T) = \begin{cases} 0 & \text{at} \quad T \le 0, \\ f(T) & \text{at} \quad T > 0, \end{cases} \quad (5.2)$$

where $T = (y + h_1)/c, h_1$ is the distance from the Ox-axis to the outermost point of the obstacle, and the function $f(T)$ can have the first-kind discontinuities in the general case.

Thus, the initial conditions take the form

$$\varphi|_{t=0} = \left.\frac{\partial\varphi}{\partial t}\right|_{t=0} = 0. \quad (5.3)$$

In the cases of acoustics and electrodynamics, the boundary conditions yield the Dirichlet, Neumann or Leontovich conditions, i.e.

$$\varphi|_{R_i} = 0, \quad \left.\frac{\partial\varphi}{\partial \boldsymbol{n}}\right|_{R_i} = 0, \quad \left(\frac{\partial\varphi}{\partial \boldsymbol{n}} + \alpha\frac{\partial\varphi}{\partial t}\right)_{R_i} = 0 \quad (i = 1, 2). \quad (5.4)$$

These conditions are inhomogeneous since φ contains the addend φ_i. In the elastic-theory case, Eq. (5.1) is replaced by a system of two equations of the same type while the boundary conditions are formulated according to the first, second and mixed boundary-value problems. Moreover, in the case of stationary diffraction problems, Sommerfeld radiation conditions should be satisfied.

The difficulties associated with such problems arise mainly because boundary conditions should be satisfied both on the plane boundary of the half-space and on the curvilinear boundary of the obstacle, the boundaries being associated with different coordinate systems. Thus, the classical separation of variables is inapplicable.

5.2 Method of Images in the case of Finite Bodies

The idea of the method is to find an auxiliary solution for an imaginary obstacle obtained by symmetric mapping of the real obstacle with respect to the flat boundary (Fig. 5.1). Each point of the mapping,

$$M(x, y, z) \in H(-\infty < x < \infty,\ -\infty < y < \infty,\ -\infty < z < \infty), \tag{5.5}$$

is associated with its specular reflection with respect to the plane $y = h$, i.e.

$$\begin{gathered} M^*\left(x^*, y^*, z^*\right) \in H^*\left(-\infty < x^* < \infty,\ -\infty < y^* < \infty,\ -\infty < z^* < \infty\right): \\ x^* \equiv x, \quad z^* \equiv z, \quad y^* \equiv -y + 2h. \end{gathered} \tag{5.6}$$

The points M and M^* are in one-to-one correspondence.

In a similar manner, we can introduce the pivot and reflection spaces of differential operators and consider the relevant physical fields that are determined by the solution of the problem for an infinite domain. For the diffraction problems this is the field diffracted by the obstacle. Solutions in H and in H^* are of similar forms differing in the asterisks over the arguments. The sum and difference of the fundamental and imaginary fields yield the total field that satisfies some boundary conditions on the flat surface $y = h$ and boundary conditions on the obstacle surface other than the given conditions. Having given the mismatch of the boundary conditions on the surface, we repeat the above procedure, obtain the second mismatch, etc. There exist some ranges of variation of the characteristic parameters where the intensity of multiple fields, i.e. the mismatch, decreases as the multiplicity gets higher. In the case of dynamical wave diffraction by a body, this process is associated with multiply scattered fields.

For the sake of convenience, we pass to the space of Laplace images, i.e.

$$\Phi = \int_0^\infty e^{-st} \varphi \, dt. \tag{5.7}$$

Taking into account initial conditions (5.3) and boundary conditions (5.4), we find from (5.1) that

$$\left(\nabla^2 - \frac{s^2}{c^2}\right)\Phi = 0, \quad F \overset{\cdot}{\longrightarrow} f, \tag{5.8}$$

$$\Phi|_{R_i} = 0, \quad \left.\frac{\partial\Phi}{\partial n}\right|_{R_i} = 0, \quad \left(\frac{\partial\Phi}{\partial n} + \alpha\, s\, \Phi\right)_{R_i} = 0, \tag{5.9}$$

where the variables denoted by capital letters refer to the Laplace images of the corresponding variables in the physical space.

We assume the solution of problem (5.8)–(5.9) in the infinite domain ($i = 1$) to be known for some condition (5.9). It is rather difficult to solve each of the three problems (5.9) for a complex shape of an obstacle. Each boundary condition (5.9) can be given on the flat surface $y = h$ ($i = 2$).

Then the following propositions hold.

Proposition 1 *In the case of scalar fields, the difference of the solution in the infinite domain and its image vanishes on the undisturbed flat boundary at* $y \to h$.

Proposition 2 *In the case of vector fields, the sum of the solution in the infinite domain and its image have the following property: a vector field component normal to the undisturbed flat boundary* $y = h$ *vanishes on this boundary.*

Within the context of Propositions 1 and 2, the solution for the nonmixed problem can be written in terms of a series that is convergent under certain conditions, i.e.

$$\Phi = \sum_{k=1}^{\infty} \Phi^{(k)}, \tag{5.10}$$

where $\Phi^{(k)}$ is the diffracted field of multiplicity k.

For the Dirichlet problem according to Proposition 1 we have

$$\Phi^{(k)}(M, M^*, s) = \Phi^{(k)}(M, s) - \Phi^{*(k)}(M^*, s), \tag{5.11}$$

where $\Phi^{(k)}(M, s)$ is the known solution for the infinite domain and $\Phi^{*(k)}(M^*, s)$ is the image solution in the coordinate system specularly reflected from the flat surface (cf. Fig. 5.1).

For the Neumann problem, it is convenient to pass from the scalar field to the vector one. For example, in the acoustic problem the potential φ is associated with the velocity field $v = \nabla\varphi$. In this case, according to Proposition 2, we have for the velocity field image

$$V^{(k)}(M,M^*,s) = V^{(k)}(M,s) + V^{*(k)}(M^*,s). \tag{5.12}$$

Thus, in the case of finite bodies with canonical shape, the solution of the diffraction problem in a semi-infinite domain is reduced to the summation of known solutions for the infinite domain. Each solution exactly satisfies the boundary conditions on the scatterer surface and does not satisfy the conditions on the flat surface. A single use of the method of images ($k = 1$) yields a solution that exactly satisfies the conditions on the flat surface but no longer satisfies the boundary conditions on the scatterer surface if the scatterer is finite. If the body is infinitesimally small, this makes the end of the procedure, whereas it is continued further for the finite-size bodies. The known solutions for the point obstacles (sources) follow from (5.10)–(5.12) as special cases for $k = 1$.

It is well known that any vector field can be represented as a sum of the gradient and vector components. In some cases, there can be introduced the corresponding potentials. In particular, the problem of elastodynamics in the case of a homogeneous isotropic medium is formalised by the following two equations for the scalar potential Φ and the vector potential $\boldsymbol{\Psi}$ in the Laplace transform space:

$$\left(\nabla^2 - \frac{s^2}{c_e^2}\right)\Phi = 0, \quad \left(\nabla^2 - \frac{s^2}{c_s^2}\right)\boldsymbol{\Psi} = 0, \tag{5.13}$$

where c_e and c_s are propagation velocities of the dilatational (P) and shear (S) waves. These equations should be solved instead of Eq. (5.7) considered above. In this case, the displacement field is determined by the expression

$$\boldsymbol{U} = \nabla\Phi + \nabla\times\boldsymbol{\Psi}, \nabla\cdot\boldsymbol{\Psi} = 0. \tag{5.14}$$

Let us consider the homogeneous boundary conditions associated with the first and second boundary-value problems [2] for a scatterer and a flat boundary, i.e.

$$\boldsymbol{U}|_{R_i} = 0 \qquad (i = 1, 2), \tag{5.15}$$

$$\Sigma_{\nu k}|_{R_i} = 0 \quad (\nu, k = 1, 2, 3). \tag{5.16}$$

Here, $\boldsymbol{U}$ and $\Sigma_{\nu k}$ are Laplace images of the displacement vector and the stress tensor components. Moreover, we also consider conditions of two types on the flat surface: pinching condition, when the normal displacement and the tangential stress are equal to zero, or the free boundary condition, when the normal stress and the tangential displacement are equal to zero, namely,

$$U_2|_{y=h} = 0, \quad \Sigma_{21}|_{y=h} = \Sigma_{23}|_{y=h} = 0, \tag{5.17}$$

$$U_1|_{y=h} = U_3|_{y=h} = 0, \quad \Sigma_{22}|_{y=h} = 0. \tag{5.18}$$

Then an additional proposition given below holds.

Proposition 3 *The sum (difference) of the solution in the infinite domain and its image, the normal (tangential) component of the displacement vector and the tangential (normal) component of the stress tensor vanish.*

To prove this proposition, it is sufficient to consider an elastic medium and a solution of any multiplicity. Suppose that $U_\nu(x, h, z, s)$ are the components of the displacement vector $\boldsymbol{U}(x, y, z, s)$ at $y \to h$, and $U_\nu^*(x^*, h^*, z^*, s)$ are the components of the image $\boldsymbol{U}^*(x^*, y^*, z^*, s)$ at $y^* \to h^*$. Within the context of the symmetry properties of the image and the definition for the degenerate quantities, we have

$$U_1 = U_1^*, \quad U_3 = U_3^*, \quad U_2 = -U_2^*, \tag{5.19}$$

and hence

$$U_1 + U_1^* = 2U_1, \quad U_3 + U_3^* = 2U_3, \quad U_2 + U_2^* = 0. \tag{5.20}$$

The first and second conditions (5.19) along with the symmetry conditions yield the relations for the stress tensor $\Sigma_{\nu k}$ in the plane $y = h$, i.e.

$$\Sigma_{21} = -\Sigma_{21}^*, \quad \Sigma_{23} = -\Sigma_{23}^*, \quad \Sigma_{22} = -\Sigma_{22}^*. \tag{5.21}$$

The proof for the difference of the solution and the image is analogous.

It follows from Proposition 3 that solutions to the problem of elastodynamics with boundary conditions (5.17) or (5.18) on a flat boundary are given by the expressions similar to (5.10), with the corresponding terms of the series given by

$$\boldsymbol{U}^{(k)}(M, M^*, s) = \boldsymbol{U}^{(k)}(M, s) \pm \boldsymbol{U}^{*(k)}(M^*, s). \tag{5.22}$$

The boundary conditions for the first and second boundary-value problems of the elastic theory cannot be satisfied completely inasmuch as waves of two types are present, dilatation and distortion.

The approach to the mixed problem for the wave equation can be given as follows. We write the third boundary condition in (5.4) on the flat boundary in the form

$$\boldsymbol{U}|_{R_2} = (U_n \boldsymbol{n} + U_\tau \boldsymbol{\tau})_{y=h} = 0, \tag{5.23}$$

where U_n and U_τ are the displacements normal and tangential to the surface. We assume that in the first approximation the solution $\boldsymbol{U}_{[1]}(M, s)$ is known and satisfies condition (5.17). This solution is obtained from (5.22) and (5.10) by mapping the first-approximation space H_1^* into H_1. Then in the second approximation satisfying condition (5.18) we get

$$U_{[2]} = \sum_{k=1}^{\infty} U_{[1]}^{(k)}(M, M^*, s).$$

Then we map the second-approximation space H_2^* into H_2 and thus find the third approximation, etc. The mismatch in (5.23) decreases as the number of approximation grows, i.e. the iteration procedure is convergent. By repeating this process iteratively, we obtain a solution that corresponds to condition (5.23). In particular, the approximation of order $(q + 1)$ is given by

$$U_{[q+1]} = \sum_{k=1}^{\infty} \left[U_{[q]}^{(k)}(M, s) + (-1)^q U_{[q]}^{(k)}(M^*, s) \right] \quad (q = 0, 1, 2 \ldots). \tag{5.24}$$

The final solution is determined by the inversion formulas. For representation (5.11), we have the solution given by

$$\begin{aligned} \varphi &= \frac{1}{2\pi i} \int_{\alpha - i\infty}^{\alpha + i\infty} \sum_{k=1}^{\infty} \Phi^{(k)} e^{st} ds \\ &= \frac{1}{2\pi i} \sum_{k=1}^{\infty} \left[\int_{\alpha - i\infty}^{\alpha + i\infty} \Phi^{(k)}(M, s) e^{st} ds - \int_{\alpha - i\infty}^{\alpha + i\infty} \Phi^{*(k)}(M^*, s) e^{st} ds \right], \end{aligned} \tag{5.25}$$

provided that the Riemann-Mellin integral is uniformly convergent. Formula (5.25) shows that going over from H^* into H can be performed after the originals of the functions $\Phi^{(k)}$ and $\Phi^{*(k)}$ are determined. Solutions of the stationary diffraction problems can be found in a manner similar to the one performed in the space of Laplace images, with $i\omega$ substituted for s.

It should be noted that the real and mirror fields are given in different coordinate systems. In some cases, the mirror field can be expressed in terms of the fundamental field variables with the use of geometric transformations or addition theorems.

It is not difficult to perform the mapping of H^* into H for a single field in the case of arbitrary orthogonal coordinates. In particular, for a cylindrical obstacle (Fig. 5.2) we have

$$\begin{aligned} r^* &= \left[1 + 4\left(\frac{h}{r}\right)^2 + 4\frac{h}{r}\cos\theta \right]^{\frac{1}{2}} r, \\ \sin n\theta^* &= \sum_{k=0}^{m_1} (-1)^k C_n^{2k+1} \eta^{-n} \left(-\frac{2h}{r} - \cos\theta \right)^{n-2k-1} \sin^{2k+1}\theta, \\ \cos n\theta^* &= \sum_{k=0}^{m_2} (-1)^k C_n^{2k} \eta^{-n} \left(-\frac{2h}{r} - \cos\theta \right)^{n-2k} \sin^{2k+1}\theta, \end{aligned} \tag{5.26}$$

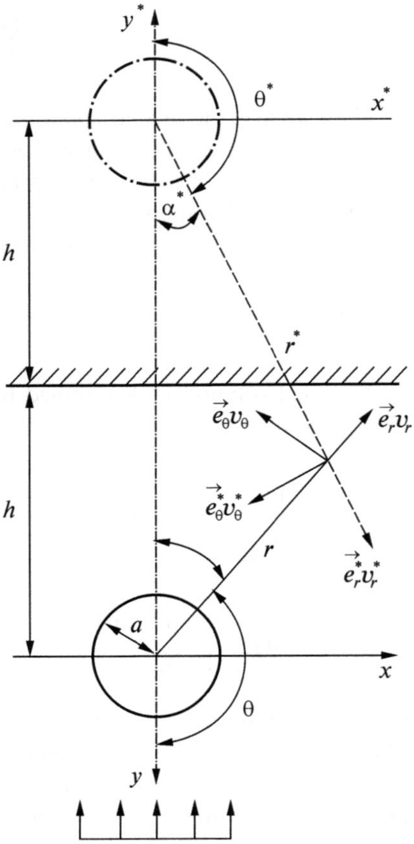

Fig. 5.2 Configuration of the problem of the plane wave scattering by a cylinder in a semi-infinite domain

where

$\eta = \frac{r^*}{r}$; $m_1 = m_2 = 0$ for $n = 0,\ 1$;

$$\frac{n-2}{2} \leq m_1 \leq \frac{n-1}{2}, \quad \frac{n-1}{2} \leq m_2 \leq \frac{n}{2} \quad \text{for} \quad n \geq 2;$$

and C_n^{2k+1} and C_n^{2k} are the binomial coefficients.

The similar formulas can be derived for the spherical and other coordinate systems.

5.3 Scattering of Plane Acoustic Waves by a Circular Cylinder

Now we consider the diffraction of acoustic waves by a cylinder in a semi-infinite domain. Suppose that plane harmonic waves are incident from infinity on a circular cylinder of radius a. The cylinder with the centre in the coordinate origin is located at distance h from a perfectly rigid boundary (Fig. 5.2). The velocity field of the incident wave propagating along the negative direction of the y-axis is given by

$$\upsilon_y = \upsilon_0 \exp[\mathrm{i}(\omega t + py)], \tag{5.27}$$

where υ_0 is the velocity amplitude, p is the wave number and ω is the oscillation frequency.

The motion is governed by the wave equation for the scalar potential φ:

$$\nabla^2\varphi - \frac{1}{c_a^2}\frac{\partial^2\varphi}{\partial t^2} = 0, \tag{5.28}$$

where $\upsilon = \nabla\varphi$ is the total velocity field (including the scattered field) and c_a is the sound velocity. This problem also concerns the scattering of surface gravity wave in the shallow-water approximation, provided that $c_{\mathrm{sh}} = \sqrt{gh}$ is substituted for c_a. In this case, g is the gravitation constant and h is the water depth.

The potential φ satisfies the boundary conditions

$$\left.\frac{\partial\varphi}{\partial r}\right|_{r=a} = 0, \quad \left.\frac{\partial\varphi}{\partial n}\right|_{y=0} = 0 \tag{5.29}$$

and the Sommerfeld condition

$$\lim_{r\to\infty} \sqrt{r}\left(\frac{\partial\varphi}{\partial r} + \mathrm{i}p\varphi\right) = 0. \tag{5.30}$$

In the case of infinite domain, the solution of Eq. (5.28) that satisfies the first boundary condition (5.29) and the Sommerfeld radiation condition (5.30) is given by

$$\varphi^{(1)} = \sum_{m=0}^{\infty}\left[f_m J_m(pr) + B_m^{(1)} H_m^{(2)}(pr)\right] \cos m\theta e^{\mathrm{i}\omega t}, \tag{5.31}$$

where $f_m = -\upsilon_0 \varepsilon_m p^{-1} \mathrm{i}^{(m-1)}$ and ε_m is the Neumann factor, i.e.

$$\varepsilon_m = \begin{cases} 1 & \text{at} \quad m = 0, \\ 2 & \text{at} \quad m = 1, 2, \ldots \end{cases}$$

In what follows we omit the factor $\exp(\mathrm{i}\omega t)$. The arbitrary constant is given by

$$B_m^{(1)} = -f_m \frac{J_m'(pa)}{H_m^{(2)\prime}(pa)}. \tag{5.32}$$

Expressions (5.31) and (5.32) describe the plane harmonic wave scattering by a cylinder in an unbounded domain.

Now we consider the field of velocities in the far region (the ratio *r/a* is large) in the Rayleigh approximation (the ratio of the wavelength to the cylinder radius is large, i.e. the value of *pa* is small). This approximation corresponds to the fields slowly varying with time. We substitute asymptotic values for the Bessel and Hankel functions in solution (5.31) [10] and thus find the scattered field in the far region to be given by

$$\upsilon_r^{(1)} \cong \mathrm{i}p\sqrt{\frac{2}{\pi}}\sqrt{\frac{a}{r}}\sqrt{pa}\sum_{m=0}^{\infty}\exp\left[\mathrm{i}\frac{\pi}{2}\left(m+\frac{1}{2}\right)\right]B_m^{(1)}\cos m\theta\exp(-\mathrm{i}pr), \tag{5.33}$$

$$\upsilon_\theta^{(1)} \cong r^{-1}\sqrt{\frac{2}{\pi}}\sqrt{\frac{a}{r}}\sqrt{pa}\sum_{m=0}^{\infty}\exp\left[\mathrm{i}\frac{\pi}{2}\left(m+\frac{1}{2}\right)\right]B_m^{(1)}m\sin m\theta\exp(-\mathrm{i}pr). \tag{5.34}$$

Here and in what follows the values of υ_r and υ_θ are normalised to the amplitude υ_0. For small values of *pa*, the functions $J_m(pa)$ and $H_m^{(2)}(pa)$ contained in $B_m^{(1)}$ are replaced by their expansions [10]. In this case the dominant coefficients are $B_0^{(1)}$ and $B_1^{(1)}$, i.e.

$$B_0^{(1)} \approx -\frac{\pi}{4}p^{-1}(pa)^2, \qquad B_1^{(1)} = \frac{\pi}{2}\mathrm{i}p^{-1}(pa)^2. \tag{5.35}$$

The field criterion in the far region that follows from the estimation of the first disregarded term is given by

$$pa\frac{r}{a} \geq \frac{4m^2-1}{8\varepsilon_e}, \tag{5.36}$$

where ε_e is the desired calculation accuracy. Expressions (5.35) are obtained with the maximum relative error

$$\frac{1}{2}(pa)^2\ln\frac{pa}{2}. \tag{5.37}$$

Now we consider a cylinder in a semi-infinite domain with a flat boundary parallel to the cylinder axis. The second boundary condition on a flat boundary can be written as

$$\upsilon_y\big|_{y=0} = 0. \tag{5.38}$$

The obstacle scatters the wave. Some portion of scattered waves goes to infinity, another portion comes to the boundary of the half-space and is reflected. Reflected and scattered fields form the primary field. Plane waves are reflected by the boundary, arrive at the cylinder and are scattered again. The secondary scattered field contains mainly the waves that arise from the reflected plane waves. The scattering continues and thus produces multiply scattered fields. Evidently, the greatest portion of energy is transported by the scattered field of the incident wave and the scattered field of the wave reflected from the flat boundary. The influence of the triply scattered field and other multiply scattered fields on the total scattered field is small and rapidly decreases as the multiplicity increases.

We write the solution that satisfies condition (5.38) in terms of a convergent series, i.e.

$$\boldsymbol{\upsilon}(r, \theta, r^*, \theta^*) = \sum_{s=1}^{\infty} \left[\boldsymbol{\upsilon}^{(s)}(r, \theta) + \boldsymbol{\upsilon}^{*(s)}(r^*, \theta^*)\right], \tag{5.39}$$

where $\boldsymbol{\upsilon}^{(s)}$ is the scattered field of multiplicity s, and $\boldsymbol{\upsilon}^{*(s)}$ is the corresponding mapping defined in the coordinate system (r^*, θ^*). The total velocity components for the scattered field of multiplicity s can be written as

$$(\boldsymbol{\upsilon}^{(s)} + \boldsymbol{\upsilon}^{*(s)})_r = V_r^{(s)} = \upsilon_r^{(s)} - \upsilon_r^{*(s)}\cos(\theta + \theta^*) + \upsilon_\theta^{*(s)}\sin(\theta + \theta^*), \tag{5.40}$$

$$(\boldsymbol{\upsilon}^{(s)} + \boldsymbol{\upsilon}^{*(s)})_\theta = V_\theta^{(s)} = \upsilon_\theta^{(s)} + \upsilon_\theta^{*(s)}\cos(\theta + \theta^*) + \upsilon_r^{*(s)}\sin(\theta + \theta^*). \tag{5.41}$$

Within the context of geometrical relations, the functions $\upsilon_r^{(s)}(r^*, \theta^*), \upsilon_\theta^{(s)}(r^*, \theta^*)$ and the trigonometric functions entering (5.40) and (5.41) can be expressed in terms of the variables r and θ. Thus, the final formulas that follow from (5.26) are given by

$$\begin{gathered}
\frac{r^*}{a} = \frac{r}{a}\eta; \quad \eta = \left[1 + 4\left(\frac{h}{r}\right)^2 + 4\frac{h}{r}\cos\theta\right]^{\frac{1}{2}}, \\
\sin\theta^* = \eta^{-1}\sin\theta; \quad \cos\theta^* = -\eta^{-1}\left(\frac{2h}{r} + \cos\theta\right), \\
\sin(\theta + \theta^*) = -\eta^{-1}\frac{2h}{r}\sin\theta; \quad \cos(\theta + \theta^*) = -\eta^{-1}\left(\frac{2h}{r}\cos\theta + 1\right).
\end{gathered} \tag{5.42}$$

We define the field at a point with coordinates r and θ as a sum of fields of two obstacles located at different distances from this point. The latter fact along with the difference between the velocities of P-waves (dilatational waves) and S-waves (shear waves) are described by the formula

$$\exp(ipr^*) = (\cos\eta_p - \mathrm{i}\sin\eta_p)\exp(ipr^*), \tag{5.43}$$

where

$$\eta_p = pr(\eta - 1).$$

We substitute (5.33) and (5.34) in the solutions (5.40) and (5.41), make use of the dependences (5.42), and perform some transformations. Thus, we obtain the formulas for the velocity vector components of the singly scattered field, i.e.

$$\begin{aligned}
\mathrm{Re}V_r^{(1)} &= 1 + 2\cos\theta - \eta^{-\frac{3}{2}}\left(\frac{2h}{r}\cos\theta + 1\right)\left[-1 + 2\eta^{-1}\left(\frac{2h}{r} + \cos\theta\right)\right](\sin\eta_p - \cos\eta_p),\\
\mathrm{Im}V_r^{(1)} &= -1 - 2\cos\theta + \eta^{-\frac{3}{2}}\left(\frac{2h}{r}\cos\theta + 1\right)\left[-1 + 2\eta^{-1}\left(\frac{2h}{r} + \cos\theta\right)\right](\sin\eta_p + \cos\eta_p),\\
\mathrm{Re}V_\theta^{(1)} &= -\eta^{-\frac{3}{2}}\frac{2h}{r}\sin\theta\left[-1 + 2\eta^{-1}\left(\frac{2h}{r} + \cos\theta\right)\right](\sin\eta_p - \cos\eta_p),\\
\mathrm{Im}V_\theta^{(1)} &= -\eta^{-\frac{3}{2}}\frac{2h}{r}\sin\theta\left[-1 + 2\eta^{-1}\left(\frac{2h}{r} + \cos\theta\right)\right](\sin\eta_p + \cos\eta_p).
\end{aligned} \tag{5.44}$$

Expressions (5.44) disregard the higher order terms. The factor $\sqrt{a/r}\,\exp(-ipr)$ is omitted in the right-hand parts of (5.44), and the left-hand parts are normalised to the factor $4(pa)^{-3/2}\pi^{-1/2}$. The first terms of the expressions for $\mathrm{Re}V_r^{(1)}$ and $\mathrm{Im}V_r^{(1)}$ correspond to the infinite domain, and all the other terms describe the influence of the boundary and have infinite numbers of zeros as h/r varies from 1 to ∞.

Analytic functions (5.44) for $1 \le h/r \le \infty$ are oscillating functions limited by constants from above and from below. At $h/r \to \infty$ all the terms containing $\eta^{-3/2}$ tend to zero, which corresponds to the passage to the infinite domain. The influence of the boundary on the scattered field decreases proportionally to $r^{-1/2}$ as the ratio h/r increases.

The incident wave field and the corresponding field of the wave reflected from the boundary are given by

$$\upsilon_y^{(1)} = \upsilon_y - \upsilon_y^* = \exp(\mathrm{i}py) - \exp[-\mathrm{i}p(y + 2h)]. \tag{5.45}$$

The secondary scattered field is the scattered field of the reflected incident wave. The velocity potential of the scattered field that is reflected from the boundary can be written as

$$\varphi^{*(2)} = \sum_{m=0}^{\infty} B_m^{(1)} H_m^{(2)}(pr^*)\cos m\theta^*.$$

We make use of the formulas

$$\theta^* = \pi - a^*, \quad r^* = r^*$$

to go over from the coordinates (r^*, θ^*) to (r^*, a^*) (Fig. 5.2). Thus, we obtain

$$\varphi^{*(2)} = \sum_{m=0}^{\infty} (-1)^m B_m^{(1)} H_m^{(2)}(pr^*) \cos ma^*.$$

In the coordinate system (r, a), the expansion of the function $H_m^{(2)}(pr^*) \cos ma^*$ in terms of cylindrical functions is determined by the Neumann addition theorem [20], i.e.

$$H_m^{(2)}(pr^*) \cos ma^* = \sum_{\nu=-\infty}^{\infty} H_{m+\nu}^{(2)}(2ph) J_\nu(pr) \cos \nu a.$$

After some transformations, the relations for the cylindrical functions yield [10]

$$H_m^{(2)}(pr^*) \cos ma^* = H_m^{(2)}(2ph) J_0(pr) + \sum_{\nu=1}^{\infty} \left[H_{\nu+m}^{(2)}(2ph) + (-1)^{2\nu-m} H_{\nu-m}^{(2)}(2ph) \right] J_\nu(pr) \cos \nu a.$$

Then, having passed again to the variable θ, we have

$$\varphi^{*(2)} = \sum_{m=0}^{\infty} (-1)^m \Big\{ H_m^{(2)}(2ph) J_0(pr) \sum_{\nu=1}^{\infty} (-1)^\nu \left[H_{\nu+m}^{(2)}(2ph) + (-1)^{2\nu-m} H_{\nu-m}^{(2)}(2ph) \right] J_\nu(pr) \cos \nu\theta \Big\}. \tag{5.46}$$

Making use of Eq. (5.46) reduces the first constituent of the secondary diffracted field to the form

$$\varphi_1^{(2)} = \sum_{\nu=0}^{\infty} \left[G_\nu^{(2)} J_\nu(pr) + F_\nu^{(2)} H_\nu^{(2)}(pr) \right] \cos \nu\theta, \tag{5.47}$$

where

$$G_\nu^{(2)} = \sum_{m=0}^{\infty} (-1)^m B_m^{(1)} D_{\nu m}^{(2)}; \quad F_\nu^{(2)} = -\frac{J_\nu'(pa)}{H_\nu^{(2)\prime}(pa)} G_\nu^{(2)},$$

$$D_{0m}^{(2)} = H_m^{(2)}(2ph); \quad D_{\nu m}^{(2)} = (-1)^\nu [H_{\nu+m}^{(2)}(2ph) + (-1)^{2\nu-m} H_{\nu-m}^{(2)}(2ph)].$$

The velocity potential of the plane wave (5.45) reflected from the boundary is given by

$$\varphi^{*(1)} = e^{-i2ph} \sum_{m=0}^{\infty} f_m^* J_m(pr) \cos m\theta. \tag{5.48}$$

We employ (5.48) to rewrite the second constituent of the secondary diffracted field in the form

$$\varphi_2^{(2)} = \sum_{m=0}^{\infty} [f_m^* J_m(pr) e^{-i2ph} + B_m^{(2)} H_m^{(2)}(pr)] \cos m\theta, \tag{5.49}$$

where

$$f_m^* = -\upsilon_0 \varepsilon_m p^{-1} \mathrm{i}^{-(m+1)}, \quad B_m^{(2)} = \mathrm{i}^{-2m} B_m^{(1)} e^{-\mathrm{i}2ph}.$$

The secondary field is equal to the sum of these fields; hence, it is given by the formula

$$\varphi^{(2)} = \sum_{\nu=0}^{\infty} [K_\nu^{(2)} J_\nu(pr) + L_\nu^{(2)} H_\nu^{(2)}(pr)] \cos \nu\theta, \tag{5.50}$$

where

$$K_\nu^{(2)} = f_\nu^* e^{-i2ph} + G_\nu^{(2)}, \quad L_\nu^{(2)} = B_\nu^{(2)} + F_\nu^{(2)}.$$

In the Rayleigh approximation, all the expressions are simplified. In our approximation, we can put with high accuracy that

$$\varphi^{(2)} = \sum_{\nu=0}^{\infty} L_\nu^{(2)} H_\nu^{(2)}(pr) \cos \nu\theta. \tag{5.51}$$

This implies that the influence of the multiply scattered fields is weak by virtue of rapid convergence of series (5.39).

The expressions for the velocity components can be written as

$$\begin{aligned}
\mathrm{Re}V_r^{(2)} &= \overline{\mathrm{Re}V_r^{(2)}} \cos 2ph + \overline{\mathrm{Im}V_r^{(2)}} \sin 2ph, \\
\mathrm{Im}V_r^{(2)} &= -\overline{\mathrm{Re}V_r^{(2)}} \sin 2ph + \overline{\mathrm{Im}V_r^{(2)}} \cos 2ph, \\
\mathrm{Re}V_\theta^{(2)} &= \overline{\mathrm{Re}V_\theta^{(2)}} \cos 2ph + \overline{\mathrm{Im}V_\theta^{(2)}} \sin 2ph, \\
\mathrm{Im}V_\theta^{(2)} &= -\overline{\mathrm{Re}V_\theta^{(2)}} \sin 2ph + \overline{\mathrm{Im}V_\theta^{(2)}} \cos 2ph,
\end{aligned} \tag{5.52}$$

where we introduced the following functions:

$$\overline{\mathrm{Re}V_r^{(2)}} = 1 - 2\cos\theta - \eta^{-\frac{3}{2}}\left(\frac{2h}{r}\cos\theta + 1\right)\left[1 + 2\eta^{-1}\left(\frac{2h}{r} + \cos\theta\right)\right](\sin\eta_p - \cos\eta_p),$$
$$\overline{\mathrm{Im}V_r^{(2)}} = 1 - 2\cos\theta - \eta^{-\frac{3}{2}}\left(\frac{2h}{r}\cos\theta + 1\right)\left[1 + 2\eta^{-1}\left(\frac{2h}{r} + \cos\theta\right)\right](\sin\eta_p + \cos\eta_p),$$
$$\overline{\mathrm{Re}V_\theta^{(2)}} = \eta^{-\frac{3}{2}}\frac{2h}{r}\sin\theta\left[1 + 2\eta^{-1}\left(\frac{2h}{r} + \cos\theta\right)\right](\sin\eta_p - \cos\eta_p),$$
$$\overline{\mathrm{Im}V_\theta^{(2)}} = \eta^{-\frac{3}{2}}\frac{2h}{r}\sin\theta\left[1 + 2\eta^{-1}\left(\frac{2h}{r} + \cos\theta\right)\right](\sin\eta_p + \cos\eta_p).$$
$$(5.53)$$

The factor $\sqrt{a/r}\exp(-ipr)$ in the right-hand parts of (5.53) is omitted, and the left-hand parts are normalised to the factor $4(pa)^{-3/2}\pi^{-1/2}$.

Thus, disregarding the small multiply scattered fields $(s \geq 3)$, we obtain the final formulas for the approximate solution of the scattering problem, i.e.

$$\upsilon_r \cong V_r^{(1)} + V_r^{(2)}, \quad \upsilon_\theta \cong V_\theta^{(1)} + V_\theta^{(2)}. \tag{5.54}$$

We conclude with a numerical example. Let us find the velocity component υ_r associated with the scattered field for the initial parameters $pa = 0.05, r/a = 75, 1.0 \leq h/r \leq 3.0$. The results calculated by formulas (5.44) and (5.52) for the primary and secondary fields, respectively, are shown in Figs. 5.3, 5.4 and 5.5. The error is calculated with the use of expressions (5.36) and (5.37). For the given parameters, it is not greater than 10%. Approaching the flat boundary to the cylinder leads to strong oscillations of the solutions. This effect considerably changes the beam patterns (Fig. 5.5). Note that for h/r = 1.0 and θ = 180° the amplitude vanishes at the point on the boundary.

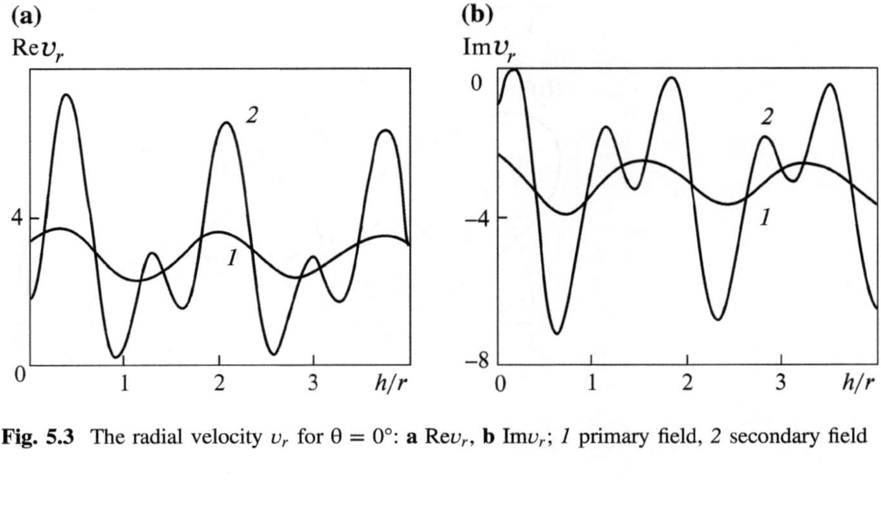

Fig. 5.3 The radial velocity υ_r for θ = 0°: **a** Reυ_r, **b** Imυ_r; *1* primary field, *2* secondary field

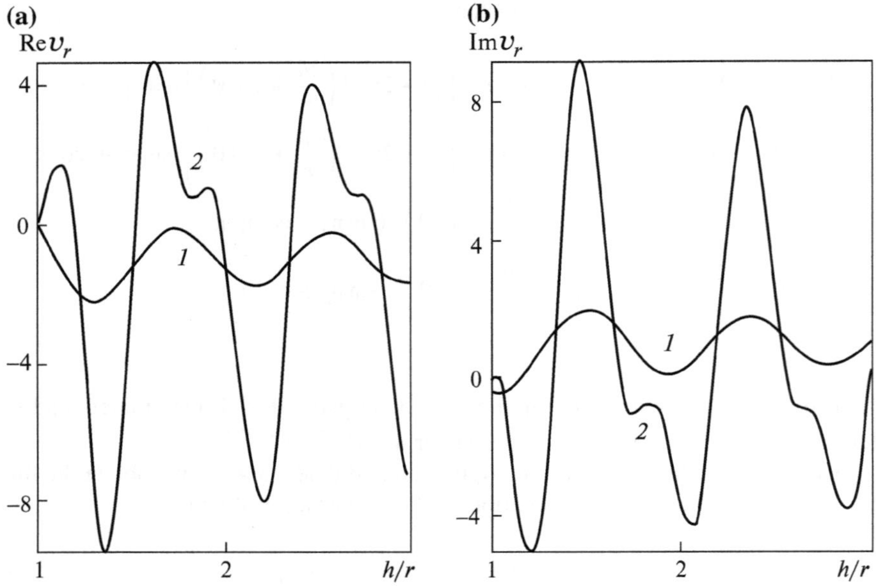

Fig. 5.4 The radial velocity v_r for $\theta = 180°$: **a** Rev_r, **b** Imv_r; *1* primary field, *2* secondary field

Fig. 5.5 Polar radiation patterns of the velocity amplitude at angles θ from 0 to 180°: **a** $h/r = 1.0$, **b** $h/r = 1.1$; *1* the field corresponding to $h/r = \infty$, *2* primary field, *3* total field

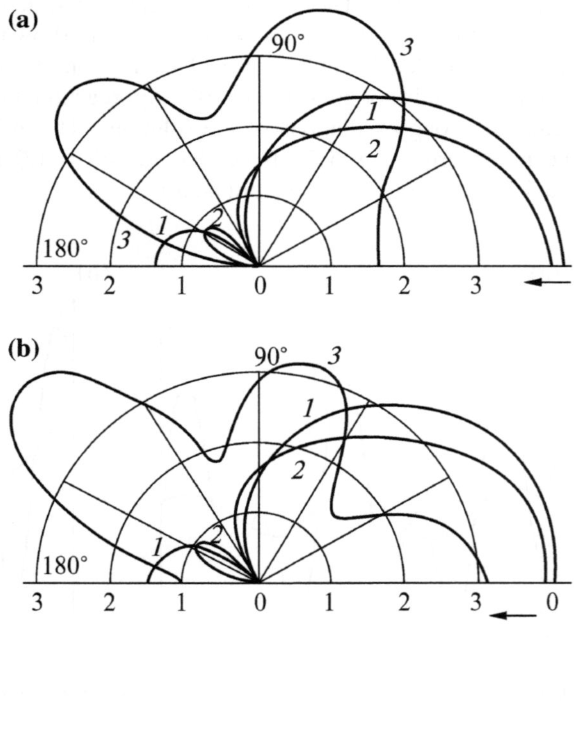

5.4 Scattering of Plane Acoustic Waves by a Sphere

We consider a system of spherical coordinates r, θ, ϕ (radial, zenith and azimuth coordinates) that corresponds to the rectangular Cartesian coordinate system x, y, z. In Fig. 5.1, we replace the y-coordinate by z, and reverse the direction of the Oy-axis.

A plane harmonic wave

$$v_z = v_0 \exp[\mathrm{i}(\omega t - kz)] \tag{5.55}$$

is incident on a perfectly rigid stationary sphere located at distance h from the flat boundary. We have to find the solution to the equation

$$(c_0^2 \nabla^2 - \partial_t^2)\Phi = 0, \quad v = -\nabla\Phi \tag{5.56}$$

with the boundary conditions

$$\left.\frac{\partial\Phi}{\partial r}\right|_{r=a} = 0, \quad \left.\frac{\partial\Phi}{\partial \boldsymbol{n}}\right|_{z=h} = 0 \tag{5.57}$$

and under the conditions of radiation and limitedness at infinity.

We write the solution of Eq. (5.56) that satisfies the first condition of (5.57) and the Sommerfeld radiation conditions as given by

$$\Phi^{(1)} = \sum_{m=0}^{\infty} \left[f_m j_m(kr) + B_m^{(1)} h_m^{(2)}(kr)\right] P_m(\cos\theta) \exp(\mathrm{i}\omega t). \tag{5.58}$$

Here, $f_m = k^{-1} v_0 (2m+1)\mathrm{i}^{-(m+1)}$, $j_m(kr)$ and $h_m^{(2)}(kr)$ are spherical Bessel and Hankel functions. In what follows we omit the factor $\exp(\mathrm{i}\omega t)$ for the sake of convenience. The coefficients $B_m^{(1)}$ are determined by the first condition (5.57). Thus, we have

$$B_m^{(1)} = -f_m \frac{j_m'(ka)}{h_m^{(2)\prime}(ka)}.$$

For large wavelengths ($ka \ll 1$), the dominant contribution in the scattered field is given by the first and second terms with the coefficients

$$B_0^{(1)} \cong \frac{1}{3}\frac{v_0}{k}(ka)^3, \quad B_1^{(1)} \cong \frac{1}{3}\mathrm{i}\frac{v_0}{k}(ka)^3. \tag{5.59}$$

For the scattered field in the far region, (5.56) and (5.58) yield

$$v_r^{(1)} = \frac{\mathrm{i}}{r} \sum_{m=0}^{\infty} B_m^{(1)} \exp\left[\mathrm{i}\frac{\pi}{2}(m+1)\right] P_m(\cos\theta) \exp(-\mathrm{i}kr), \tag{5.60}$$

$$v_\theta^{(1)} = \frac{\mathrm{i}}{kr^2} \sum_{m=0}^{\infty} B_m^{(1)} \exp\left[\mathrm{i}\frac{\pi}{2}(m+1)\right] P_m'(\cos\theta) \sin\theta \exp(-\mathrm{i}kr). \tag{5.61}$$

As before, the solution to Eq. (5.56) that satisfies all the boundary conditions of (5.57) is given by

$$\boldsymbol{v}(r,\theta,r^*,\theta^*) = \sum_{s=1}^{\infty} \left[\boldsymbol{v}^{(s)}(r,\theta) + \boldsymbol{v}^{*(s)}(r^*,\theta^*)\right] = \sum_{s=1}^{\infty} \boldsymbol{V}^{(s)}(r,\theta,r^*,\theta^*). \tag{5.62}$$

Setting $s = 1$ in formula (5.62) (the case of a singly scattered field) and taking into account (5.59)–(5.61) yields an approximate solution

$$\begin{aligned}
\mathrm{Re}V_r^{(1)} &= -2 + 3\cos\theta + \eta^{-2}\left(\frac{2h}{r}\cos\theta - 1\right)\cos\eta_p\left[2 - 3\eta^{-1}\left(\frac{2h}{r} - \cos\theta\right)\right],\\
\mathrm{Im}V_r^{(1)} &= -\eta^{-2}\left(\frac{2h}{r}\cos\theta - 1\right)\sin\eta_p\left[2 - 3\eta^{-1}\left(\frac{2h}{r} - \cos\theta\right)\right],\\
\mathrm{Re}V_\theta^{(1)} &= \eta^{-2}\frac{2h}{r}\sin\theta\cos\eta_p\left[2 - 3\eta^{-1}\left(\frac{2h}{r} - \cos\theta\right)\right],\\
\mathrm{Im}V_\theta^{(1)} &= -\eta^{-2}\frac{2h}{r}\sin\theta\cos\eta_p\left[2 - 3\eta^{-1}\left(\frac{2h}{r} - \cos\theta\right)\right].
\end{aligned} \tag{5.63}$$

Solutions (5.63) possess the property that the terms with η^{-2} tend to zero at $h/r \to \infty$, and the effect of the boundary decreases as r^{-1} when h/r increases.

The secondary scattered field consists of the scattered field of the primary scattered waves reflected from the boundary and the scattered field of the reflected incident wave. In terms of the variables r^* and θ^*, the scattered field reflected from the boundary is given by

$$\Phi^{*(2)} = \sum_{m=0}^{\infty} \bar{B}_m^{(1)} h_m^{(2)}(kr^*) P_m(\cos\theta^*). \tag{5.64}$$

We employ the addition theorem for the spherical wave functions [2], expand field (5.64) in terms of spherical functions and thus find the first constituent of the secondary diffracted field. After some transformations we have

$$\Phi_1^{(2)} = \sum_{\nu=0}^{\infty} \left[C_\nu^{(2)} j_\nu(kr) + D_\nu^{(2)} h_\nu^{(2)}(kr)\right] P_\nu(\cos\theta), \tag{5.65}$$

where

$$C_\nu^{(2)} = \sum_{m=0}^{\infty} \sum_p \bar{a}_m^{(1)} (2\nu+1)\hat{a}(0,0;\, p,\, m,\, \nu) h_p^{(2)}(2kh), \; h<r<2h,$$

$$D_\nu^{(2)} = -\frac{C_\nu^{(2)} j_\nu'(ka)}{h_\nu^{(2)\prime}(ka)}, \; \bar{B}_m^{(1)} = -\bar{f}_m \frac{j_m'(ka)}{h_m^{(2)\prime}(ka)}, \; \bar{f}_m = k^{-1} \upsilon_0 (2m+1) \mathrm{i}^{m-1},$$

and $\hat{a}(0,0;\, p,\, m,\, \nu)$ is the coefficient arising in the expansion of the product of two Legendre functions:

$$P_n^m(\cos\theta) P_\nu^\mu(\cos\theta) = \sum_p \hat{a}(|m|,|\mu|;\, p,\, n,\, \nu) P_p^{m+\mu}(\cos\theta).$$

Its explicit form is given in Ref. [2].

The expansion of the plane wave reflected from the boundary is given by

$$\Phi^{*(1)} = \exp(-\mathrm{i}2kh) \sum_{m=0}^{\infty} \bar{f}_m j_m(kr) P_m(\cos\theta). \tag{5.66}$$

Within the context of (5.66), we obtain the second constituent of the secondary diffracted field, i.e.

$$\Phi_2^{(2)} = \sum_{m=0}^{\infty} \left[\bar{f}_m j_m(kr) \exp(-\mathrm{i}2kh) + B_m^{(2)} h_m^{(2)}(kr) \right] P_m(\cos\theta), \tag{5.67}$$

where

$$B_m^{(2)} = \mathrm{i}^{2m} B_m^{(1)} \exp(-\mathrm{i}2kh).$$

The final expression for the secondary scattered field follows from (5.65) and (5.67):

$$\Phi^{(2)} \cong \sum_{\nu=0}^{\infty} \left(B_\nu^{(2)} + D_\nu^{(2)} \right) h_\nu^{(2)}(kr) P_\nu(\cos\theta). \tag{5.68}$$

The triply scattered field and higher order scattered fields can be found in the similar manner. The effect of multiply scattered fields is small by virtue of rapid convergence of series (5.62). For example, in the case of long wavelengths, it is sufficient to put $s = 1$ and $s = 2$ to obtain an adequate expression for the field in the far region. In this case, expressions for the secondary scattered field are as follows:

$$\begin{aligned}
\mathrm{Re}V_r^{(2)} &= \overline{\mathrm{Re}V_r^{(2)}}\cos 2kh + \overline{\mathrm{Im}V_r^{(2)}}\sin 2kh,\\
\mathrm{Im}V_r^{(2)} &= \overline{\mathrm{Im}V_r^{(2)}}\cos 2kh - \overline{\mathrm{Re}V_r^{(2)}}\sin 2kh,\\
\mathrm{Re}V_\theta^{(2)} &= \overline{\mathrm{Re}V_\theta^{(2)}}\cos 2kh + \overline{\mathrm{Im}V_\theta^{(2)}}\sin 2kh,\\
\mathrm{Im}V_\theta^{(2)} &= \overline{\mathrm{Im}V_\theta^{(2)}}\cos 2kh - \overline{\mathrm{Re}V_\theta^{(2)}}\sin 2kh.
\end{aligned} \tag{5.69}$$

The expressions for $\overline{\mathrm{Re}V_r^{(2)}}$, $\overline{\mathrm{Im}V_r^{(2)}}$, $\overline{\mathrm{Re}V_\theta^{(2)}}$ and $\overline{\mathrm{Im}V_\theta^{(2)}}$ follow from (5.63), provided that we substitute 2 for the superscript 1 and change the signs of the second terms within the square brackets. Formulas (5.63) and (5.69) are derived with the relative error $(kr)^{-1}$, the distance criterion is determined by the inequality $kr \gg 1/\varepsilon_e$. The factor $\exp(-\mathrm{i}kr)(kr)^{-1}$ is omitted in the right-hand parts of these expressions, the left-hand parts are normalised to the factor $1/6\,(ka)^3 \upsilon_0$.

The approximate solution is given by series (5.62) with the first and second terms:

$$\upsilon_r \cong V_r^{(1)} + V_r^{(2)}, \quad \upsilon_\theta \cong V_\theta^{(1)} + V_\theta^{(2)}.$$

As an example, we found the scattered field υ_r and υ_θ for $ka = 0.2, r/a = 50$, and $0 \le \theta \le 180°$. Figure 5.6 shows the real and imaginary parts of υ_r and υ_θ. The point $h/r = 1.0$, $\theta = 0°$ lies on the boundary, and hence the amplitude vanishes in this case. Similar to the case of a circular cylinder, a flat boundary considerably influences the scattered field and produces strong oscillations and phase changes.

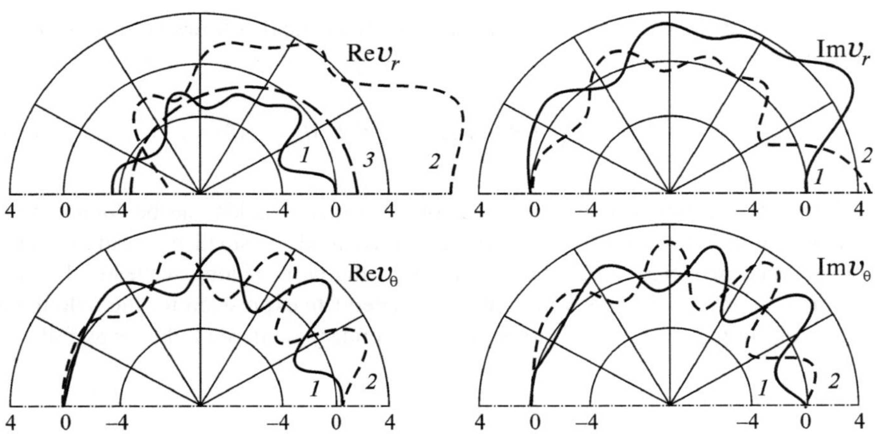

Fig. 5.6 Real and imaginary parts of the scattered field υ_r and υ_θ for *1* $h/r = 1.0$, *2* $h/r = 1.1$, and *3* $h/r = \infty$

5.5 Scattering of Electromagnetic Waves by a Circular Cylinder

The problem of electromagnetic wave scattering by a cylinder in a semi-infinite domain is studied by means of the method of images. Explicit solutions for long wavelengths and far-region field are found. Suppose a Cartesian and cylindrical coordinate systems, (x, y, z) and (r, θ, z), are related to the cylinder of radius a as shown in Fig. 5.7. The cylinder surface $r = a$ and the boundary $y = h$ are perfect conductors. A plane harmonic wave is incident from infinity. It is polarised in a way that the magnetic field intensity component is parallel to the z-axis, i.e.

$$H_z = H_0 e^{\mathrm{i}(ky-\omega t)}. \tag{5.70}$$

In this case, the equation of motion is given by

$$\nabla^2 \boldsymbol{E} - \frac{1}{c^2}\frac{\partial^2 \boldsymbol{E}}{\partial t^2} = 0, \tag{5.71}$$

and the relevant boundary conditions are

$$E_\theta|_{r=a} = \frac{\mathrm{i}}{\varepsilon_0\omega}\frac{\partial H_z}{\partial r}\bigg|_{r=a} = 0, \quad E_x = \frac{\partial H_z}{\partial y}\bigg|_{y=h} = 0. \tag{5.72}$$

The unknown functions should satisfy the boundary conditions and Sommerfeld radiation conditions.

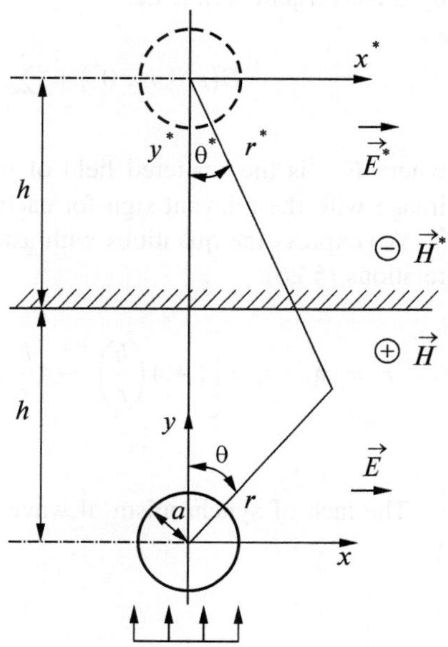

Fig. 5.7 Configuration of the problem

The solution to Eq. (5.71) for an infinite domain that satisfies condition (5.72) can be written as

$$E_\theta = \sum_{m=0}^{\infty} B_m^{(1)} H_m^{(1)}(kr) \cos m\theta \, e^{-\mathrm{i}\omega t}. \tag{5.73}$$

In what follows, we omit the factor $\exp(-\mathrm{i}\omega t)$ for convenience. The coefficients $B_m^{(1)}$ are determined by the first boundary condition in (5.72).

In the case of scattered field in the far region and large wavelengths, expression (5.73) reduces to

$$E_\theta = \frac{\sqrt{\pi}}{4a\omega\varepsilon_0}(ka)^{\frac{5}{2}}(1+\mathrm{i})(1-2\cos\theta)\sqrt{\frac{a}{r}}\, e^{\mathrm{i}kr} \tag{5.74}$$

with the relative error $1/2(ka)^2 \ln(ka/2)$ and the remoteness criterion

$$kr \geq \frac{3}{8\varepsilon_e}, \tag{5.75}$$

where ε_e is some given error.

In a system with a flat boundary, scattered waves partially propagate to infinity and partially arrive at the boundary and are reflected. These scattered and reflected fields form the primary field. The incident plane wave is reflected from the boundary, arrives at the cylinder and is scattered again, thus providing the dominant contribution in the secondary scattered field. The process continues and induces multiply scattered fields whose intensity decreases as the multiplicity increases.

The solution of Eq. (5.71) that satisfies the boundary conditions (5.72) is given by a convergent series, i.e.

$$\boldsymbol{E}\left(r,\theta,r^*,\theta^*\right) = \sum_{s=1}^{\infty}\left[\boldsymbol{E}^{(s)}(r,\theta) + \boldsymbol{E}^{*(s)}\left(r^*,\theta^*\right)\right], \tag{5.76}$$

where $\boldsymbol{E}^{(s)}$ is the scattered field of multiplicity s, and $\boldsymbol{E}^{*(s)}$ is the corresponding image with the relevant sign for each case.

We express the quantities with asterisks in terms of r and θ in accordance with relations (5.26):

$$r^* = r\eta, \quad \eta = \left[1 + 4\left(\frac{h}{r}\right)^2 - 4\frac{h}{r}\cos\theta\right]^{1/2}, \quad \cos\theta^* = \left(\frac{2h}{r} - \cos\theta\right)\eta^{-1}. \tag{5.77}$$

The lack of synchronism of wave incidence is described by the formulas

$$e^{ikr^*} = (\cos\eta_p + \mathrm{i}\sin\eta_p)e^{ikr}, \quad \eta_p = kr(\eta - 1). \tag{5.78}$$

An approximate solution for a singly scattered field, $s = 1$, follows from (5.76) with (5.74), (5.77), and (5.78) taken into account:

$$\mathrm{Re}E_\theta^{(1)} = 1 - 2\cos\theta + \eta^{-\frac{1}{2}}\left[1 - 2\eta^{-1}\left(\frac{2h}{r} - \cos\theta\right)\right](\sin\eta_p - \cos\eta_p), \tag{5.79}$$

$$\mathrm{Im}E_\theta^{(1)} = 1 - 2\cos\theta - \eta^{-\frac{1}{2}}\left[1 - 2\eta^{-1}\left(\frac{2h}{r} - \cos\theta\right)\right](\sin\eta_p + \cos\eta_p). \tag{5.80}$$

The factor $\sqrt{a/r}\exp(ikr)$ is omitted in the right-hand parts of (5.79) and (5.80), the left-hand parts are normalised to $4\,\omega\varepsilon_0 a(ka)^{-5/2}\pi^{-1/2}$. In the above solutions, the first and second terms in the expressions for $\mathrm{Re}E_\theta^{(1)}$ and $\mathrm{Im}E_\theta^{(1)}$ are associated with the infinite domain. All the other terms describe the influence of the boundary and have infinite number of zeros as h/r varies from 1 to ∞. Analytic functions (5.79) and (5.80) possess the following properties in the range $1 \le h/r < \infty$: (i) they are oscillating functions limited by constants from above and from below; (ii) all the terms containing $\eta^{-1/2}$ tend to zero at $h/r \to \infty$, which corresponds to the transition to the infinite domain; and (iii) the influence of the boundary on the scattered field decreases proportionally to $r^{-1/2}$ as h/r increases.

The field of the incident wave and its reflection from the boundary are given by

$$E_x^{(1)} = E_x - E_x^* = e^{iky} - e^{ik(2h-y)}. \tag{5.81}$$

The secondary scattered field consists of the scattered field of the reflected primarily scattered waves and the scattered field of the reflected incident wave. The scattered field reflected from the boundary is given by

$$E_\theta^{*(2)} = \sum_{m=0}^{\infty} B_m^{(1)} H_m^{(1)}(kr^*)\cos m\theta^*. \tag{5.82}$$

We make use of the Neumann addition theorem [6, 20] to expand field (5.82) in terms of cylindrical functions and thus find the first constituent of the secondary diffracted field. After some transformations we have

$$\begin{gathered}\left(E_\theta^{(2)}\right)_1 = \sum_{\nu=0}^{\infty}\left[G_\nu^{(2)} J_\nu(kr) + F_\nu^{(2)} H_\nu^{(1)}(kr)\right]\cos\nu\theta^*,\\ G_\nu^{(2)} = \sum_{m=0}^{\infty} B_m^{(1)} D_{\nu m}^{(2)}, \quad F_\nu^{(2)} = -\frac{J_\nu'(ka)}{H_\nu^{(1)\prime}(ka)} G_\nu^{(2)},\\ D_{0m}^{(2)} = H_m^{(1)}(2kh), \quad D_{\nu m}^{(2)} = H_{\nu+m}^{(1)}(2kh) + (-1)^{2\nu-m} H_{\nu-m}^{(1)}(2kh).\end{gathered} \tag{5.83}$$

We expand the reflected plane wave in terms of cylindrical functions and thus obtain the second constituent of the diffracted field:

$$\left(E_\theta^{(2)}\right)_2 = \sum_{\nu=0}^{\infty} \left[f_\nu^* J_\nu(kr)e^{i2kh} + B_\nu^{(2)} H_\nu^{(1)}(kr)\right] \cos \nu\theta, \tag{5.84}$$

where

$$B_\nu^{(2)} = -i^{-2\nu} e^{i2kh} B_\nu^{(1)}.$$

As follows from (5.83) and (5.84), the secondary scattered field is given by

$$E_\theta^{(2)} = \sum_{\nu=0}^{\infty} \left(B_\nu^{(2)} + F_\nu^{(2)}\right) H_\nu^{(1)}(kr) \cos \nu\theta. \tag{5.85}$$

The expressions for the fields of the third and higher multiplicities are constructed in the similar manner. It can be shown that the effect of the multiply scattered fields is small by virtue of rapid convergence of the series (5.76). In the case of long wavelengths for the far-region field, it is sufficient to keep the terms with $s = 1$ and $s = 2$ in (5.76) in order to stay within the limits of the error introduced by formula (5.75). In this case, the secondary scattered field is described by the expressions

$$\begin{aligned} \mathrm{Re}E_\theta^{(2)} &= (\cos 2kh - \sin 2kh)\left\{-1 - 2\cos\theta + \eta^{-1/2}\cos\eta_p\left[1 + 2\eta^{-1}\left(\frac{2h}{r} - \cos\theta\right)\right]\right\} \\ &\quad - (\cos 2kh + \sin 2kh)\eta^{-1/2}\sin\eta_p\left[1 + 2\eta^{-2}\left(\frac{2h}{r} - \cos\theta\right)\right], \end{aligned} \tag{5.86}$$

$$\begin{aligned} \mathrm{Im}E_\theta^{(2)} &= (\cos 2kh - \sin 2kh)\eta^{-1/2}\sin\eta_p\left[1 + 2\eta^{-1}\left(\frac{2h}{r} - \cos\theta\right)\right] \\ &\quad + (\cos 2kh + \sin 2kh)\left\{-1 - 2\cos\theta + \eta^{-1/2}\cos\eta_p\left[1 + 2\eta^{-1}\left(\frac{2h}{r} - \cos\theta\right)\right]\right\}. \end{aligned} \tag{5.87}$$

Thus, the approximate solution is equal to the sum of (5.79), (5.80), (5.86), and (5.87):

$$E_\theta \cong E_\theta^{(1)} + E_\theta^{(2)}. \tag{5.88}$$

As an example, we find the scattered field E_θ for $ka = 0.3$, $r/a = 12.5, 0 \le \theta \le 180°$. The primary field is calculated by formulas (5.79) and (5.80), and the secondary field is determined by formulas (5.86) and (5.87), the relative error being not greater than 10%. Figure 5.8 shows the polar radiation

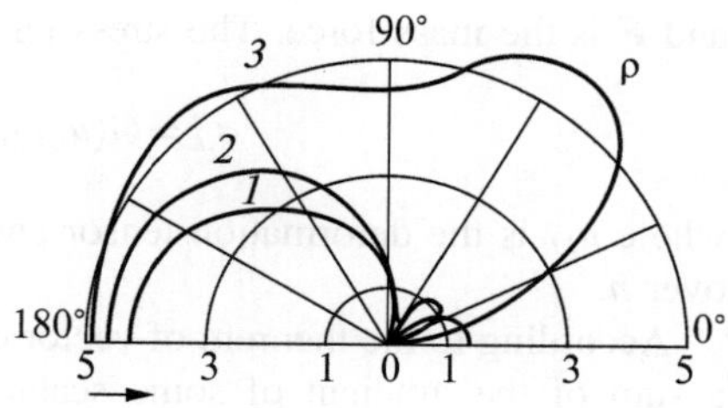

Fig. 5.8 Polar radiation patterns for the variation of the field E_θ: *1* the field corresponding to $h/r = \infty$, *2* primary field for $h/r = 1$, *3* total field for $h/r = 1$

patterns for the amplitude ρ of the field E_θ for the angles varying from 0 to 180° and the case $h/r = 1$. The point $h/r = 1$ and $\theta = 0°$ lies on the boundary, and hence the corresponding amplitude vanishes. The polar amplitude considerably changes as distance h between the scatterer and the flat boundary varies within a fixed circle of radius r. This effect is produced by strong oscillations of the functions contained in solution (5.79), (5.80), (5.86) and (5.87).

5.6 Diffraction of Elastic Waves by a Cylinder

Now we consider a stationary problem of the scattering of a plane monochromatic wave by a cylinder (Fig. 5.9) in the case when the displacement vector vanishes on the cylinder surface while its normal component vanishes on the flat boundary (sliding).

The motion of the elastic medium is described by the equation for the displacement vector $\boldsymbol{u}$, i.e.

$$G\nabla^2\boldsymbol{u} + (\lambda + G)\,\nabla(\nabla\cdot\boldsymbol{u}) + \rho\boldsymbol{F} = \rho\frac{\partial^2\boldsymbol{u}}{\partial t^2}, \tag{5.89}$$

where G and λ are Lamé constants expressed in terms of the Young modulus E and Poisson coefficient ν by the formulas

$$G = \frac{E}{2(1+\nu)}, \quad \lambda = \frac{E\nu}{(1+\nu)(1-2\nu)}$$

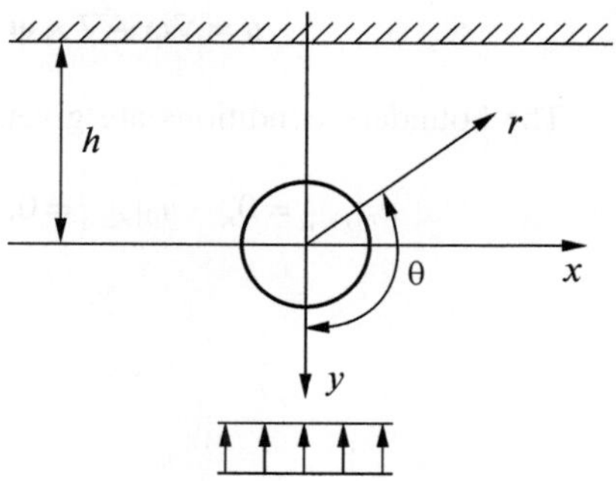

Fig. 5.9 Configuration of the problem

and $\boldsymbol{F}$ is the mass force. The stress tensor component is given by

$$\sigma_{ik} = G(u_{i,k} + u_{k,i}) + \lambda u_{n,n}\delta_{ik},$$

where $u_{i,k}$ is the deformation tensor and the term with $u_{n,n}$ implies the summation over n.

According to the theorem of vector calculus, the vector field $\boldsymbol{u}$ can be written as a sum of the gradient of some scalar φ and the curl of some vector $\boldsymbol{\psi}$ whose divergence is equal to zero:

$$\boldsymbol{u} = \nabla\varphi + \nabla \times \boldsymbol{\psi}, \quad \nabla \cdot \boldsymbol{\psi} = 0.$$

The field φ is called the scalar potential and the field $\boldsymbol{\psi}$ is called the vector potential. The mass force is given by a similar expression: $\boldsymbol{F} = \nabla\Phi + \nabla \times \boldsymbol{\Psi}$.

Thus, the initial vector equation (5.89) reduces to the two independent wave equations given by

$$\begin{aligned} c_e^2 \nabla^2 \varphi - \frac{\partial^2 \varphi}{\partial t^2} &= -\Phi, \\ c_s^2 \nabla^2 \boldsymbol{\psi} - \frac{\partial^2 \boldsymbol{\psi}}{\partial t^2} &= -\boldsymbol{\Psi}. \end{aligned} \tag{5.90}$$

The first equation describes the propagation of a dilatational wave that is characterised by volume changes without shape distortions. This wave propagates with the velocity $c_e = \sqrt{(\lambda + 2G)/\rho}$. The second equation describes the propagation of a shear wave that is characterised by shape distortions. The shear wave velocity is $c_s = \sqrt{G/\rho}$.

For a semi-infinite domain, the problem is formulated as follows. Within the context of (5.90), the equations of motion of an elastic medium without taking into account the mass forces are given by

$$\left(\nabla^2 - \frac{1}{c_e^2}\frac{\partial^2}{\partial t^2}\right)\varphi = 0, \quad \left(\nabla^2 - \frac{1}{c_s^2}\frac{\partial^2}{\partial t^2}\right)\psi = 0. \tag{5.91}$$

The displacement vector is

$$\boldsymbol{u} = \nabla\varphi + \nabla \times \boldsymbol{\psi}, \quad \boldsymbol{\psi} = \boldsymbol{e}_z\psi, \quad \nabla \cdot \boldsymbol{\psi} = 0. \tag{5.92}$$

The boundary conditions are given by

$$u_r\big|_{r=a} = 0, \quad u_\theta\big|_{r=a} = 0, \quad u_y\big|_{y=-h} = 0, \quad \sigma_{xy}\big|_{y=-h} = 0. \tag{5.93}$$

The equation for the incident wave is given by

$$u_y = u_0 e^{\mathrm{i}(py+\omega t)}.$$

The solutions should satisfy the Sommerfeld radiation conditions.

Solutions of Eqs. (5.91)–(5.92) that satisfy the first and second conditions in (5.93) that correspond to an infinite domain are given by

$$\varphi^{(1)} = \sum_{n=0}^{\infty} \left[f_n J_n(pr) + A_n^{(1)} H_n^{(2)}(pr)\right] \cos n\theta, \quad \psi^{(1)} = \sum_{n=0}^{\infty} B_n^{(1)} H_n^{(2)}(qr) \sin n\theta, \tag{5.95}$$

where

$$f_n = \varepsilon_n u_0 p^{-1} \mathrm{i}^{(n-1)}, \quad p = \frac{\omega}{c_e}, \quad q = \frac{\omega}{c_s},$$

and ε_n is the Neumann factor, i.e.

$$\varepsilon_n = \begin{cases} 1 & \text{at} \quad n = 0, \\ 2 & \text{at} \quad n = 1, 2, \ldots \end{cases}$$

The boundary conditions yield the coefficients $A_n^{(1)}$ and $B_n^{(1)}$:

$$\begin{aligned} A_n^{(1)} &= \Delta_n^{-1} f_n \left[-n^2 J_n(pa)\, H_n^{(2)}(qa) + paqa\, J_n'(pa) H_n^{(2)\prime}(qa)\right], \\ B_n^{(1)} &= \Delta_n^{-1} f_n\, npa \left[J_n(pa)\, H_n^{(2)\prime}(pa) - J_n'(pa) H_n^{(2)}(pa)\right], \\ \Delta_n &= n^2 H_n^{(2)}(pa)\, H_n^{(2)}(qa) - paqa\, H_n^{(2)}(pa) H_n^{(2)}(qa). \end{aligned} \tag{5.96}$$

Thus, for the field in the far region $r/a \gg 1$, we have

$$\begin{aligned} \frac{u_r^{(1)}}{u_0} &\cong -\mathrm{i}\sqrt{\frac{2}{\pi}\frac{a}{r}}\, pa \sum_{n=0}^{\infty} e^{\frac{\mathrm{i}\pi}{2}\left(n+\frac{1}{2}\right)} A_n^{(1)} \cos n\theta\, e^{-\mathrm{i}pr}, \\ \frac{u_\theta^{(1)}}{u_0} &\cong -\mathrm{i}\sqrt{\frac{2}{\pi}\frac{a}{r}}\, qa \sum_{n=0}^{\infty} e^{\frac{\mathrm{i}\pi}{2}\left(n+\frac{1}{2}\right)} B_n^{(1)} \sin n\theta\, e^{-\mathrm{i}qr}. \end{aligned} \tag{5.97}$$

In the Rayleigh approximation, we have $pa \ll 1$, $qa \ll 1$ and the cylindrical functions in the coefficients $A_n^{(1)}$ and $B_n^{(1)}$ can be substituted by expansions. It can be show that the dominant contribution comes from the coefficients $A_1^{(1)}$ and $B_1^{(1)}$:

$$
\begin{aligned}
A_1^{(1)} &\cong -0.5\pi^2 (pa)^{-1} b_2 b_1^{-1} - \mathrm{i}\pi (qa)^{-1} b_1^{-1}, \\
B_1^{(1)} &\cong -0.5\pi^2 \frac{q}{p}^{-1} b_2 b_1^{-2} - \mathrm{i}\,\pi\, b_1^{-1}, \\
b_1 &= \frac{p}{q}\ln\frac{pa}{2} + \frac{q}{p}\ln\frac{qa}{2}, \quad b_2 = \left(\frac{p}{q}\right)^2 + 1.
\end{aligned}
\tag{5.98}
$$

The field criteria in the far region are

$$
\begin{aligned}
&(8/3\,pr)^{-1} \le \varepsilon_e \text{ for } u_r, \\
&\left(pr\sqrt{pa\,qa}\right)^{-1} \le \varepsilon_e \text{ for } u_\theta, \\
&b_2 |b_1|^{-1} \frac{q}{p} \le \varepsilon_e.
\end{aligned}
$$

These conditions determine the relative error ε_e of calculating the corresponding coefficients.

The solution in the semi-infinite domain that satisfies all the boundary conditions in (5.93) is given by a convergent series

$$
\boldsymbol{u}(r,\theta,r^*,\theta^*) = \sum_{s=1}^{\infty} \left[\boldsymbol{u}^{(s)}(r,\theta) + \boldsymbol{u}^{*(s)}(r^*,\theta^*)\right], \tag{5.99}
$$

where $\boldsymbol{u}^{(s)}$ is the diffracted displacement field of multiplicity s and $\boldsymbol{u}^{*(s)}$ is the corresponding mapping that should be expressed in terms of the coordinates r, θ.

The total components of the displacement vector for the scattered field of multiplicity s are determined by the formulas

$$
\begin{aligned}
(\boldsymbol{u}^{(s)} + \boldsymbol{u}^{*(s)})_r &= U_r^{(s)} = u_r^{(s)} - u_r^{*(s)} \cos(\theta+\theta^*) + u_\theta^{*(s)} \sin(\theta+\theta^*), \\
(\boldsymbol{u}^{(s)} + \boldsymbol{u}^{*(s)})_\theta &= U_\theta^{(s)} = u_\theta^{(s)} + u_\theta^{*(s)} \cos(\theta+\theta^*) + u_r^{*(s)} \sin(\theta+\theta^*).
\end{aligned}
\tag{5.100}
$$

We express all the quantities with asterisks in (5.100) in terms of the variables r and θ as given by the relations

$$
\begin{aligned}
&\frac{r^*}{a} = \frac{r}{a}\eta, \ \eta = \left[1 + 4\left(\frac{h}{r}\right)^2 + 4\frac{h}{r}\cos\theta\right]^{\frac{1}{2}}, \ \cos(\theta+\theta^*) = -\eta^{-1}\left(\frac{2h}{r}\cos\theta + 1\right), \\
&\sin\theta^* = \eta^{-1}\sin\theta, \ \sin(\theta^*+\theta^*) = -\eta^{-1}\frac{2h}{r}\sin\theta, \\
&\cos\theta^* = -\eta^{-1}\left(\frac{2h}{r} + \cos\theta\right).
\end{aligned}
\tag{5.101}
$$

Nonsimultaneous arrival of P- and S-waves and the difference of the distances from the point (r, θ) to the real and imaginary bodies are taken into account by the formulas

$$\exp \alpha_j = (\cos \eta_j - \mathrm{i} \sin \eta_j) \exp(ipr), \quad j = 0, 1, 2,$$
$$\eta_j = pr\left[\left(\frac{q}{p}\right)^{j(j-2)+1} \eta^{0.5j(3-j)} - 1\right], \ \alpha_0 = -iqr, \ \alpha_1 = -ipr^*, \ \alpha_2 = -iqr^*. \tag{5.102}$$

We substitute expressions (5.97) into (5.100) and take into account relations (5.98), (5.101), and (5.102). After some transformations, we find the displacement-vector components of the singly scattered field to be given by

$$\begin{aligned}
\mathrm{Re}U_r^{(1)} \cong & -\cos\theta + \left(\cos\eta_1 + \frac{B_2}{B_1}\sin\eta_1\right)\left(\frac{2h}{r} + \cos\theta\right)\left(\frac{2h}{r}\cos\theta + 1\right)\eta^{-\frac{5}{2}} \\
& - \left(\cos\eta_2 + \frac{B_2}{B_1}\sin\eta_2\right)(paqa)^{\frac{1}{2}}\frac{q}{p}\frac{2h}{r}\eta^{-\frac{5}{2}}\sin^2\theta, \\
\mathrm{Im}U_r^{(1)} \cong & -\frac{B_2}{B_1}\cos\theta + \left(\frac{B_2}{B_1}\cos\eta_1 - \sin\eta_1\right)\left(\frac{2h}{r} + \cos\theta\right) \\
& \times\left(\frac{2h}{r}\cos\theta + 1\right)\eta^{-\frac{5}{2}} - \left(\frac{B_2}{B_1}\cos\eta_2 - \sin\eta_2\right)(paqa)^{\frac{1}{2}}\frac{q}{p}\frac{2h}{r}\eta^{-\frac{5}{2}}\sin^2\theta,
\end{aligned} \tag{5.103}$$

$$\begin{aligned}
\mathrm{Re}U_\theta^{(1)} \cong & \left[\left(\cos\eta_0 + \frac{B_2}{B_1}\sin\eta_0\right)(paqa)^{\frac{1}{2}}\frac{q}{p}\right. \\
& - \left(\cos\eta_2 + \frac{B_2}{B_1}\sin\eta_2\right)\left(\frac{2h}{r}\cos\theta + 1\right)(paqa)^{\frac{1}{2}}\frac{q}{p}\eta^{-\frac{5}{2}} \\
& \left. - \left(\cos\eta_1 + \frac{B_2}{B_1}\sin\eta_1\right)\left(\frac{2h}{r} + \cos\theta\right)\frac{2h}{r}\eta^{-\frac{5}{2}}\right]\sin\theta, \\
\mathrm{Im}U_\theta^{(1)} \cong & \left[\left(\frac{B_2}{B_1}\cos\eta_0 - \sin\eta_0\right)(paqa)^{\frac{1}{2}}\frac{q}{p}\right. \\
& - \left(\frac{B_2}{B_1}\cos\eta_2 - \sin\eta_2\right)\left(\frac{2h}{r}\cos\theta + 1\right)(paqa)^{\frac{1}{2}}\frac{q}{p}\eta^{-\frac{5}{2}} \\
& \left. - \left(\frac{B_2}{B_1}\cos\eta_1 - \sin\eta_1\right)\left(\frac{2h}{r} + \cos\theta\right)\frac{2h}{r}\eta^{-\frac{5}{2}}\right]\sin\theta.
\end{aligned} \tag{5.104}$$

Here $B_{1,2} = 0.5\pi^{3/2}b_2 \pm \pi^{1/2}pq^{-1}b_1$. The factor $\sqrt{a/r}\exp(-ipr)$ is omitted in the right-hand parts of (5.103) and (5.104), and the left-hand parts are normalised to the factor $(pa)^{1/2}b_1^2B_1^{-1}$.

The first terms of the solutions thus obtained are associated with the infinite domain. The other terms describe the influence of the flat boundary and have infinite number of zeros for h/r varying from 1 to ∞. Analytic functions (5.103) and (5.104) possess the following properties within the range $1 \le h/r < \infty$: (i) they are oscillating functions limited by constants from above and from below; (ii) all the terms containing $\eta^{-5/2}$ tend to zero at $h/r \to \infty$, which corresponds to the passage to the infinite domain; and (iii) the influence of the flat boundary on the scattered field decreases proportionally to $(r/a)^{-1/2}$ as h/r increases.

The field of the incident wave and the corresponding field of the wave scattered from the boundary are given by

$$u_y^{(1)} = u_y - u_y^* = \exp(ipy) - \exp[-ip(y+2h)]. \tag{5.105}$$

The singly diffracted field is determined by the relevant sum of solutions (5.103)–(5.105).

The main part of the secondary scattered field is simplified in the longwave approximation, i.e.

$$\begin{aligned} \mathrm{Re}U_r^{(2)} &= -\left(\mathrm{Re}U_r^{(1)} \cos 2ph + \mathrm{Im}U_r^{(1)} \sin 2ph\right), \\ \mathrm{Im}U_r^{(2)} &= \mathrm{Re}U_r^{(1)} \sin 2ph - \mathrm{Im}U_r^{(1)} \cos 2ph. \end{aligned} \tag{5.106}$$

Expressions for $\mathrm{Re}U_\theta^{(2)}$ and $\mathrm{Im}U_\theta^{(2)}$ are determined by the same formulas with the subscript θ being substituted for r. The remaining part of the secondary scattered field can be written as

$$\begin{aligned} \varphi^{(2)} &\cong \sum_{n=0}^{\infty} \left[D_n^{(2)} J_n(pr) + A_n^{(2)} H_n^{(2)}(pr)\right] \cos n\theta, \\ \psi^{(2)} &\cong \sum_{n=0}^{\infty} \left[E_n^{(2)} J_n(qr) + B_n^{(2)} H_n^{(2)}(qr)\right] \sin n\theta. \end{aligned} \tag{5.107}$$

Here, $D_n^{(2)}$ and $E_n^{(2)}$ are the known coefficients to be found in terms of the Neumann addition theorem [8], $A_n^{(2)}$ and $B_n^{(2)}$ should be obtained from boundary conditions (5.93). This constituent of the scattered field is much smaller than the first one; hence, we regard formulas (5.106) to be an approximation for the secondary scattered field. The secondary diffracted field is given by the superposition of fields (5.105) and (5.106). The small multiply scattered fields $(s \ge 3)$ being ignored, relations (5.99), (5.100), (5.102)–(5.106) lead to an approximate solution given by

$$\boldsymbol{u} \cong \sum_{s=1,\,2} \boldsymbol{U}^{(s)}. \tag{5.108}$$

Figures 5.10 and 5.11 show $\mathrm{Re}u_r$ and $\mathrm{Im}u_r$ as functions of h/r for $\theta = 0°$ and $\theta = 180°$. The scattered field u_r was calculated with the accuracy $\varepsilon_e < 0.1$ for the Poisson coefficient $\nu = 0.25$ and the parameters $pa = 0.001$; $r/a = 3100$;

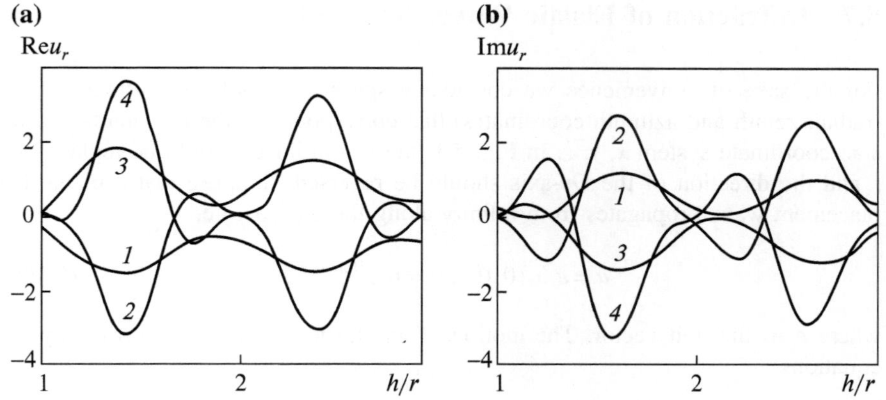

Fig. 5.10 The real part Reu_r (**a**) and imaginary part Imu_r (**b**) of the scattered field for $\theta = 0°$ (*curves 1* correspond to the primary field and *curves 2* correspond to the total field) and for $\theta = 180°$ (*curves 3* correspond to the primary field and *curves 4* correspond to the total field)

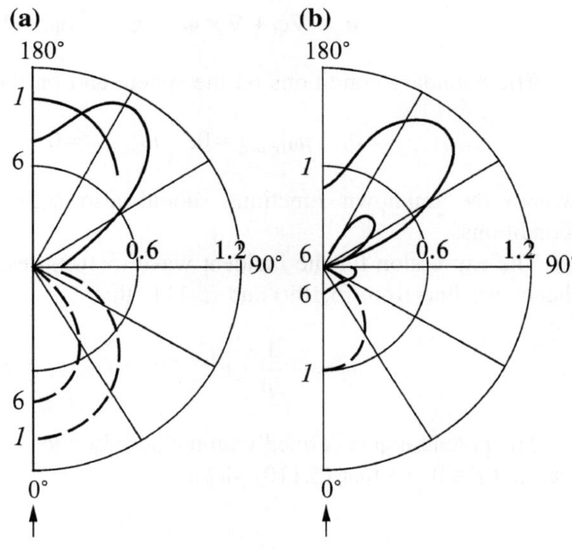

Fig. 5.11 Real parts of the scattered field Reu_r for $h/r = \infty$ (*curves 1*) and 1.12 (*curves 6*): **a** primary field, **b** total field

$0.9 \le h/r \le 3.0$. In this case, the last terms in formulas (5.103), (5.104) can be disregarded since $(pr\sqrt{pa\,qa})^{-1} \le \varepsilon_e$ for u_θ and $b_2|b_1|^{-1}q/p \le \varepsilon_e$. It can be seen from Fig. 5.10 that strong oscillations arise even under moderate changes of the distance *h*/*r* between the scatterer and the flat boundary from 1 to 3. Figure 5.11 demonstrates the polar radiation pattern of the field Reu_r.

5.7 Diffraction of Elastic Waves by a Sphere

For the sake of convenience we consider a spherical coordinate system r, θ, ϕ (radial, zenith and azimuth coordinates) that corresponds to the rectangular Cartesian coordinate system x, y, z. In Fig. 5.1, the y-coordinate should be replaced by z and the direction of the Oy-axis should be reversed. Suppose that a plane displacement wave propagates from infinity along the Oz-axis, i.e.

$$\boldsymbol{u} = \boldsymbol{e}_z u_z(0, 0, z, t) = \boldsymbol{u}_0 e^{\mathrm{i}(\omega t - pz)}, \tag{5.109}$$

where $\boldsymbol{e}_z$ is the unit vector. The motion of an elastic medium is governed by the equations

$$\left(\nabla^2 - \frac{1}{c_e^2}\frac{\partial^2}{\partial t^2}\right)\varphi = 0, \quad \left(\nabla^2 - \frac{1}{c_s^2}\frac{\partial^2}{\partial t^2}\right)\psi = 0, \tag{5.110}$$

while the displacement vector is determined by the formula

$$\boldsymbol{u} = \nabla\varphi + \nabla \times \boldsymbol{\psi}, \quad \boldsymbol{\psi} = \boldsymbol{e}_z \psi, \quad \nabla \cdot \boldsymbol{\psi} = 0. \tag{5.111}$$

The boundary conditions on the sphere and on the flat boundary are given by

$$u_r|_{r=a} = 0, \quad u_\theta|_{r=a} = 0, \quad u_z|_{z=h} = 0, \quad \sigma_{xz}|_{z=h} = 0, \tag{5.112}$$

where the unknown functions should also satisfy the Sommerfeld radiation conditions.

The expression for the incident wave (5.109) describes dilatational waves, and hence we find from (5.109) and (5.111) that

$$\varphi = -\frac{1}{\mathrm{i}p} u_0 e^{\mathrm{i}(\omega t - pz)} + f(x, z, 0, t) + \text{const}. \tag{5.113}$$

The potential φ is defined with the accuracy to an arbitrary function. Hence, we can put $f \equiv 0$, so that (5.110) yields

$$\left(\frac{\partial^2}{\partial z^2} - \frac{1}{c_e^2}\frac{\partial^2}{\partial t^2}\right)\varphi(z, t) = 0. \tag{5.114}$$

Thus, taking into account (5.113), we obtain $p = \omega/c_e$, $q = \omega/c_s$.

In what follows we apply the formulas in terms of the spherical coordinates, i.e.

$$\nabla^2 \boldsymbol{\psi} = \boldsymbol{e}_r \left[\nabla^2 \psi_r - \frac{2}{r^2} \psi_r - \frac{2}{r^2 \sin\theta} \frac{\partial}{\partial \theta} (\sin\theta\, \psi_\theta) - \frac{2}{r^2 \sin\theta} \frac{\partial \psi_\phi}{\partial \phi} \right]$$
$$+ \boldsymbol{e}_\theta \left[\nabla^2 \psi_\theta - \frac{\psi_\theta}{r^2 \sin^2\theta} + \frac{2}{r^2} \frac{\partial \zeta_r}{\partial \theta} - \frac{2\cos\theta}{r^2 \sin^2\theta} \frac{\partial \psi_\phi}{\partial \phi} \right]$$
$$+ \boldsymbol{e}_\phi \left[\nabla^2 \psi_\phi - \frac{\psi_\phi}{r^2 \sin^2\theta} + \frac{2}{r^2 \sin\theta} \frac{\partial \psi_r}{\partial \phi} + \frac{2\cos\theta}{r^2 \sin^2\theta} \frac{\partial \psi_\theta}{\partial \phi} \right],$$
$$\nabla \varphi = \boldsymbol{e}_r \frac{\partial \varphi}{\partial r} + \frac{\boldsymbol{e}_\theta}{r} \frac{\partial \varphi}{\partial \theta} + \frac{\boldsymbol{e}_\phi}{r \sin\theta} \frac{\partial \varphi}{\partial \phi},$$
$$\nabla \times \boldsymbol{\psi} = \frac{\boldsymbol{e}_r}{r \sin\theta} \left[\frac{\partial}{\partial \theta} \left(\sin\theta\, \psi_\phi \right) - \frac{\partial \psi_\theta}{\partial \phi} \right] + \frac{\boldsymbol{e}_\theta}{r} \left[\frac{1}{\sin\theta} \frac{\partial \psi_r}{\partial \phi} - \frac{\partial}{\partial r} \left(r \psi_\phi \right) \right]$$
$$+ \frac{\boldsymbol{e}_\phi}{r} \left[\frac{\partial}{\partial r} (r \psi_\theta) - \frac{\partial \psi_r}{\partial \theta} \right].$$

The shear deformations are possible only in the planes normal to the surfaces of constant ϕ and are produced by the vector $\boldsymbol{e}_\phi \psi_\phi$. This implies that the potential $\boldsymbol{\psi}$ has only one nontrivial component ψ_ϕ, i.e.

$$\boldsymbol{\psi} = \boldsymbol{e}_r 0 + \boldsymbol{e}_\theta 0 + \boldsymbol{e}_\phi \psi_\phi.$$

Therefore, we have

$$\nabla^2 \boldsymbol{\psi} = \boldsymbol{e}_\phi \left(\nabla^2 \psi_\phi - \frac{\psi_\phi}{r^2 \sin^2\theta} \right)$$
$$= \boldsymbol{e}_\phi \left[\frac{1}{r^2} \frac{\partial}{\partial r} \left(r^2 \frac{\partial \psi_\phi}{\partial r} \right) + \frac{1}{r^2 \sin\theta} \frac{\partial}{\partial \theta} \left(\sin\theta \frac{\partial \psi_\phi}{\partial \theta} \right) - \frac{\psi_\phi}{r^2 \sin^2\theta} \right]. \tag{5.115}$$

Next, we introduce the function $\xi = \partial \psi_\phi / \partial \theta$ and obtain

$$\nabla^2 \boldsymbol{\psi} = \boldsymbol{e}_\phi \frac{\partial}{\partial \theta} \nabla^2 \xi, \tag{5.116}$$

where we took into account that

$$\nabla^2 \xi = \frac{1}{r^2} \frac{\partial}{\partial r} \left(r^2 \frac{\partial \xi}{\partial r} \right) + \frac{1}{r^2 \sin\theta} \frac{\partial}{\partial \theta} \left(\sin\theta \frac{\partial \xi}{\partial \theta} \right) + \frac{1}{r^2 \sin\theta} \frac{\partial^2 \xi}{\partial \phi^2}.$$

Thus, a scalar wave equation for ξ can be employed instead of the equation for $\boldsymbol{\psi}$:

$$\left(\nabla^2 - \frac{1}{c_s^2} \frac{\partial^2}{\partial t^2} \right) \xi = 0. \tag{5.117}$$

After the separation of variables, this equation can be reduced to the Legendre and Bessel equations.

The solutions for the functions φ and ξ in the infinite domain can be written as

$$\varphi=\sum_{m=0}^{\infty}\left[f_m j_m(pr)+B_m h_m^{(2)}(pr)\right]P_m(\cos\theta),$$
$$\xi=\sum_{m=0}^{\infty}b_m h_m^{(2)}(qr)\frac{\partial}{\partial\theta}P_m(\cos\theta), \tag{5.118}$$

where $f_m=-(2m+1)\,u_0 p^{-1}\mathrm{i}^{-(m+1)}$, $j_m(pr)$ and $h_m^{(2)}(pr)$ are spherical Bessel and Hankel functions. For example,

$$j_m(\zeta)=J_{m+\frac{1}{2}}(\zeta)\sqrt{\frac{\pi\zeta}{2}}.$$

The coefficients B_m and b_m can be found from boundary conditions (5.112), i.e.

$$\left[\frac{\partial\varphi}{\partial r}+\frac{1}{r\sin\theta}\frac{\partial}{\partial\theta}\left(\sin\theta\,\psi_\phi\right)\right]_{r=a}=0,\quad \left[\frac{1}{r}\frac{\partial\varphi}{\partial\theta}+\frac{1}{r}\left(-\frac{\partial}{\partial r}r\psi_\phi\right)\right]_{r=a}=0. \tag{5.119}$$

Thus, we obtain

$$B_m=f_m\Delta_m^{-1}\{m(m+1)j_m(pa)h_m(qa)-pa\,j'_m(pa)[h_m(qa)+qah'_m(qa)]\},$$
$$b_m=f_m\Delta_m^{-1}pa\{h_m(pa)j'_m(pa)-h'_m(pa)j_m(pa)\}, \tag{5.120}$$
$$\Delta_m=pa\,h'_m(pa)[h_m(qa)+qa\,h'_m(qa)]-m(m+1)\,h_m(pa)\,h_m(qa).$$

The approximate solutions for the field in the far region $r/a\gg 1$ are given by

$$u_r\cong\sum_{m=0}^{\infty}B_m\mathrm{i}^m\frac{1}{r}e^{-\mathrm{i}pr}P_m(\cos\theta),\quad u_\theta\cong-\sum_{m=0}^{\infty}b_m\mathrm{i}^m\frac{1}{r}e^{-\mathrm{i}qr}\frac{\partial}{\partial\theta}P_m(\cos\theta). \tag{5.121}$$

In the Rayleigh approximation $pa\ll 1, qa\ll 1,$ the dominant coefficients are

$$B_1\cong\mathrm{i}\,3a\left[1+2\left(\frac{q}{p}\right)^2\right]^{-1},\quad b_1\cong\mathrm{i}\,3a\left[1+2\left(\frac{q}{p}\right)^2\right]^{-1}\left(\frac{q}{p}\right)^2.$$

The field criterion in the far region is

$$pr\geq\frac{q}{p}\frac{2}{\varepsilon_e}.$$

The maximum error in calculating the coefficients is given by

$$\frac{1}{3}\left[1+2\left(\frac{q}{p}\right)^3\right]\left[1+2\left(\frac{q}{p}\right)^2\right]^{-1} pa \leq \varepsilon_e.$$

We write the solution that satisfies conditions (5.112) in the form

$$\boldsymbol{u}(r,\theta,r^*,\theta^*) = \sum_{s=1}^{\infty}\left[\boldsymbol{u}^{(s)}(r,\theta)+\boldsymbol{u}^{*(s)}(r^*,\theta^*)\right], \tag{5.122}$$

where the displacement components of the total scattered field of multiplicity s are given by

$$\begin{aligned}(\boldsymbol{u}^{(s)}+\boldsymbol{u}^{*(s)})_r &= U_r^{(s)} = u_r^{(s)} - u_r^{*(s)}\cos(\theta+\theta^*) + u_\theta^{*(s)}\sin(\theta+\theta^*),\\ (\boldsymbol{u}^{(s)}+\boldsymbol{u}^{*(s)})_\theta &= U_\theta^{(s)} = u_\theta^{(s)} + u_\theta^{*(s)}\cos(\theta+\theta^*) + u_r^{*(s)}\sin(\theta+\theta^*).\end{aligned} \tag{5.123}$$

The difference of the distances from the real and imaginary obstacles to some point r, θ and nonsimultaneous arrival of the P- and S-waves are taken into account in terms of the formulas given by

$$\begin{aligned}&\exp(-\mathrm{i}\alpha_j) = (\cos\eta_j - \mathrm{i}\sin\eta_j)\exp(-ipr), \quad \alpha_0 = qr, \quad \alpha_1 = pr^*, \quad \alpha_2 = qr^*,\\ &\eta_0 = pr\left(\frac{q}{p}-1\right), \quad \eta_1=\eta_p = pr(\eta-1), \quad \eta_2=\eta_q = pr\left(\frac{q}{p}\eta-1\right).\end{aligned} \tag{5.124}$$

Taking into account formulas (5.121)–(5.124) and performing some transformations, we come to the following expressions for the singly scattered field:

$$\begin{aligned}U_r^{(1)} \cong& -\cos\theta + \left(\frac{2h}{r}\cos\theta - 1\right)\left(\frac{2h}{r} - \cos\theta\right)(\cos\eta_p - \mathrm{i}\sin\eta_p)\eta^{-3}\\ &+\left(\frac{q}{p}\right)^2\left(\cos\eta_q - \mathrm{i}\sin\eta_q\right)\frac{2h}{r}\eta^{-3}\sin^2\theta,\\ U_\theta^{(1)} \cong& \left[-\left(\frac{q}{p}\right)^2(\cos\eta_0 - \mathrm{i}\sin\eta_0) - \left(\frac{2h}{r}\cos\theta - 1\right)\left(\cos\eta_q - \mathrm{i}\sin\eta_q\right)\eta^{-3}\left(\frac{q}{p}\right)^2\right.\\ &\left.-\left(\frac{2h}{r} - \cos\theta\right)(\cos\eta_p - \mathrm{i}\sin\eta_p)\frac{2h}{r}\eta^{-3}\right]\sin\theta.\end{aligned} \tag{5.125}$$

The factor $(a/r)\exp(ipr)$ is omitted in the right-hand parts of expressions (5.125), the left-hand parts are normalised to the factor $3\,[1+2(q/p)^2]^{-1}$, and the displacements are normalised to u_0.

The first terms of the solutions are associated with the infinite domain. The other terms decrease proportionally to r^{-1} as h/r increases and the terms containing η^{-3} tend to zero at $h/r \to \infty$.

The incident wave field and the corresponding field of the wave reflected from the boundary are given by

$$u_y = u_y - u_y^* = \exp(ipy) - \exp[i(py - 2ph)]. \tag{5.126}$$

Therefore, the singly diffracted field is described by formulas (5.125) and (5.126).

The secondary diffracted field contains two constituents. For the first constituent, the incident field can be presented in terms of waves scattered by the imaginary obstacle and arriving at the real obstacle. Making use of the addition theorem for the spherical wave functions, we obtain

$$\begin{aligned}\varphi^{(2)} &\cong \sum_{m=0}^{\infty}\left[D_m^{(2)}h_{m+q}^{(2)}(pr) + B_m^{(2)}h_m^{(2)}(pr)\right]P_m(\cos\theta),\\ \xi^{(2)} &\cong \sum_{m=0}^{\infty}\left[E_m^{(2)}h_{m+q}^{(2)}(qr) + b_m^{(2)}h_m^{(2)}(qr)\right]\frac{\partial}{\partial\theta}P_m(\cos\theta),\end{aligned} \tag{5.127}$$

where $D_m^{(2)}$ and $E_m^{(2)}$ are the known but rather cumbersome coefficients and

$$\left(B_m^{(2)}, b_m^{(2)}\right) = i^{2m}\exp(-i2ph)\left(B_m^{(1)}, b_m^{(1)}\right). \tag{5.128}$$

The second constituent of the secondary field is described by the formulas similar to (5.127).

In the longwave approximation, the scattered fields as well as the fields of higher multiplicities ($k \geq 3$) can be ignored within the limits of the adopted computational error in the range $h < r < 2h$. Then we find the secondary scattered field to be given by

$$\left(U_r^{(2)}, U_\theta^{(2)}\right) \cong -\left(U_r^{(1)}, U_\theta^{(1)}\right)e^{-i2ph}.$$

The approximate solution of the scattering problem is given by

$$u_r \cong U_r^{(1)} + U_r^{(2)}; \quad u_\theta \cong U_\theta^{(1)} + U_\theta^{(2)}.$$

As an example, we calculated the displacements u_r and u_θ for the Poisson coefficient $\nu = 0.25$ and for $pa = 0.18, r/a = 200$. The relative computational error was no more than 10%. Calculations are carried out for the points separated by the step $\pi/36$. The corresponding results are shown in Fig. 5.12.

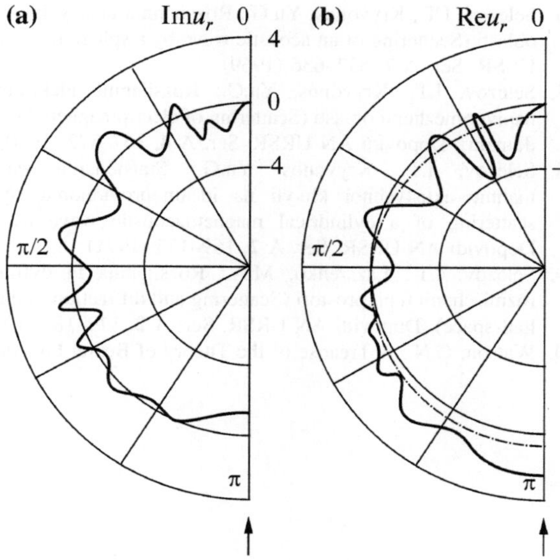

Fig. 5.12 Variation of Imu_r and Reu_r of the scattered field for $h/r = 1.0$ (*solid curve*) and $h/r = \infty$ (*dot-dashed curve*)

References

1. Borovikov, V.A.: Difraktsiya na mnogougolnikakh i mnogogrannikakh (Diffraction by Polygons and Polyhedra). Nauka, Moscow (1966)
2. Friedman, B., Russek, J.: Addition theorem for spherical waves. Quart. Appl. Math. **12**(1), 13–23 (1954)
3. Grinchenko, V.T., Vovk, I.V.: Volnovye zadachi rasseyaniya zvuka na uprugikh obolochkakh (Wave Problems of Sound Scattering on Elastic Shells). Naukova Dumka, Kiev (1986)
4. Guz, A.N., Kubenko, V.D., Cherevko, M.A.: Difraktsiya uprugikh voln (Diffraction of Elastic Waves). Naukova Dumka, Kiev (1987)
5. Hönl, H., Maue, A.W., Westpfahl, K.: Theorie der Beugung. Springer, Berlin (1961)
6. Ivanov, E.A.: Difraktsiya elektromagnitnykh voln na dvukh telakh (Diffraction of Electromagnetic Waves by Two Bodies). Nauka i Tekhnika, Minsk (1968)
7. Jackson, J.D.: Classical Electrodynamics. Wiley, New York (1962)
8. Katsenelenbaum, B.Z.: High-Frequency Electrodynamics. Wiley-VCH, Weinheim (2006)
9. King, R.W.P., Wu, T.T.: The Scattering and Diffraction of Waves. Harvard University Press, Cambridge (1959)
10. Kratzer, A., Franz, W.: Transzendente Funktionen. In: Akademische Verlagsgesellschaft Geest & Portig, Leipzig (1960)
11. Nikolskii, V.V.: Teoriya elektromagnitnogo polya (Electromagnetic Field Theory). Vysshaya Shkola, Moscow (1961)
12. Rushchitsky, J.: Theory of Waves in Materials. Ventus Publishing ApS, Copenhagen (2011)
13. Selezov, I.T.: Diffraction of waves by convex bodies in semibounded regions. Sov. Appl. Mech. **6**(3), 258–263 (1970)
14. Selezov, I.T.: Dyfraktsiia khvyl na tsylindri v pruzhnomu pivprostori (Wave diffraction by a cylinder in an elastic half-space). Dopovidi AN URSR. Ser. A **3**, 259–264 (1970)
15. Selezov, I.T., Kryvonos, Yu.G.: Rasseyanie akusticheskikh voln na tsilindre v poluogra-nichennoi oblasti (Scattering of acoustic waves by a cylinder in a semibounded domain). Gidromekhanika **15**, 99–106 (1969)

16. Selezov, I.T., Kryvonos, Yu.G.: Rozsiiannia akustychnoi khvyli na sferi v napivobmezhenii oblasti (Scattering of an acoustic wave by a sphere in a semibounded domain). Dopovidi AN URSR. Ser. A **7**, 632–636 (1969)
17. Selezov, I.T., Kryvonos, Yu.G.: Rozsiiannia elektromahnitnykh khvyl na tsylindri v napivobmezhenii oblasti (Scattering of electromagnetic waves by a cylinder in a semibounded domain). Dopovidi AN URSR. Ser. A **4**, 367–372 (1970)
18. Selezov, I.T., Kryvonos, Yu.G.: Statsionarna zadacha rozsiiannia tsylindrychnoi mahnito-akustychnoi khvyli na idealnoprovidnomu tsylindri (A stationary problem of scattering of a cylindrical magnetoacoustic wave by an ideally conducting cylinder). Dopovidi AN URSR. Ser. A **2**, 169–173 (1971)
19. Selezov, I.T., Lazarenko, M.A.: Rozsiiannia ta dyfraktsiia pruzhnykh khvyl na sferi, rozmischenii u pivprostori (Scattering and diffraction of elastic waves by a sphere located in a half-space). Dopovidi AN URSR. Ser. A **2**, 179–182 (1966)
20. Watson, G.N.: A Treatise of the Theory of Bessel Functions. Macmillan, Cambridge (1945)

Chapter 6
Propagation and Evolution of Transient Water Waves

6.1 Generation of Tsunami Waves by Underwater Earthquakes

The generation and propagation of tsunami waves have long been a subject of extensive studies. The main theoretical and applied aspects of this problem are well described in Refs. [37, 51, 58, 90]. The stages of tsunami wave evolution are well known: generation, propagation, transformation and run-up on the coast. Nevertheless, predicting the tsunami wave run-up on the coast within an acceptable time still remains to be a crucial task that should allow the timely evacuation of the regions under the risk and mitigation of the destructive effect of tsunamis.

The indeterminacy of the problem of the tsunami wave generation is mainly attributed to the mechanism of underwater earthquakes that represent the major source of tsunami waves. The corresponding initial-boundary-value problem also remains to be nondetermined, in view of the unknown exact initial moment of tsunami generation [68]. Another problematic issue is the indeterminacy in the tsunami wave initiation in each particular case. Some underwater earthquakes of the similar power do generate the tsunami waves and some do not. A typical example is the tsunami that took place in Southeast Asia (December 2004, 9.3 magnitude) [31], whereas the recurring earthquakes (March 2005, 6.3 magnitude) did not result in the generation of any tsunami wave.

In this regard, special attention must be given to some characteristic examples. One should recall the tsunami event that took place on the coast of Kamchatka in 1737, when 70 m waves were running up on the coastline. They followed one another at intervals of 10 min. Other big tsunami events include those in Kamchatka (1952), Chile (1960, 9 magnitude), Mexico City (1985, 8.5 magnitude) and Hokkaido island (1993, 7.5 magnitude, 30 m). Smaller tsunami waves of up to 2 m were even observed in the Black and Azov seas (1959, Kerch Strait). Modelling the tsunami waves generated by the landslide of Mt. Mayuyama in 1792 showed that such a landslide could generate 20 m high tsunami waves [2]. Three large tsunami

I.T. Selezov et al., *Wave Propagation and Diffraction*,
Foundations of Engineering Mechanics, DOI 10.1007/978-981-10-4923-1_6

waves (10 m elevation above the sea level) were observed in a lagoon in Papua New Guinea in 1998. They were generated by an underwater earthquake caused by earthquake shocks. After the destructive tsunami in South-East Asia in December 2004 and January 2005, there was an earthquake in the Crimea near Sudak. These events could probably be connected, the latter being the result of the solitary wave propagation in accordance with the new model of wave-type earthquake proposed in Ref. [85]. This model is based on the concept of the hydrodynamic flow of geomass along the tectonic ray tube, and the evolution of the damaged medium is described by the kinetic equation for the damage coefficient $\lambda \in [0, 1]$. The analysis of the dispersion relation showed that the problem of propagation of soliton-like disturbances could be solved for narrow pencils of waves around Morse's critical points. A singular degeneration takes place at $\lambda \to 1$, and it can result in the jumps of structural parameters that can cause an earthquake. The observations of the tsunami in the Pacific Ocean dated 28 March 1964 showed that its wavelength could exceed the fluid depth fivefold [97]. Therefore, such a tsunami can be described by the theory of shallow-water waves.

Despite big efforts, the problem of earthquake prediction as the main source of tsunamis and tsunami waves as their consequence remains open until now. The possible trigger mechanisms include the repolarisation of elastic waves that can lead to a disturbance of a frictional contact interface [95] and resonance effects in the elastic waveguide [12, 14, 28, 46, 54]. Various theoretical approaches to solving the problem of earthquake prediction were developed in Refs. [11, 15, 18, 26, 43, 56, 59, 60, 67, 70, 83, 85, 87]. Certain computational techniques were used in the form of cluster programmes in Ref. [27] to define the earthquake forerunners and to compile the data on large-magnitude earthquake prediction.

Of fundamental importance is the problem of the tsunami wave suppression. The data on the tsunami that took place in South-East Asia in 2014–2015 revealed some local areas along the coastline where the tsunami run-up was weak, although there was a huge run-up observed in other areas. Therefore, the bottom relief in the near-coastal zone has a profound effect on the run-up phenomenon. These effects are similar to the wave energy focusing and defocusing in the course of water wave refraction in the coastal zone [68].

There is good reason to place some bottom obstacles along the direction of wave propagation in the coastal zone prior to the wave run-up on the coast. The tsunami breakwater calculations were made in Ref. [92]. Rubble mound breakwaters were evaluated in terms of the tsunami impact in Ref. [6]. The performance of the tsunami breakwater built on the Sanriku Coast in Japan in 1968 was analysed in Ref. [32]. The run-up of a solitary wave on a sloping coast is considered in Ref. [38]. A generalised Boussinesq model with high-order nonlinear effects taken into account was proposed in Ref. [44] in relation to the water wave research in the surf zone. It would be practical to place a sea wall in some coastal zones exposed to extreme waves [40].

Here, we consider the problem of the surface gravity wave excitation in the ocean caused by an underwater source (earthquake). The simplest axisymmetric problem is solved in order to estimate the effect of the local bottom elevation on the generation of tsunami waves. The corresponding initial-boundary-value problem is solved directly with the use of the Laplace transform with respect to time t and the Hankel transform with respect to the radial coordinate r. As a result, the problem is reduced to the boundary-value problem for an ordinary differential equation in the space of the Laplace–Hankel transforms that can be solved analytically. The inverse Hankel transform leads to a solution in the integral form in the space of the Laplace transforms. Going over to the original space is performed using the numerical inversion of the Laplace transform.

Consider the propagation of surface gravity waves generated by a local axisymmetric bottom disturbance. Fluid is assumed to be incompressible and inviscid. Motion is assumed to be irrotational allowing us to introduce the velocity potential $\varphi(r, z, t)$ for the potential flow. The axisymmetric problem is considered in the cylindrical coordinate system r, θ, z for the fluid of finite depth H_2. With these assumptions, the fluid flow is described by the Laplace equation for the velocity potential $\varphi(r, z, t)$:

$$\frac{\partial^2\varphi}{\partial z^2} + \frac{1}{r}\frac{\partial}{\partial r}\left(r\frac{\partial\varphi}{\partial r}\right) = 0, \quad -H_2 \le z \le 0, \ r > 0, \ t > 0. \tag{6.1}$$

The boundary condition at the free surface $z = 0$ is written as

$$\left(\frac{1}{g}\frac{\partial^2\varphi}{\partial t^2} + \frac{\partial\varphi}{\partial z}\right)_{z=0} = 0. \tag{6.2}$$

The boundary condition at the bottom $z = -H_2$ expresses the equality of the vertical velocity $V_z = \partial\varphi/\partial z$ and the lift speed $\partial\xi/\partial t$ of the bottom, namely,

$$\frac{\partial\varphi}{\partial z}\Big|_{z=-H_2} = \frac{\partial\xi(r, t)}{\partial t}. \tag{6.3}$$

The system is assumed to be at rest at the initial moment $t = 0$, so that the zero initial conditions hold true:

$$\varphi\,|_{t=0} = \frac{\partial\varphi}{\partial t}\Big|_{t=0} = \xi\,|_{t=0} = 0. \tag{6.4}$$

The free surface deviation is written as

$$\eta = -\frac{1}{g}\frac{\partial\varphi}{\partial t}\Big|_{z=0}. \tag{6.5}$$

The initial-boundary-value problem (6.1)–(6.5) describes the generation of water waves by a vertical bottom disturbance. Introduce the dimensionless variables using the following formulas (asterisks are omitted henceforth):

$$\left(r^*, z^*, H_2^*, \eta^*\right) = \left(r, z, H_2, \eta\right)/gT^2, \quad t^* = t/T, \quad \varphi^* = \varphi/g^2T^3, \quad \xi_0^* = \xi_0/gT^2. \tag{6.6}$$

Going over to variables (6.6) is equivalent to setting $g = 1$.
Let us define the function $\xi(r, t)$ as follows:

$$\xi(r, t) = \xi_0\, \xi_1(t)\, \xi_2(r). \tag{6.7}$$

Subsequently, we introduce the Laplace transform with respect to time t,

$$f_L(r, s) = \int_0^\infty f(r, t)\, e^{-st} \mathrm{d}t \tag{6.8}$$

and the Hankel transform with respect to the radial coordinate r,

$$f_H(k, t) = \int_0^\infty f(r, t)\, rJ_0(kr)\, \mathrm{d}r. \tag{6.9}$$

The original problem (6.1)–(6.5) and (6.7) can be reduced to the following one using transforms (6.8) and (6.9):

$$\frac{\mathrm{d}^2\varphi_{LH}}{\mathrm{d}z^2} - k^2\varphi_{LH} = 0, \quad -H_2 \le z \le 0, \tag{6.10}$$

$$\left(s^2\varphi_{LH} + \frac{\mathrm{d}\varphi_{LH}}{\mathrm{d}z}\right)_{z=0} = 0, \tag{6.11}$$

$$\left.\frac{\mathrm{d}\varphi_{LH}}{\mathrm{d}z}\right|_{z=-H_2} = s\xi_0\xi_{1L}(s)\,\xi_{2H}(k). \tag{6.12}$$

A solution to problem (6.10)–(6.12) is written as

$$\varphi_{LH} = -\frac{s}{2k}\,\xi_0\xi_{1L}(s)\,\xi_{2H}(k)\,\frac{(s^2+k)\,e^{-kz} - (s^2-k)\,e^{kz}}{s^2\cosh kH_2 + k\sinh kH_2}. \tag{6.13}$$

After the analytical inversion of the Hankel transform, we obtain a solution in the Laplace transform space for the free surface deviation in the form of an integral:

$$\eta_L = -\frac{1}{g}\frac{\partial\varphi}{\partial t}\Big|_{z=0} = s^2\xi_0\xi_{1L}(s) \times \int_0^\infty \frac{\xi_{2H}(k)\;\, kJ_0\,(kr_1)}{s^2\cosh kH_2 \,+\, k\sinh kH_2}\,\mathrm{d}k, \tag{6.14}$$

the integrand including the Laplace transform parameter s.

Going over from Laplace transform (6.14) to the original functions is performed using the numerical inversion in terms of symmetric sine functions described in Sect. 1.5. The functions ξ_1 and ξ_2 are specified as follows. The function $\xi_1(t)$ linearly increases from $t=0$ to $t=t_0$ and then exponentially decreases, and the function $\xi_2(r)$ quickly decreases with increasing r:

$$\xi_1(t) = \begin{cases} t/t_0, & 0 < t \leq t_0, \\ \exp[-\alpha(t-t_0)], & t \geq t_0, \end{cases} \quad (6.15)$$

$$\xi_2(r) = \xi\left(\xi^2 + r^2\right)^{-3/2}, \quad \xi > 0. \quad (6.16)$$

Figures 6.1, 6.2, 6.3, 6.4 and 6.5 demonstrate the results of calculating the deviation of the free surface η depending on time t obtained at various distances from the epicentre. The free surface deviation is transformed into diverging cylindrical waves with decreasing amplitudes as the distance from the epicentre increases. Figures 6.1, 6.2 and 6.3 show that a decrease in ξ results in an increase in $\xi_2(r, \xi)$, i.e. in a greater localisation of the initial bottom height and, consequently, in a larger deviation of the free surface. Figures 6.4 and 6.5 demonstrate that increasing the disturbance rate in time ($t_0 = 0.3,\ 0.6,\ 0.9$) along with its decrease rate ($\alpha = 1.5,\ 2.5,\ 3.5$) leads to a substantial increase in the elevation amplitude of the free surface.

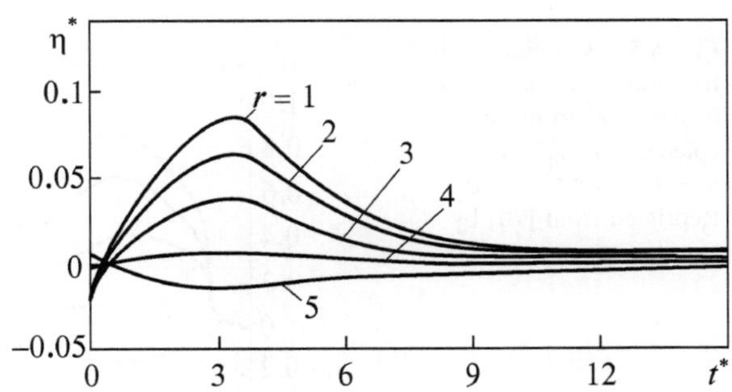

Fig. 6.1 Deviation of the free surface η^* at various distances r^* from the epicentre for $t_0^* = 0.3$, $\alpha = 1.5$, and $\xi^* = 1.5$. Reprinted from [70] by permission

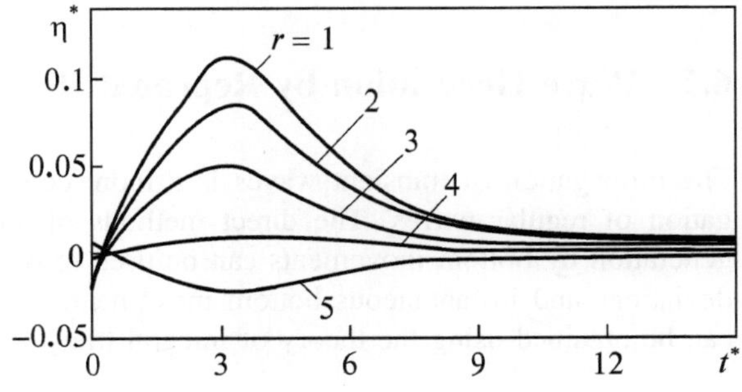

Fig. 6.2 Deviation of the free surface η^* at various distances r^* from the epicentre for $t_0^* = 0.3$, $\alpha = 1.5$, and $\xi^* = 1.0$. Reprinted from [70] by permission

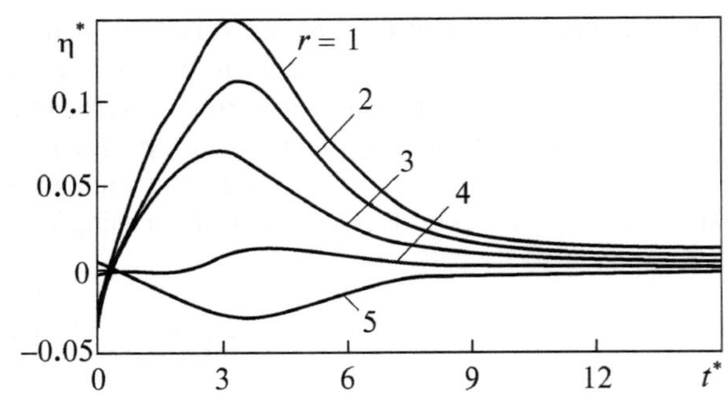

Fig. 6.3 Deviation of the free surface η^* at various distances r^* from the epicentre for $t_0^* = 0.3$, $\alpha = 1.5$, and $\xi^* = 0.5$. Reprinted from [70] by permission

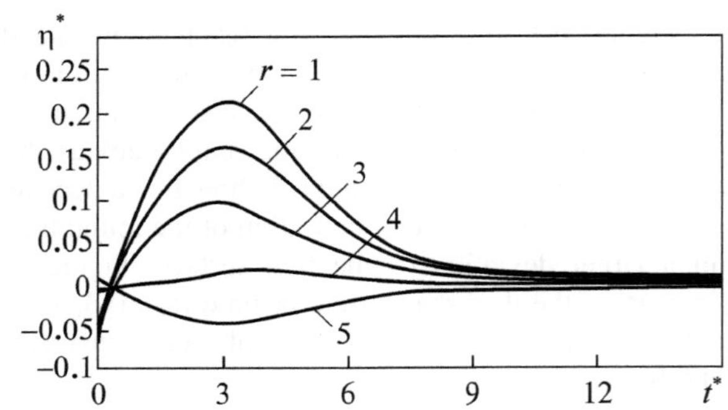

Fig. 6.4 Deviation of the free surface η^* at various distances r^* from the epicentre for $t_0^* = 0.6$, $\alpha = 2.5$, and $\xi^* = 1.0$. Reprinted from [70] by permission

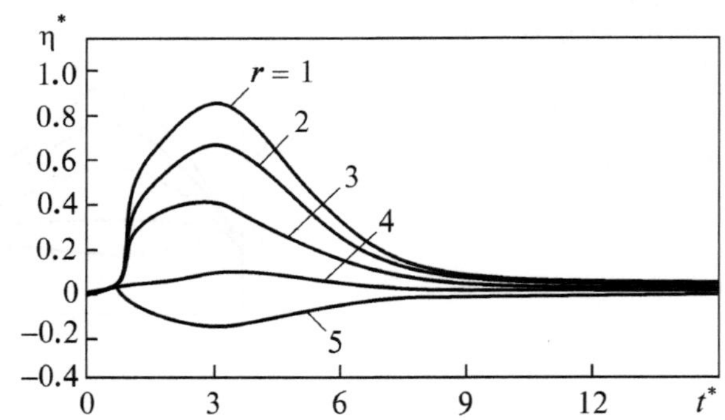

Fig. 6.5 Deviation of the free surface η^* at various distances r^* from the epicentre for $t_0^* = 0.9$, $\alpha = 3.5$, and $\xi^* = 1.0$. Reprinted from [70] by permission

6.2 Wave Generation by Repeated Disturbances

The propagation of transient waves is a more complex problem than the propagation of regular waves. The direct methods of solving the problems of wave generation by bottom movements can be used to obtain solutions only for simple deviations and instantaneous bottom movement. In more general cases, solutions can be obtained using the theory of integral transforms [63, 84].

Here, we investigate the free surface deviations caused by two repeated bottom disturbances in the framework of the finite-depth fluid model. We consider two bottom disturbances that follow one another, with the first disturbance taking place at the initial moment $t = 0$ and the second disturbance taking place at some later moment $t = t_d$. This situation corresponds to the generation of tsunami waves by an underwater earthquake during aftershocks. In this case, the first main shock is followed (after some time interval) by another shock that can differ from the first shock by power, lag time, and increase and decrease rates. The problem is solved in the linear formulation, so that the superposition principle for the solutions holds true.

For simplicity, we assume the both shocks to be identical. The velocity of the localised elevation of the bottom surface with amplitude $\eta^b(r,t)$ is supposed to be known at the initial moment $t = 0$. It is assumed to increase in time to its maximum and then to decrease exponentially. We consider a domain filled with inviscid incompressible fluid with density ρ in the cylindrical coordinate system (r, θ, z). This fluid is assumed to be bounded by the free surface $z = 0$ from the top and the bottom surfaces $z = -H_0$. The gravity force is assumed to be directed opposite to the Oz axis. The fluid is supposed to be at rest prior to the initial moment $t = 0$, when the first bottom disturbance takes place. The fluid motion is assumed to be irrotational, allowing us to introduce the velocity potential φ determined as $\boldsymbol{v} = \nabla\varphi$, where $\boldsymbol{v}$ is the velocity vector and ∇ is the gradient operator. In this case, the problem is reduced to the Laplace equation for the potential $\varphi(r,z,t)$ with the dynamic and kinematic boundary conditions at the free surface and the kinematic boundary condition at the bottom [70]:

$$\frac{\partial^2\varphi}{\partial r^2} + \frac{1}{r}\frac{\partial\varphi}{\partial r} + \frac{\partial^2\varphi}{\partial z^2} = 0, \quad -H_0 \le z \le 0, \quad r > 0, \quad t > 0, \tag{6.17}$$

$$\left(\frac{1}{g}\frac{\partial^2\varphi}{\partial t^2} + \frac{\partial\varphi}{\partial z}\right)\bigg|_{z=0} = 0, \quad \eta = -\frac{1}{g}\frac{\partial\varphi}{\partial t}\bigg|_{z=0}, \quad \frac{\partial\varphi(r,z,t)}{\partial z}\bigg|_{z=-H_0} = \frac{\partial\eta^b(r,t)}{\partial t}, \tag{6.18}$$

where η is the deviation of the free surface, η^b is the deviation of the bottom surface and g is the acceleration due to gravity. The corresponding initial conditions at the moment $t = 0$ (first disturbance) are as follows:

$$\varphi(r,z,t)\big|_{t=0} = \frac{\partial\varphi(r,z,t)}{\partial t}\bigg|_{t=0} = \eta^b(r,t)\big|_{t=0} = 0. \tag{6.19}$$

Solving the initial-boundary-value problem requires the fluid velocity v to vanish at infinity at any t. The functions η_1^b and η_2^b corresponding to the bottom disturbance amplitude for the first shock ($t = 0$) and the second shock ($t = t_d$), respectively, are specified as [81]

$$\eta_1^b = \eta_0 \psi(r) f(t), \quad \eta_2^b = \eta_0^d \psi^d(r) f^d(t). \tag{6.20}$$

Next, we introduce the dimensionless variables using the following formulas (asterisks are omitted henceforth):

$$r^* = \frac{r}{r_0}, \quad r_0^* = 1, \quad z^* = \frac{z}{H_0}, \quad t^* = t\frac{c_{\text{sh}}}{r_0}, \quad \eta^* = \frac{\eta}{H_0}, \quad \varphi^* = \frac{\varphi}{r_0 c_{\text{sh}}}, \quad \beta = \frac{r_0}{H_0}, \tag{6.21}$$

where r_0 is some characteristic radius of the bottom disturbance and $c_{\text{sh}} = \sqrt{gH_0}$ is the wave speed on shallow water (the limiting value in the long-wave approximation). In terms of dimensionless variables, the initial-boundary-value problem (6.17)–(6.19) is rewritten as

$$\frac{\partial^2 \varphi}{\partial r^2} + \frac{1}{r}\frac{\partial \varphi}{\partial r} + \beta^2 \frac{\partial^2 \varphi}{\partial z^2} = 0, \quad -1 \le z \le 0, \quad r > 0, \quad t > 0, \tag{6.22}$$

$$\left(\frac{\partial^2 \varphi}{\partial t^2} + \beta^2 \frac{\partial \varphi}{\partial z} \right)\bigg|_{z=0} = 0, \quad \eta = -\frac{1}{g}\frac{\partial \varphi}{\partial t}\bigg|_{z=0}, \quad \beta^2 \frac{\partial \varphi}{\partial z}\bigg|_{z=-1} = \frac{\partial \eta^b}{\partial t}. \tag{6.23}$$

Initial conditions (6.19) remain the same in the dimensionless form.

To solve this problem, we use the integral Laplace transform with respect to time t [84]:

$$\varphi^L(r, z, s) = \int_0^\infty \varphi(r, z, t)\, e^{-st}\, \mathrm{d}t, \tag{6.24}$$

where s is the Laplace transform parameter. After applying transform (6.24) to (6.22), (6.23) with taking into account initial conditions (6.19), we go over to the problem formulated in the Laplace transform space:

$$\frac{\partial^2 \varphi^L}{\partial r^2} + \frac{1}{r}\frac{\partial \varphi^L}{\partial r} + \beta^2 \frac{\partial^2 \varphi^L}{\partial z^2} = 0, \quad -1 \le z \le 0, \quad r > 0, \tag{6.25}$$

$$\left(s^2 \varphi^L + \beta^2 \frac{\partial \varphi^L}{\partial z} \right)\bigg|_{z=0} = 0, \quad \beta^2 \frac{\partial \varphi^L}{\partial z}\bigg|_{z=-1} = s\eta_0 \psi^d(r) f^{dL}(s). \tag{6.26}$$

Next, we use the integral Hankel transform with respect to the radial coordinate r:

$$\varphi^{LH}(k, z, s) = \int_0^\infty \varphi^L(k, z, s) r J_0(kr)\, \mathrm{d}r, \tag{6.27}$$

where k is the Hankel transform parameter. After applying transform (6.27) to problem (6.25), (6.26), we come to the following problem in the Laplace–Hankel transform space:

$$\frac{d^2\varphi^{LH}}{dz^2} - \left(\frac{k}{\beta}\right)^2 \varphi^{LH} = 0, \quad -1 \le z \le 0, \tag{6.28}$$

$$\left(s^2\varphi^{LH} + \beta^2 \frac{d\varphi^{LH}}{dz} \right)\Bigg|_{z=0} = 0, \quad \beta^2 \frac{d\varphi^{LH}}{dz}\Bigg|_{z=-1} = s\eta_0\psi^{dH}(k)f^{dL}(s). \tag{6.29}$$

Solving Eqs. (6.28) and (6.29) yields an expression for the velocity potential:

$$\varphi^{LH}(k,z,s) = -\frac{1}{2}\frac{s}{\beta k}\eta_0\psi^{dH}(k)f^{dL}(s)\frac{(s^2+\beta k)e^{-\frac{k}{\beta^2}z} - (s^2-\beta k)e^{\frac{k}{\beta^2}z}}{s^2\cosh\frac{k}{\beta} - \beta k \sinh\frac{k}{\beta}}. \tag{6.30}$$

We assume that the disturbance is generated by the same axisymmetric elevations of the horizontal bottom [70]:

$$\psi(r) = \psi^d(r) = \xi(\xi^2 + r^2)^{-3/2}, \quad \xi > 0,$$

with two successive disturbances $f_n(t)$, $n = 1, 2$ taken into consideration:

$$\begin{aligned} f_1(t) &= f(t) = te^{-\alpha t}, & t &\ge 0, \\ f_2(t) &= f^d(t) = te^{-\alpha t}H(t-t_d), & t &\ge t_d, \end{aligned}$$

where $H(t)$ is the Heaviside step function. The bottom disturbance function $\psi(r)$ is defined in the infinite domain $r > 0$ and decreases asymptotically with increasing radial coordinate r.

Thus, the problem is solved using the integral Hankel transform with respect to the radial coordinate and the Laplace transform with respect to time, followed by the numerical inversion. The inversion technique is described in detail in Sect. 1.5 and also in Ref. [66]. After the inversion of Hankel transform (6.27), we get an expression for the free surface deviation in the Laplace transform space:

$$\eta_n^L = s^2\eta_0 f_n^L(s) \int_0^\infty \frac{e^{-\xi\lambda}\lambda J_0(\lambda k r)}{s^2\cosh\lambda + \lambda\sinh\lambda}\, d\lambda, \tag{6.31}$$

where [24]

$$f_1^L = \frac{1}{(s+\alpha)^2}, \quad f_2^L = \frac{1}{(s+\alpha)^2}\Gamma(2, t_d(s+\alpha)),$$

$\Gamma(x)$ being the incomplete gamma function.

To calculate the original functions η_n in terms of the image functions η_n^L, we perform the numerical inversion of the Laplace transform with the use the Fourier sine series, according to the technique described in Ref. [57]. It requires the knowledge of the image values only at the equidistant points $s = (2n + 1)\sigma$, where $\sigma > 0$ is an arbitrary number greater than zero and $n = 0, 1, \ldots$. At smaller t the parameter σ is selected sufficiently large, and at larger t the parameter σ is selected to be relatively small. Thus, to go over from the Laplace transform space to the original functions, we calculate the integral in (6.31) by varying the Laplace transform parameter s and selecting appropriately the correction parameter σ appearing in the inversion algorithm.

We investigated the free surface deviation η/η_0 at different distances from the epicentre $r = 0$. The model parameters were selected as $\alpha = 2.5$; $\xi = 0.5, 1.0$. Figures 6.6 and 6.7 show the curves corresponding to the distances from the epicentre equal to 0, 1, 2, 3 and 4. Figure 6.6 demonstrates the deviations η/η_0 caused by only the first disturbance, which took place at the moment $t = 0$. Figure 6.7 shows the same curves in the case of the superposition of the both disturbances,

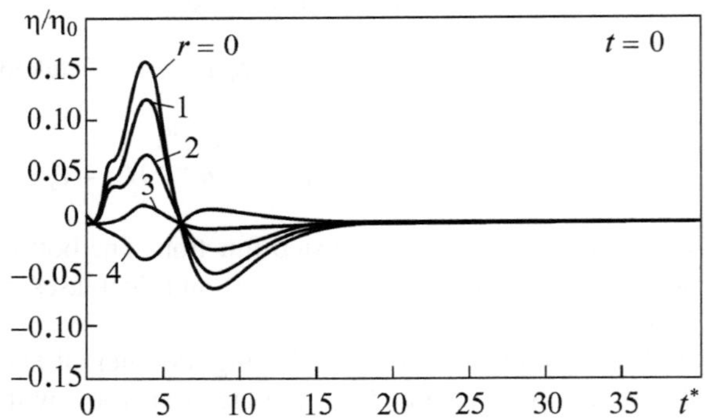

Fig. 6.6 The free surface deviation with only the first bottom disturbance taken into consideration, the distances from the epicentre being $r/r_0 = 0, 1, 2, 3, 4$. Reprinted from [81] by permission

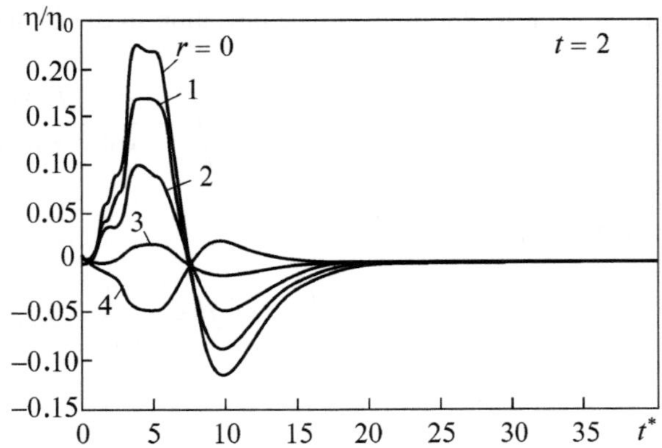

Fig. 6.7 The free surface deviation with both disturbances ($t = 0$ and $t = 2$) taken into consideration. Reprinted from [81] by permission

after the second disturbance took place at the moment $t=2$. Comparing Figs. 6.6 and 6.7, one can see that the free surface deviations η/η_0 are larger in the case of the inclusion of the second disturbance. In the latter case, the amplitudes in the epicentre $r=0$ and at $r=1$ are larger by approximately 20%, and this difference becomes even more distinct (about 30%) at twice larger distances from the epicentre ($r=2$).

6.3 Evolution of Long Water Waves Over a Disturbed Bottom

Studying the nonlinear dispersive wave effects during the propagation of tsunamis is a problem of great significance. In the ocean, tsunami waves propagate in accordance with the linear shallow-water theory. However, nonlinear effects are of importance for the wave propagation in the shelf zone, in view of the possibility for the formation of solitary waves [64, 65, 98]. Here, we consider the three-dimensional problem of the propagation of nonlinear dispersive waves over an inhomogeneous disturbed bottom. The problem is solved by the power series method, which was first successfully used in constructing the degenerate hyperbolic models of wave propagation in plates and shells [65, 69] and then for constructing the degenerate models of wave propagation on finite-depth water [98]. According to this approach, the velocity potential is represented as a power series in small vertical coordinate. Next, this power series is substituted into the equations of motion to reduce the original three-dimensional problem to the two-dimensional one. This approach leads to an infinite system of equations from which weakly nonlinear dispersive evolution equations can be obtained as the limiting linear equations with the use of certain asymptotic estimates.

Consider an ideal incompressible fluid in the Cartesian coordinate system x, y, z. The undisturbed fluid occupies the domain $\Omega=\{(x, y, z): -\infty<x, y<\infty, z\in[0, -h_0]\}$. The problem of nonlinear wave propagation over an inhomogeneous moving bottom is formulated in terms of the velocity potential $\varphi(x, y, z, t)$ as

$$\beta\,\nabla^2\varphi+\varphi_{zz}=0 \text{ in } \Omega, \tag{6.32}$$

$$\eta_t+\alpha\,\nabla\varphi\nabla\eta=\frac{1}{\beta}\,\varphi_z \text{ at } z=\alpha\,\eta, \tag{6.33}$$

$$\eta+\varphi_t+\frac{\alpha}{2\beta}\,\varphi_z^2+\frac{\alpha}{2}\,(\nabla\varphi)^2=0 \text{ at } z=\alpha\,\eta, \tag{6.34}$$

$$\gamma(\xi_t+\alpha\,\nabla\varphi\nabla\xi)-\alpha\,\nabla\varphi\nabla h_0=\frac{\alpha}{\beta}\,\varphi_z \text{ at } z=-h_0(x,y)+\gamma\,\xi(x,y,t) \tag{6.35}$$

where ∇^2 and ∇ are the horizontal operators, $\varphi_z \equiv \partial\varphi/\partial z, \eta_t \equiv \partial\eta/\partial t, \eta$ is the elevation of the free surface, h_0 is the water depth and ξ is the bottom disturbance that is assumed to be a function of horizontal coordinates and time. Equations (6.32)–(6.35) are written in dimensionless form in terms of the variables (asterisks have been omitted)

$$(x^*, y^*) = (x, y)/l, \ (z^*, h_0^*) = (z, h_0)/h_0, \ \xi^* = \xi/\xi_0, \ \eta^* = \eta/a,$$
$$\varphi^* = \varphi\sqrt{g h_0}/g l a, \ t^* = t\sqrt{g h_0}/l,$$

where l and h_0 are the characteristic length and depth, a and ξ_0 are the amplitudes of the free surface and bottom disturbances, respectively. Hence, the system of Eqs. (6.32)–(6.35) is characterised by the following dimensionless parameters: the nonlinearity parameter $\alpha = a/h_0$, the dispersion parameter $\beta = (h_0/l)^2$ and the bottom nonstationarity parameter $\gamma = \xi_0/h_0$.

The approximation of the long-wave theory for the finite but small wave amplitudes implies that $\alpha \ll 1, \beta \ll 1, \gamma = O(\alpha)$. Neglecting the $O(\alpha, \beta)$ terms as compared to the $O(1)$ terms in (6.32)–(6.35), we obtain

$$\frac{\partial^2\varphi}{\partial z^2} + \beta\nabla^2\varphi = 0, \tag{6.36}$$

$$\beta\frac{\partial^2\varphi}{\partial t^2} + \frac{\partial\varphi}{\partial z} = 0, \ \eta = -\frac{\partial\varphi}{\partial t} \ \text{at } z = \alpha\,\eta, \tag{6.37}$$

$$\beta\left[\frac{\partial\xi}{\partial t} - \nabla\varphi\,\nabla h_0\right] = \frac{\partial\varphi}{\partial z}, \ z = -h_0 + \gamma\,\xi. \tag{6.38}$$

In general, ξ is an arbitrary function, but in the case of elastic bottom it is equal to the vertical displacement of the bottom. Equations (6.32)–(6.35) can be used to study the weakly nonlinear and weakly dispersive waves on the surface of a fluid of variable depth. Neglecting the terms of order α, we obtain the well-known linear system of equations for water waves.

Next, restricting our attention to the case of constant undisturbed depth $H = 1$, we write the function φ as a power series:

$$\varphi = \sum_{n=0}^{\infty} \varphi_n(x, y, t)\,(z+1)^n. \tag{6.39}$$

Substituting expansion (6.39) into (6.36)–(6.38), we obtain

$$\beta\nabla^2\varphi_n + (n+1)(n+2)\varphi_{n+2} = 0, \tag{6.40}$$

$$\sum_{n=0}^{\infty}\left[\beta\frac{\partial^2\varphi_n}{\partial t^2} + (n+1)\varphi_{n+1}\right](1+\alpha\eta)^n = 0, \tag{6.41}$$

$$\sum_{n=0}^{\infty}(n+1)\varphi_{n+1}(\gamma\xi)^n = \beta\frac{\partial\xi^n}{\partial t}. \tag{6.42}$$

Equations (6.36)–(6.38) have thereby been reduced to an infinite system of equations that includes recurrence relations (6.40). Taking into account these recurrence relations, Eqs. (6.41) and (6.42) can be reduced to the following evolution equations, correct to the $O(\alpha, \beta, \gamma)$ terms [80]:

$$\frac{\partial^2\varphi_0}{\partial t^2} - c_0^2(\eta, \xi)\,\nabla^2\varphi_0 - \frac{\beta}{2}\frac{\partial^2\nabla^2\varphi_0}{\partial t^2} + \frac{\beta}{6}\nabla^4\varphi_0 = \frac{\partial F}{\partial t}, \tag{6.43}$$

$$F = -\xi - \beta\frac{\partial^2\xi}{\partial t^2} + \frac{\beta}{2}\nabla^2\xi, \quad \eta_0 = -\frac{\partial\varphi_0}{\partial t}, \quad c_0^2(\eta, \xi) = 1 + \alpha\eta_0 - \gamma\xi. \tag{6.44}$$

In the absence of dispersion ($\alpha \to 0$, $\beta \to 0$), the system of Eqs. (6.43), (6.44) can be reduced to a single equation in the linear approximation of shallow water with variable depth:

$$\nabla(h_0\nabla\eta_0) - \frac{\partial^2\eta_0}{\partial t^2} = -\frac{\partial^2\xi}{\partial t^2}. \tag{6.45}$$

Equation (6.45) implies that the motion of the bottom gives rise to a disturbance force and results in the change of propagation speed c_0.

In the case of nonrigid bottom within the framework of the planar problem at $h_0 = 1$, Eq. (6.45) can be written as

$$\frac{\partial^2\eta_0}{\partial x^2} - \frac{\partial^2\eta_0}{\partial t^2} = -\frac{\partial^2\xi}{\partial t^2}. \tag{6.46}$$

The elastic nonrigid bottom under consideration is described as

$$\xi = \frac{1}{\mu}\eta_0, \tag{6.47}$$

where μ is the constant foundation modulus. This simplest one-parameter model is usually referred to as the Winkler foundation model. Substituting (6.47) into (6.46), we obtain

$$\frac{\partial^2\eta_0}{\partial x^2} - \frac{1}{\tilde{c}^2}\frac{\partial^2\eta_0}{\partial t^2} = 0, \tag{6.48}$$

where

$$\hat{c} = \sqrt{\mu/(\mu - 1)}. \tag{6.49}$$

As may be inferred from expression (6.49), no solution exists at $\mu \leq 1$. Therefore, μ changes within the interval

$$1 < \mu < \infty. \tag{6.50}$$

When $\mu \to \infty$, we have $\xi = 0$ and $\hat{c} = 1 = c_{sh}^{*}$, which corresponds to a rigid bottom. When $\mu \to 1$, we have $\xi = \eta_0$, which corresponds to the resonance behaviour. The propagation speed $\hat{c}$ increases as μ goes down from ∞ to 1. Thus, the propagation speed $\hat{c}$ increases in the case of nonrigid bottom.

In the case of a more general two-parameter model (Pasternak foundation),

$$\eta_0 = \mu\xi - G\frac{\partial^2 \xi}{\partial x^2}, \tag{6.51}$$

the following expression for the phase velocity can be obtained:

$$c_p = \left[1 + \frac{G}{\mu}\left(\frac{2\pi}{\lambda}\right)^2\right]^{1/2} \left[1 - \frac{1}{\mu} + \frac{G}{\mu}\left(\frac{2\pi}{\lambda}\right)^2\right]^{-1/2}. \tag{6.52}$$

The velocity $\hat{c}$ described by formula (6.49) can be obtained from expression (6.52) as the limiting case $G \to 0$ or at large wavelengths ($\lambda \to \infty$).

6.4 Diffraction of Cylindrical Waves by a Radial Inhomogeneity

We assume that the fluid motion can described by the equations of the long-wave approximation and that the fluid is at rest at the initial moment $t = 0$. The free fluid surface is supposed to deflect from the horizontal position as follows:

$$\eta' = \eta_0(r_1)\, U(b - r_1), \tag{6.53}$$

where $U(b - r_1)$ is the Heaviside step function.

In the domain of variable depth $H(r)$, the deviation of the free surface is described by the following equation:

$$\nabla[H(r)\nabla\eta_1] - \frac{\partial^2 \eta_1}{\partial t^2} = 0, \quad 0 < r \leq a, \quad t > 0. \tag{6.54}$$

In the domain of constant depth H_2, the corresponding equation is written as

$$\nabla^2\eta_2 - \frac{\partial^2\eta_2}{\partial t^2} = 0, \quad \eta_2 = \eta_s + \eta_i, \quad r \geq 0, \quad t > 0. \tag{6.55}$$

The unknown functions should meet the matching conditions that take into account the continuity of the velocity and free surface elevation:

$$\left.\frac{\partial\eta_1}{\partial r}\right|_{r=0} = \left.\frac{\partial\eta_2}{\partial r}\right|_{r=0}, \quad \eta_1|_{r=0} = \eta_2|_{r=0}. \tag{6.56}$$

The following initial conditions should be satisfied as well:

$$\eta_t|_{t=0} = \eta_s|_{t=0} = \left.\frac{\partial\eta_1}{\partial t}\right|_{t=0} = \left.\frac{\partial\eta_s}{\partial t}\right|_{t=0} = \left.\frac{\partial\eta_i}{\partial t}\right|_{t=0}, \quad \eta_i|_{t=0} = \eta'. \tag{6.57}$$

Here, η_i and η_s are the free surface deviations of the incident and scattered waves, respectively. In addition, in the domain of variable depth the unknown functions should meet the regularity condition at $r=0$ and the function η_s should meet the Sommerfeld radiation conditions.

Here, we use the following dimensionless variables:

$$\left(r^*, r_1^*\right) = \frac{1}{a}(r, r_1), \; t^* = \frac{\sqrt{gH_2}}{a}t,$$
$$\left[\eta_1^*, \eta_2^*, H^*\left(r^*\right), H_2^*\right] = \frac{1}{H^2}[\eta_1, \eta_2, H(r), H_2].$$

By setting $\eta_0(r_1) = d\left(1 - r_1^2/b^2\right)$, the radiation problems (6.53), (6.55), (6.57) can be solved in the Laplace transform space yielding the following expression for the free surface deviation of the incident wave:

$$\eta_i = \begin{cases} d\left[\frac{1}{s} - \frac{r_1^2}{b^2 s} - \frac{4}{b^2 s^3} + \frac{2K_0(sb)I_0(sr_1)}{s} + \frac{4K_1(sb)I_0(sr_1)}{bs^2}\right], & r_1 \leq b, \\ d\left[I_0(sb) - \frac{2I_1(sb)}{bs}\frac{2K_0(sr_1)}{s}\right], \; r_1 \geq b, \end{cases} \tag{6.58}$$

where $I_m(z)$ and $K_m(z)$ are the modified Bessel functions.

After applying the Neumann's addition theorem in the (r, θ, z) coordinate system associated with the submerged inhomogeneity, solution (6.58) can be written as

$$\eta_i = \frac{F}{s}\sum_{m=0}^{\infty} \varepsilon_m I_m(sr)K_m(ki_2)\cos m\theta, \quad r_1 > b, \tag{6.59}$$

where

$$F = 2d\left[I_0(sb) - \frac{2I_1(sb)}{bs}\right], \quad \varepsilon_m = \begin{cases} 1, & m=0, \\ 2, & m>0. \end{cases}$$

For the scattered wave in the image space, we have the following solution:

$$\eta_s = \sum_m^\infty a_m K_m(sr)\cos m\theta, \quad r>a. \tag{6.60}$$

Consider the case when the depth changes according to a parabolic pattern, $H(r) = h + r^2(1-h)$. Then, taking into account the regularity of solutions at $r=0$ and using the power series method described in Sect. 1.2, we come to the following expansion in the domain of variable depth that satisfies Eq. (6.54):

$$\eta_i = \sum_{m=0}^\infty b_m R_m(r)\cos m\theta, \tag{6.61}$$

where

$$R_m(r) = r^m \sum_{n=0}^\infty a_n r^n, \quad a_0 = \frac{1}{2^m m}, \quad a_{2n+1} = 0,$$

$$a_{2n} = \frac{(1-h)\left[(2n-1)^2 - 1 + 2m(2n-1) - s^2\right]}{4nh(m+n)} a_{2n-2}, \quad h>0.5.$$

Substituting expressions (6.59)–(6.61) in the matching condition (6.56), we find the unknown coefficients:

$$a_m = \frac{\varepsilon_m F K_m(sl_2)}{s} \frac{sI'_m(sa)R_m(a) - R'_m(a)I_m(sa)}{R'_m(a)K_m(sa) - sR_m(a)K'_m(sa)},$$

$$b_m = \varepsilon_m \frac{FK_m(sl_2)[I'_m(sa)R_m(sa) - I_m(sa)K'_m(sa)]}{R'_m(a)K_m(sa) - sR_m(a)K'_m(sa)}.$$

The free surface deviation η/d was calculated as a function of x/a for the cases of protrusion ($h_0 > 1$) and depression ($h_0 < 1$) in the bottom. The calculations were performed with formula (6.61) at $n = 9$ with $l_2 = 1$ and $b = 0.5$ using the numerical inversion of the Laplace transform described in Sect. 1.5. Figure 6.8 shows the contour lines of the free surface in the plane $y = 0$. The numerical results demonstrate that there are some points at which the free surface deviation does not depend on the depth of a submerged obstacle (the so-called self-similarity points).

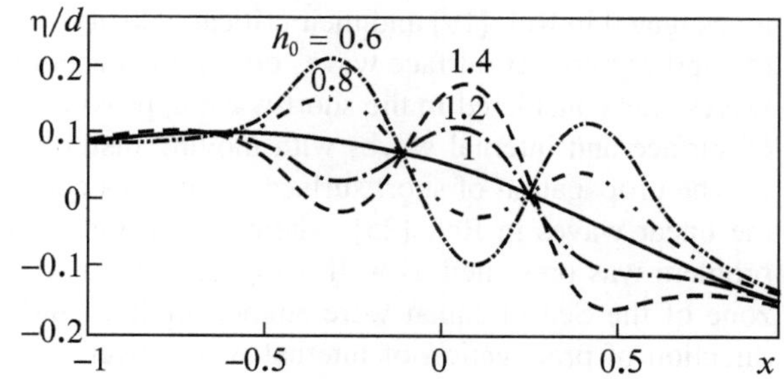

Fig. 6.8 The shape of the free surface η/d at various h_0

6.5 Wave Evolution in a Two-Layer Fluid

Finite-amplitude internal waves have attracted the attention of many researchers [7–9, 13, 16, 17, 20–22, 48, 55]. The majority of theoretical studies of finite-amplitude internal waves deal with the analysis of wave motion in the systems where the internal waves are weakly nonlinear and long as compared to the fluid depth [1, 10, 45, 49, 52, 62, 91, 93, 98, 99]. The Korteweg–de Vries equation that describes the evolution of the wave motion under the balance between nonlinearity and dispersion is well studied. The methods of finding the exact solutions for arbitrary initial conditions were also developed [61].

For the first time, the propagation and stability of wave trains on the interface between semi-infinite fluids was investigated by Nayfeh [53]. The stability of internal Stokes waves in a two-layer fluid was analysed in Ref. [100] using the method of multiscale asymptotic expansions. The propagation of nonlinear waves in two- and three-layer fluids was investigated in Refs. [78, 79], and the excitation of nonlinear waves in the case of obstacles located on the interface was studied in Ref. [82]. The so-called half-space/half-space, half-space/layer and layer/layer systems as well as some associated peculiarities of the method of multiscale expansions were considered in Refs. [3, 4, 71–75].

Studying the interaction of internal and surface waves is of particular interest as well. The generation of internal waves by surface waves was studied in Ref. [96]. The nonlinear wave propagation in a two-layer fluid in the presence of an upper free boundary was considered in Ref. [50]. A nonlinear evolution equation was derived with the help of the asymptotic expansion in terms of a small parameter without referring to the shallow-water approximation. The asymptotic analysis of interface waves in a two-layer fluid was performed in Ref. [42] in terms of a small parameter equal to the difference between the densities of the upper and lower fluids. The resonant excitation of weakly nonlinear waves at the interface between the layers during the wave propagation on the free surface was investigated in Ref. [30] by the method of multiscale expansions. The propagation of solitary interface waves and a resonance between short and slow waves were studied in Ref. [23] in the case of a two-layer fluid in the presence of an upper free boundary. Internal capillary-gravity waves in a two-layer fluid with free surface over a bottom obstruction were investigated in Ref. [19] and their critical velocity was determined. The propagation of capillary-gravity surface waves during the generation of large-amplitude internal waves was considered in the short-wave approximation in Ref. [5]. The interaction of surface and internal waves with moving disturbances was studied in Ref. [35].

The propagation of short surface waves was considered using the ray theory for the linear waves in Ref. [25], where the fully nonlinear numerical solution of the problem was presented as well. Nonlinear internal waves and solitons in the shelf zone of the Sea of Japan were studied in Ref. [88]. The basic regularities in the direction of propagation of internal waves over the shelf and near the continental slope were revealed. Second-order solutions describing internal and surface waves in a two-fluid system were obtained with the use of perturbation method in Ref. [47]. The transfer function and the influence of second-order effects on the wave profile were considered. Note that the mass transfer is caused by nonlinear effects that result in nonclosed particle trajectories in wave flows [86, 94].

The interaction between short surface waves and long internal waves in a two-layer fluid was studied in Ref. [39] by the method multiscale expansions. A stationary solution for the wave envelope was obtained in the case when the group velocity of surface waves was equal to the phase velocity of internal waves. Such an interaction was proven to have a strong effect on the envelope. This problem was also discussed in Refs. [23, 41, 89]. The nonlinear wave motion in the gravity force field in a two-layer fluid stratified by density was considered in Ref. [33] in the case of finite thickness of the upper (lighter) layer, and the nonlinear internal resonance interaction of the gravity waves generated by the free surface of the upper layer and the interface was studied. A new nonlinear problem of the wave train propagation in the system layer-with-rigid-bottom/layer-with-free-surface was considered in Ref. [77]. The first three linear approximations of the nonlinear problem were obtained using the multiscale expansion method. Here, we demonstrate the presence of nonoscillatory terms in the second approximation, which corresponds to the calculations given in Refs. [29, 34, 36].

6.5.1 Problem Formulation and Solution Method

Consider the wave train propagation in the hydrodynamic system layer-with-rigid-bottom/layer-with-free-surface. The upper and lower fluids are assumed to be ideal and incompressible. The surface tension at the interface (T) and on the free surface (T_0) is taken into consideration. The wave motion is supposed to be potential and is characterised by a small but finite amplitude (Fig. 6.9). The force of gravity, which is perpendicular to the interface, is directed opposite to the z-axis.

Introduce the dimensionless variables using the characteristic length L, the maximum free surface deviation a, the characteristic time $(L/g)^{1/2}$ and the lower fluid density ρ_1, where g is the acceleration due to gravity:

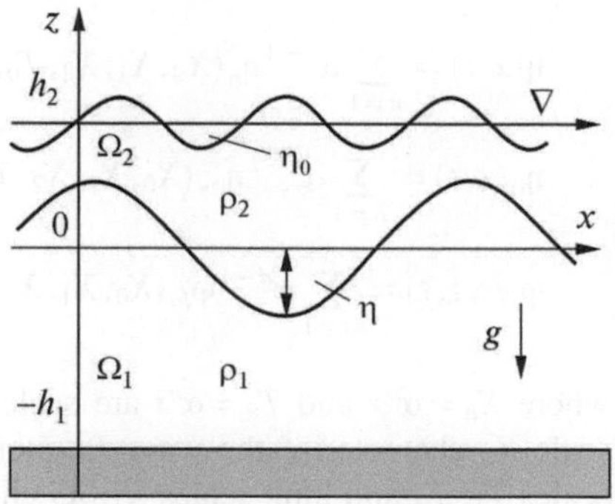

Fig. 6.9 Problem formulation

$$(x,z)=L(x^*,\ z^*),\quad t=\sqrt{\tfrac{L}{g}}\,t^*,\quad \rho_2=\rho_1\rho^*,\quad (\eta,\eta_0)=a(\eta^*,\eta_0^*),$$
$$\varphi=\tfrac{La}{\sqrt{L/g}}\varphi^*,\quad (T,T_0)=L^2\rho_1 g(T^*,T_0^*). \tag{6.62}$$

Then the mathematical formulation of the problem is as follows (asterisks are omitted henceforth):

$$\frac{\partial^2\varphi_j}{\partial x^2}+\frac{\partial^2\varphi_j}{\partial z^2}=0 \text{ in } \Omega_j,\quad j=1,2, \tag{6.63}$$

$$\frac{\partial\eta}{\partial t}-\frac{\partial\varphi_j}{\partial z}=-\alpha\frac{\partial\eta}{\partial x}\frac{\partial\varphi_j}{\partial x}\ \text{ at } z=\alpha\eta\,(x,t),\quad j=1,2, \tag{6.64}$$

$$\frac{\partial\eta_0}{\partial t}-\frac{\partial\varphi_2}{\partial z}=-\alpha\frac{\partial\eta_0}{\partial t}\frac{\partial\varphi_2}{\partial z}\ \text{ at } z=\alpha\eta_0(x,t), \tag{6.65}$$

$$\begin{aligned}&\frac{\partial\varphi_1}{\partial t}-\rho\frac{\partial\varphi_2}{\partial t}+(1-\rho)\eta+\frac{1}{2}\alpha\left[(\nabla\varphi_1)^2-\rho(\nabla\varphi_2)^2\right]\\ &-T\left(1+\left(\alpha\frac{\partial\eta}{\partial x}\right)^2\right)^{-\frac{3}{2}}\frac{\partial^2\eta}{\partial x^2}=0 \text{ at } z=\alpha\eta(x,t),\end{aligned} \tag{6.66}$$

$$\frac{\partial\varphi_2}{\partial t}+\eta+\frac{1}{2}\alpha(\nabla\varphi_2)^2-T_0\left(1+\left(\alpha\frac{\partial\eta_0}{\partial x}\right)^2\right)^{-\frac{3}{2}}\frac{\partial^2\eta_0}{\partial x^2}=0 \text{ at } z=\alpha\eta_0(x,t), \tag{6.67}$$

$$\frac{\partial\varphi_1}{\partial z}=0\quad \text{at } z=-h_1=-\frac{h_1}{L}, \tag{6.68}$$

where the nonlinearity factor α was introduced as the ratio $a/L=\alpha$.

To find the approximate solution for small but finite amplitudes, we use the multiscale expansion method [53, 55]:

$$\eta(x,t) = \sum_{n=1}^{3} \alpha^{n-1}\eta_n(X_0, X_1, X_2, T_0, T_1, T_2) + O(\alpha^3),$$
$$\eta_0(x,t) = \sum_{n=1}^{3} \alpha^{n-1}\eta_{0n}(X_0, X_1, X_2, T_0, T_1, T_2) + O(\alpha^3), \tag{6.69}$$
$$\varphi_j(x,z,t) = \sum_{n=1}^{3} \alpha^{n-1}\varphi_{jn}(X_0, X_1, X_2, z, T_0, T_1, T_2) + O(\alpha^3), \quad j = 1, 2,$$

where $X_n = \alpha^n x$ and $T_n = \alpha^n t$ are scaled variables. The fast scale T_0 and the short scale X_0 characterise the wave frequency and wavelength, respectively. The slow scales T_1, T_2 and long scales X_1, X_2 characterise the temporal and spatial variations of the wave phase and amplitude.

6.5.2 Linear Approximations

Substituting expansions (6.69) in the system of Eqs. (6.63)–(6.68) and collecting the coefficients with the same powers of α leads to three linear problems for the unknown functions η_1, η_{01}, φ_{11}, φ_{21}, η_2, η_{02}, φ_{12}, φ_{22}, η_3, η_{03}, φ_{31}, φ_{32} that appear in the multiscale expansions of potentials φ_1, φ_2, interface deviation η, and free surface deviation η_0.

The first linear approximation of problem (6.63)–(6.68) (the terms proportional to α) is written as follows:

$$\frac{\partial^2\varphi_{11}}{\partial X_0^2} + \frac{\partial^2\varphi_{11}}{\partial z^2} = 0 \text{ in } \Omega_1, \tag{6.70}$$

$$\frac{\partial^2\varphi_{21}}{\partial X_0^2} + \frac{\partial^2\varphi_{21}}{\partial z^2} = 0 \text{ in } \Omega_2, \tag{6.71}$$

$$\frac{\partial\eta_1}{\partial T_0} - \frac{\partial\varphi_{11}}{\partial z} = 0 \text{ at } z = 0, \tag{6.72}$$

$$\frac{\partial\eta_2}{\partial T_0} - \frac{\partial\varphi_{21}}{\partial z} = 0 \text{ at } z = 0, \tag{6.73}$$

$$\frac{\partial\eta_{01}}{\partial T_0} - \frac{\partial\varphi_{21}}{\partial z} = 0 \text{ at } z = h_2, \tag{6.74}$$

$$\frac{\partial\varphi_{11}}{\partial T_0} - \rho\frac{\partial\varphi_{21}}{\partial T_0} + (1-\rho)\,\eta_1 - T\frac{\partial^2\eta_1}{\partial X_0^2} = 0 \text{ at } z = 0, \tag{6.75}$$

$$\frac{\partial\varphi_{21}}{\partial T_0}+\eta_{01}-T\frac{\partial^2\eta_{01}}{\partial X_0^2}=0 \text{ at } z=h_2, \tag{6.76}$$

$$\frac{\partial\varphi_{11}}{\partial z}=0 \text{ at } z=-h_1. \tag{6.77}$$

The solution to Eqs. (6.70) and (6.71) is

$$\eta_1=Ae^{i\theta}+A^*e^{-i\theta}, \tag{6.78}$$

$$\varphi_{11}=-\frac{i\omega}{k}(Ae^{i\theta}-A^*e^{-i\theta})\frac{\cosh(k(h_1+z))}{\sinh(kh_1)}, \tag{6.79}$$

$$\varphi_{21}=\frac{i\omega}{k}\left(\frac{\omega^2\sinh(k(h_2-z))-(k+T_0k^3)\cosh(k(h_2-z))}{\omega^2\cosh(kh_2)-(k+T_0k^3)\sinh(kh_2)}\right)(Ae^{i\theta}-A^*e^{-i\theta}), \tag{6.80}$$

$$\eta_{01}=\left(\frac{\omega^2}{\omega^2\cosh(kh_2)-(k+T_0k^3)\sinh(kh_2)}\right)(Ae^{i\theta}+A^*e^{-i\theta}), \tag{6.81}$$

where $\theta=kx_0-\omega t_0$, k is the wave number of the wave train centre, ω is the frequency of the wave train centre and A^* is the complex conjugate of the internal wave train envelope A. After substituting these solutions into Eq. (6.75), we obtain a dispersion equation that defines the relationship between the frequency and wave number of the wave train centre:

$$\omega^2\coth(kh_1)+\rho\omega^2\left(\frac{\omega^2-(k+T_0k^3)\coth(kh_2)}{\omega^2\coth(kh_2)-(k+T_0k^3)}\right)=(1-\rho)k+Tk^3. \tag{6.82}$$

The interface is linearly stable or unstable depending on whether k is larger or smaller than the critical wave number $k_c=[(1-\rho)/T]^{1/2}$.

The relationship between the internal wave train envelope A and the surface wave train envelope $A^{(0)}$ follows from formula (6.82):

$$A^{(0)}=\frac{\omega^2}{\lambda}A, \tag{6.83}$$

where

$$\lambda=\omega^2\cosh(kh_2)-(k+T_0k^3)\sinh(kh_2).$$

The second linear approximation of problem (6.63)–(6.68) (the terms proportional to α^2) is written as

$$
\begin{gathered}
\frac{\partial^2\varphi_{12}}{\partial X_0^2}+\frac{\partial^2\varphi_{12}}{\partial z^2}+2\frac{\partial^2\varphi_{11}}{\partial X_0\partial X_1}=0 \text{ in } \Omega_1,\\
\frac{\partial^2\varphi_{22}}{\partial X_0^2}+\frac{\partial^2\varphi_{22}}{\partial z^2}+2\frac{\partial^2\varphi_{21}}{\partial X_0\partial X_1}=0 \text{ in } \Omega_2,\\
\frac{\partial\eta_1}{\partial T_1}+\frac{\partial\eta_2}{\partial T_0}-\frac{\partial\varphi_{12}}{\partial z}-\eta_1\frac{\partial^2\varphi_{11}}{\partial z^2}=-\frac{\partial\eta_1}{\partial X_0}\frac{\partial\varphi_{11}}{\partial X_0} \text{ at } z=0,\\
\frac{\partial\eta_1}{\partial T_1}+\frac{\partial\eta_2}{\partial T_0}-\frac{\partial\varphi_{22}}{\partial z}-\eta_1\frac{\partial^2\varphi_{21}}{\partial z^2}=-\frac{\partial\eta_1}{\partial X_0}\frac{\partial\varphi_{21}}{\partial X_0} \text{ at } z=0,\\
\frac{\partial\eta_{01}}{\partial T_1}+\frac{\partial\eta_{02}}{\partial T_0}-\frac{\partial\varphi_{22}}{\partial z}-\eta_{01}\frac{\partial^2\varphi_{21}}{\partial z^2}=-\frac{\partial\eta_{01}}{\partial X_0}\frac{\partial\varphi_{21}}{\partial X_0} \text{ at } z=h_2,\\
\frac{\partial\varphi_{11}}{\partial T_1}+\frac{\partial\varphi_{12}}{\partial T_0}+\eta_1\frac{\partial^2\varphi_{11}}{\partial T_0\partial z}-\rho\left(\frac{\partial\varphi_{21}}{\partial T_1}+\frac{\partial\varphi_{22}}{\partial T_0}+\eta_1\frac{\partial^2\varphi_{21}}{\partial T_0\partial z}\right)+(1-\rho)\eta_2\\
+0.5\left(\frac{\partial\varphi_{11}}{\partial X_0}\right)^2+0.5\left(\frac{\partial\varphi_{11}}{\partial z}\right)^2-\rho\left(0.5\left(\frac{\partial\varphi_{21}}{\partial X_0}\right)^2+0.5\left(\frac{\partial\varphi_{21}}{\partial z}\right)^2\right)\\
-T\left(\frac{\partial^2\eta_2}{\partial X_0^2}+2\frac{\partial^2\eta_1}{\partial X_0\partial X_1}\right)=0 \text{ at } z=0,\\
\frac{\partial\varphi_{21}}{\partial T_1}+\frac{\partial\varphi_{22}}{\partial T_0}+\eta_{01}\frac{\partial^2\varphi_{21}}{\partial T_0\partial z}+\eta_{02}+0.5\left(\frac{\partial\varphi_{21}}{\partial X_0}\right)^2+0.5\left(\frac{\partial\varphi_{21}}{\partial z}\right)^2\\
-T_0\left(\frac{\partial^2\eta_{02}}{\partial X_0^2}+2\frac{\partial^2\eta_{01}}{\partial X_0\partial X_1}\right)=0 \text{ at } z=h_2,\\
\frac{\partial\varphi_{12}}{\partial z}=0 \text{ at } z=-h_1.
\end{gathered}
\tag{6.84}
$$

Equations (6.84) have the following solution:

$$
\begin{aligned}
\varphi_{12}&=B_{11}(z+h_1)\sinh(k(z+h_1))e^{i\theta}+B_{21}\cosh(k(z+h_1))e^{i\theta}\\
&\quad+B_{22}\cosh(2k(z+h_1))e^{2i\theta}+c.c.,\\
\varphi_{22}&=(C_{10}+C_{11}z)e^{i\theta+k(h_2-z)}+C_{20}e^{2i\theta+2k(h_2-z)}\\
&\quad+(E_{10}+E_{11}z)e^{i\theta-k(h_2-z)}+E_{20}e^{2i\theta-2k(h_2-z)}+c.c.,\\
\eta_2&=D_0+D_1e^{i\theta}+D_2e^{2i\theta}+c.c.,\\
\eta_{02}&=F_0+F_1e^{i\theta}+F_2e^{2i\theta}+c.c.,
\end{aligned}
\tag{6.85}
$$

where B_{ij}, C_{ij}, E_{ij}, D_i and F_i are the undetermined coefficients, *c.c.* denotes the complex conjugate of the preceding terms.

Substituting expressions (6.85) and the solutions (6.78)–(6.81) of the first linear approximation into the first two equations of system (6.84), we obtain the following relations:

$$
\begin{aligned}
B_{11}&=-\frac{\omega}{k\sinh(kh_1)}A_{x_1},\\
C_{11}&=-\frac{\omega}{2k}\left(\frac{\omega^2-(k+T_0k^3)}{\lambda}\right)A_{x_1},\\
E_{11}&=-\frac{\omega}{2k}\left(\frac{\omega^2+(k+T_0k^3)}{\lambda}\right)A_{x_1},
\end{aligned}
\tag{6.86}
$$

where the indices next to A designate the corresponding partial derivatives. Other coefficients are calculated in the same way from other equations of system (6.84) leading to two independent systems of equations with respect to the remaining unknown coefficients.

The first system of equations with respect to the unknown coefficients B_{10}, C_{10}, E_{10}, D_1 and F_1 is obtained by equating the coefficients at the $e^{i\theta}$ terms. This system contains inconsistent equations, and its solvability condition is

$$\begin{aligned}
&\left(\frac{(k+T_0k^3)\cosh(kh_2)-\omega^2\sinh(kh_2)}{k\omega}-\frac{2(k+T_0k^3)\omega}{k\lambda}\right)A_{t_1}\\
&-\lambda\frac{\omega\coth(kh_1)+(k+k\rho+Tk^3)}{k\rho\,\omega^3}A_{t_1}+\frac{(1+T_0k)\cosh^2(kh_2)}{k}A_{x_1}\\
&-\cosh(kh_2)\left(\frac{\coth(kh_1)(1-kh_1\coth(kh_1))+2k^3T-\omega^2kh_1}{k^2\rho}\right)A_{x_1}+\\
&+\left(1+T_0k^2\right)\left(\frac{\coth(kh_1)\sinh(kh_2)(1-kh_1\coth(kh_1))}{k\omega\rho}\right)A_{x_1}\\
&-\left(\frac{\omega^2\sinh(kh_2)}{k^2}+\frac{2T_0k\omega^2}{\lambda}+\left(1+T_0k^2\right)\frac{\omega^2+kh_2(k+T_0k^3)}{k\lambda}\right)A_{x_1}\\
&+\left(1+T_0k^2\right)\frac{\sinh(kh_2)(2k^2T-\omega^2h_1)}{\rho\,\omega^2}A_{x_1}=0.
\end{aligned}\tag{6.87}$$

If condition (6.87) is met and $D_1=0$, the other unknown coefficients can be found from the following equations:

$$\begin{aligned}
k\sinh(kh_1)B_{10}&=A_{t_1}+\tfrac{\omega(1-kh_1\coth(kh_1))}{k}A_{x_1},\\
ke^{kh_2}C_{10}-ke^{-kh_2}E_{10}&=-A_{t_1}-\tfrac{\omega}{k}A_{x_1},\\
kC_{10}-kE_{10}-i\omega F_1&=-\tfrac{\omega^2}{\lambda}A_{t_1}-\tfrac{\omega}{k\lambda}(\omega^2+kh_2(k+T_0k^3))A_{x_1}\\
-\omega\cosh(kh_1)B_{10}+\rho\,\omega\,e^{kh_2}C_{10}+\rho\,\omega\,e^{-kh_2}E_{10}&=\tfrac{1-\rho+Tk^2}{\omega}A_{t_1}+\left(2kT-\tfrac{\omega^2h_1}{k}\right)A_{x_1}.
\end{aligned}$$

The second system of equations with respect to the unknown coefficients B_{20}, C_{20}, E_{20}, D_2 and F_2 is obtained by equating the coefficients at the $e^{2i\theta}$ terms:

$$\begin{aligned}
-2i\,\omega D_2-2k\sinh(2kh_1)B_{20}&=-2i\,k\omega\coth(kh_1)A^2,\\
-2i\,\omega D_2+2ke^{2kh_2}C_{20}-2ke^{-2kh_2}E_{20}&=2i\,k\tfrac{(1-\rho)k-Tk^3-\omega^2\coth(kh_1)}{\rho\,\omega}A^2,\\
-2i\,\omega F_2+2kC_{20}-2kE_{20}&=-\tfrac{2i\,\omega^3\left(k+T_0k^3\right)}{\lambda^2}A^2,\\
(1+4T_0k^2)F_2-2i\,\omega C_{20}-2i\,\omega E_{20}&=\tfrac{1.5\omega^6-0.5\omega^2\left(k+T_0k^3\right)^2}{\lambda^2}A^2,\\
(1-\rho+2Tk^2)D_2-2i\,\omega\cosh(2kh_1)B_{20}&+2i\,\rho\,\omega\,e^{2kh_2}C_{20}+2i\,\rho\,\omega\,e^{-2kh_2}E_{20}\\
&=1.5\,\omega^2A^2+\left(1.5\,\rho\,\omega^2-0.5\,\omega\coth^2(kh_1)\right)A^2\\
&+\left(0.5\rho\omega\tfrac{\left(\omega^2\sinh(kh_2)-\left(k+T_0k^3\right)\cosh(kh_2)\right)^2}{\lambda^2}\right)A^2.
\end{aligned}$$

This is a consistent system of equations, which can easily be solved using symbolic mathematical processors. Its explicit solution is quite cumbersome and will not be presented here.

The third linear approximation of problem (6.63)–(6.68) (the terms proportional to α^3) is written as follows:

$$
\begin{aligned}
&\frac{\partial^2\varphi_{13}}{\partial X_0^2}+\frac{\partial^2\varphi_{11}}{\partial X_1^2}+\frac{\partial^2\varphi_{13}}{\partial z^2}+2\frac{\partial^2\varphi_{12}}{\partial X_0\partial X_1}+2\frac{\partial^2\varphi_{11}}{\partial X_0\partial X_2}=0 \text{ in } \boldsymbol{\Omega}_1,\\
&\frac{\partial^2\varphi_{23}}{\partial X_0^2}+\frac{\partial^2\varphi_{21}}{\partial X_1^2}+\frac{\partial^2\varphi_{23}}{\partial z^2}+2\frac{\partial^2\varphi_{22}}{\partial X_0\partial X_1}+2\frac{\partial^2\varphi_{21}}{\partial X_0\partial X_2}=0 \text{ in } \boldsymbol{\Omega}_2,\\
&\frac{\partial\eta_2}{\partial T_1}+\frac{\partial\eta_1}{\partial T_2}+\frac{\partial\eta_3}{\partial T_0}-\frac{\partial\varphi_{13}}{\partial z}-\eta_1\frac{\partial^2\varphi_{12}}{\partial z^2}-\eta_2\frac{\partial^2\varphi_{11}}{\partial z^2}-0.5\eta_1^2\frac{\partial^3\varphi_{11}}{\partial z^3}\\
&\quad=-\frac{\partial\eta_1}{\partial X_1}\frac{\partial\varphi_{11}}{\partial X_0}-\frac{\partial\eta_2}{\partial X_0}\frac{\partial\varphi_{11}}{\partial X_0}-\frac{\partial\eta_1}{\partial X_0}\frac{\partial\varphi_{11}}{\partial X_1}-\frac{\partial\eta_1}{\partial X_0}\frac{\partial\varphi_{12}}{\partial X_0}-\eta_1\frac{\partial\eta_1}{\partial X_0}\frac{\partial^2\varphi_{11}}{\partial X_0\partial z} \text{ at } z=0,\\
&\frac{\partial\eta_2}{\partial T_1}+\frac{\partial\eta_1}{\partial T_2}+\frac{\partial\eta_3}{\partial T_0}-\frac{\partial\varphi_{23}}{\partial z}-\eta_1\frac{\partial^2\varphi_{22}}{\partial z^2}-\eta_2\frac{\partial^2\varphi_{21}}{\partial z^2}-0.5\eta_1^2\frac{\partial^3\varphi_{21}}{\partial z^3}\\
&\quad=-\frac{\partial\eta_1}{\partial X_1}\frac{\partial\varphi_{21}}{\partial X_0}-\frac{\partial\eta_2}{\partial X_0}\frac{\partial\varphi_{21}}{\partial X_0}-\frac{\partial\eta_1}{\partial X_0}\frac{\partial\varphi_{21}}{\partial X_1}-\frac{\partial\eta_1}{\partial X_0}\frac{\partial\varphi_{22}}{\partial X_0}-\eta_1\frac{\partial\eta_1}{\partial X_0}\frac{\partial^2\varphi_{21}}{\partial X_0\partial z} \text{ at } z=0,\\
&\frac{\partial\eta_{02}}{\partial T_1}+\frac{\partial\eta_{01}}{\partial T_2}+\frac{\partial\eta_{03}}{\partial T_0}-\frac{\partial\varphi_{23}}{\partial z}-\eta_{01}\frac{\partial^2\varphi_{22}}{\partial z^2}-\eta_{02}\frac{\partial^2\varphi_{21}}{\partial z^2}-0.5\eta_{01}^2\frac{\partial^3\varphi_{21}}{\partial z^3}=-\frac{\partial\eta_{01}}{\partial X_1}\frac{\partial\varphi_{21}}{\partial X_0}-\frac{\partial\eta_{02}}{\partial X_0}\frac{\partial\varphi_{21}}{\partial X_0}\\
&\quad-\frac{\partial\eta_{01}}{\partial X_0}\frac{\partial\varphi_{21}}{\partial X_1}-\frac{\partial\eta_{01}}{\partial X_0}\frac{\partial\varphi_{22}}{\partial X_0}-\eta_{01}\frac{\partial^2\varphi_{21}}{\partial X_0\partial z}\frac{\partial\eta_{01}}{\partial X_0} \text{ at } z=h_2,\\
&\frac{\partial\varphi_{11}}{\partial T_2}+\frac{\partial\varphi_{12}}{\partial T_1}+\frac{\partial\varphi_{13}}{\partial T_0}+\eta_2\frac{\partial^2\varphi_{11}}{\partial T_0\partial z}+\eta_1\frac{\partial^2\varphi_{11}}{\partial T_1\partial z}+0.5\eta_1^2\frac{\partial^3\varphi_{11}}{\partial T_0\partial z^2}\\
&\quad-\rho\left(\frac{\partial\varphi_{21}}{\partial T_2}+\frac{\partial\varphi_{22}}{\partial T_1}+\frac{\partial\varphi_{23}}{\partial T_0}+\eta_2\frac{\partial^2\varphi_{21}}{\partial T_0\partial z}+\eta_1\frac{\partial^2\varphi_{21}}{\partial T_1\partial z}+0.5\eta_1^2\frac{\partial^3\varphi_{21}}{\partial T_0\partial z^2}\right)\\
&\quad+\eta_1\frac{\partial^2\varphi_{12}}{\partial T_0\partial z}+(1-\rho)\eta_3+\frac{\partial\varphi_{11}}{\partial X_0}\frac{\partial\varphi_{11}}{\partial X_1}+\frac{\partial\varphi_{11}}{\partial X_0}\frac{\partial\varphi_{12}}{\partial X_1}+\eta_1\frac{\partial^2\varphi_{11}}{\partial X_0\partial z}\frac{\partial\varphi_{11}}{\partial X_0}\\
&\quad+\frac{\partial\varphi_{11}}{\partial z}\frac{\partial\varphi_{12}}{\partial z}+\eta_1\frac{\partial\varphi_{11}}{\partial z}\frac{\partial^2\varphi_{11}}{\partial z^2}-\rho\eta_1\frac{\partial^2\varphi_{22}}{\partial T_0\partial z}-\rho\frac{\partial\varphi_{21}}{\partial X_0}\frac{\partial\varphi_{21}}{\partial X_1}-\rho\frac{\partial\varphi_{21}}{\partial X_0}\frac{\partial\varphi_{22}}{\partial X_1}\\
&\quad-\rho\eta_1\frac{\partial^2\varphi_{21}}{\partial X_0\partial z}\frac{\partial\varphi_{21}}{\partial X_0}-\rho\frac{\partial\varphi_{21}}{\partial z}\frac{\partial\varphi_{22}}{\partial z}-\rho\eta_1\frac{\partial\varphi_{21}}{\partial z}\frac{\partial^2\varphi_{21}}{\partial z^2}-T\frac{\partial^2\eta_3}{\partial X_0^2}-T\frac{\partial^2\eta_1}{\partial X_1^2}\\
&\quad-2T\frac{\partial^2\eta_2}{\partial X_0\partial X_1}-2T\frac{\partial^2\eta_1}{\partial X_0\partial X_2}-1.5T\left(\frac{\partial\eta_1}{\partial X_0}\right)^2\frac{\partial^2\eta_1}{\partial X_0^2}=0 \text{ at } z=0,\\
&\frac{\partial\varphi_{21}}{\partial T_2}+\frac{\partial\varphi_{22}}{\partial T_1}+\frac{\partial\varphi_{23}}{\partial T_0}+\eta_{02}\frac{\partial^2\varphi_{21}}{\partial T_0\partial z}+\eta_{01}\frac{\partial^2\varphi_{21}}{\partial T_1\partial z}+0.5\eta_{01}^2\frac{\partial^3\varphi_{21}}{\partial T_0\partial z^2}+\eta_{01}\frac{\partial^2\varphi_{22}}{\partial T_0\partial z}+\eta_{03}\\
&\quad+\frac{\partial\varphi_{21}}{\partial X_0}\frac{\partial\varphi_{21}}{\partial X_1}+\frac{\partial\varphi_{21}}{\partial X_0}\frac{\partial\varphi_{22}}{\partial X_{01}}+\eta_{01}\frac{\partial^2\varphi_{21}}{\partial X_0\partial z}\frac{\partial\varphi_{21}}{\partial X_0}+\frac{\partial\varphi_{21}}{\partial z}\frac{\partial\varphi_{22}}{\partial z}+\eta_{01}\frac{\partial\varphi_{21}}{\partial z}\frac{\partial^2\varphi_{21}}{\partial z^2}-T_0\frac{\partial^2\eta_{03}}{\partial X_0^2}\\
&\quad-T_0\frac{\partial^2\eta_{01}}{\partial X_1^2}-2T_0\frac{\partial^2\eta_{02}}{\partial X_0\partial X_1}-T_0\frac{\partial^2\eta_{01}}{\partial X_0\partial X_2}-1.5T_0\left(\frac{\partial\eta_{01}}{\partial X_0}\right)^2\frac{\partial^2\eta_{01}}{\partial X_0^2}=0 \text{ at } z=h_2,\\
&\frac{\partial\varphi_{13}}{\partial z}=0 \text{ at } z=-h_1.
\end{aligned}
\tag{6.88}
$$

These equations are solved in the same way as the equations of the first and second linear approximations [74, 75, 77]. As in the second approximation, the homogeneous part of the third-order problem has a nontrivial solution implying that the nonhomogeneous problem has a solution if and only if the following condition is met:

$$W_1A_{t_2} + W_2A_{x_2} + W_3A_{x_1x_1} + W_4A^2\overline{A} = 0, \tag{6.89}$$

where W_i $(i=\overline{1,4})$ are the coefficients depending on k, ρ, ω, h_1, h_2, T, and T_0.

To simplify the solvability conditions for the second- and third-order problems, we rewrite the equations for the envelope amplitudes as

$$A_{t_1} + \omega' A_{x_1} = 0, \tag{6.90}$$

$$A_{t_2} + \omega' A_{x_2} - 0.5\,\mathrm{i}\,\omega'' A_{x_1x_1} = \mathrm{i}\,IA^2\overline{A}, \tag{6.91}$$

where $\omega' = \mathrm{d}\omega/\mathrm{d}k$ and I is the coefficient that depends on all the geometrical and physical parameters of the system and is calculated in explicit form but is not presented here because it is rather cumbersome. The partial derivatives A_t, A_x and A_{xx} can be represented as

$$\begin{aligned} A_t &= \sum_{n=1}^{2} \alpha^n A_{t_n} + O(\alpha^3),\\ A_x &= \sum_{n=1}^{2} \alpha^n A_{x_n} + O(\alpha^3),\\ A_{xx} &= \alpha^2 A_{x_1x_1} + O(\alpha^3). \end{aligned} \tag{6.92}$$

Multiplying Eq. (6.90) by α^2 and adding Eq. (6.91) multiplied by α with taking into account expressions (6.92), we obtain the following evolution equation for the wave train envelope at the interface:

$$A_t + \omega' A_x - 0.5\,\mathrm{i}\,\omega'' A_{xx} = \mathrm{i}\,\alpha^2 IA^2A^*. \tag{6.93}$$

It has the form of the nonlinear Schrödinger equation derived earlier in other hydrodynamic systems [76, 98, 99]. The evolution equation for the wave train envelope on the free surface is derived from Eq. (6.93) using formula (6.83):

$$A_t^{(0)} + \omega' A_x^{(0)} - 0.5\,\mathrm{i}\,\omega'' A_{xx}^{(0)} = \mathrm{i}\,\alpha^2 I_0 (A^{(0)})^2 A^{(0)*}, \tag{6.94}$$

where $I_0 = (\lambda^2/\omega^4)\,I$. Equation (6.93) has a trivial solution dependent only on time:

$$A = a\,\exp\left(\mathrm{i}\,\alpha^2 I |a|^2 t\right), \tag{6.95}$$

where a is an arbitrary amplitude. Similarly, for Eq. (6.94) we obtain

$$A^{(0)} = a\,\exp\left(\mathrm{i}\,\alpha^2 I_0\,|a|^2\,t\right). \tag{6.96}$$

Thus, the original nonlinear problem (6.63)–(6.68) was rewritten in the form of three linear approximations with the use of multiscale expansions (6.69) up to the third order. Partial solutions to the first two linear problems were found, with the dispersion equation derived and the solvability conditions for the second and third linear problems formulated. Solvability conditions (6.90) and (6.91) lead to the evolution equations in the form of nonlinear Schrödinger equation for the wave train envelopes at the interface and on the free surface. Hereafter, we will restrict our attention to the detailed analysis of the first approximation.

6.5.3 Analysis of the First Linear Problem

The first linear problem (6.70)–(6.77) has solutions (6.78)–(6.81) with dispersion equation (6.82). This dispersion equation determines the conditions of propagation of linearly stable wave trains in the hydrodynamic system under consideration (layer-with-rigid-bottom/layer-with-free-surface). It contains a multiplier

$$\left(\frac{\omega^2 - (k + T_0 k^3)\coth(kh_2)}{\omega^2 \coth(kh_2) - (k + T_0 k^3)}\right)$$

whose replacement by $\coth(kh_2)$ leads to the dispersion equation for the layer/layer system [53]. Note that the boundary conditions at the free surface and at the so-called rigid cover cannot be derived from one another by passing to any limit because they are physically incompatible.

The following biquadratic equation can be obtained from dispersion equation (6.82):

$$\begin{aligned}\omega^4(\rho + \coth(kh_1)\coth(kh_2)) - \omega^2[\coth(kh_1)(k + T_0k^3) + \rho\,\coth(kh_2)(k + T_0k^3)\\ + \coth(kh_2)(k - \rho k + Tk^3)] + (k + T_0k^3)(k - \rho k + Tk^3) = 0,\end{aligned} \tag{6.97}$$

provided that

$$\omega^2 \coth(kh_2) - (k + T_0k^3) \neq 0.$$

Equation (6.97) has two pairs of roots:

$$\omega_1^2 = \frac{-b - \sqrt{b^2 - 4ac}}{2a}, \quad \omega_2^2 = \frac{-b + \sqrt{b^2 - 4ac}}{2a}, \tag{6.98}$$

where

$$a = \rho + \coth(kh_1)\coth(kh_2); \quad c = (k + T_0k^3)(k - \rho k + Tk^3);$$
$$b = -\left[\coth(kh_1)(k + T_0k^3) + \rho\coth(kh_2)(k + T_0k^3) + \coth(kh_2)(k - \rho k + Tk^3)\right].$$

The first pair of roots ω_1^2 corresponds to a pair of roots of the dispersion equation for the half-space/half-space system, inasmuch as expression (6.98) for ω_1^2 degenerates into the corresponding dispersion relation at $h_2 \to \infty$ and $h_1 \to \infty$ [53]:

$$\omega^2 = \frac{(1 - \rho + Tk^2)k}{(1 + \rho)}.$$

The second pair of roots ω_2^2 is characteristic of similar hydrodynamic systems such as half-space with free surface, half-space/half-space, layer-with-rigid-cover/half-space and layer/layer [53, 71]. Such additional roots of the dispersion equation originate owing to the existence of two surfaces along which the wave trains can propagate: internal waves propagate along the interface and surface waves propagate along the free surface.

It follows from relations (6.98) that $\omega_1^2 = 0$ when $k = \sqrt{(\rho - 1)/T}$ and that ω_2^2 is always greater than zero. Thus, the interface of the system layer-with-rigid-bottom/layer-with-free-surface is linearly stable or unstable depending on whether k is larger or smaller than the critical value $k_c = \sqrt{(\rho - 1)/T}$, which serves as a separatrix between the linearly stable and unstable wave motion.

Restricting ourselves to the first approximation, let us analyse the ratios of the amplitude of the free surface deviation $\eta_0(x,t)$ to the amplitude of the interface deviation $\eta(x,t)$ corresponding to ω_1^2 and ω_2^2, respectively:

$$a_1 = \frac{\omega_1^2}{\omega_1^2\cosh(kh_2) - (k + T_0k^3)\sinh(kh_2)},$$
$$a_2 = \left(\frac{\omega_2^2}{\omega_2^2\cosh(kh_2) - (k + T_0k^3)\sinh(kh_2)}\right)^{-1}.$$

The parameter a_1 characterises the contribution of the wave with frequency ω_1 to the surface motion, and the parameter a_2 characterises the contribution of the wave with frequency ω_2 to the wave motion at the fluid interface.

Figure 6.10a shows a_1 as a function of the lower layer thickness for the wave numbers $k = 0.01,\ 1,\ 2,\ 3$ in the case when $h_2 = 1, T = T_0 = 0$, and $\rho = 0.6$. As the lower layer thickness increases at each fixed k, the parameter a_1 tends to some limiting value that characterises the ratio of the amplitude of the free surface deviation to the amplitude of the interface deviation for the wave with frequency ω_1. The interface deviations with frequency ω_1 are larger than the free surface deviations, which is typical for internal waves.

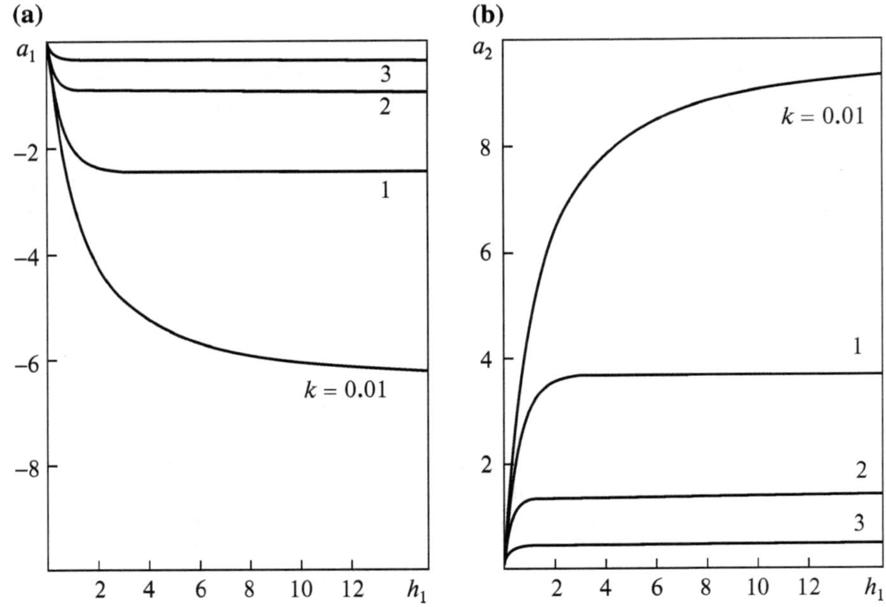

Fig. 6.10 Parameters a_1 and a_2 as functions of the lower layer thickness h_1: **a** ratio $a_1(h_1)$ of the amplitude of the free surface deviation to the amplitude of the interface deviation for the wave with frequency ω_1; **b** ratio $a_2(h_1)$ of the amplitude of the interface deviation to the amplitude of the free surface deviation for the wave with frequency ω_2

Figure 6.10b shows a_2 a function of the lower layer thickness for the wave numbers $k = 0.01,\ 1,\ 2,\ 3$ in the case when $h_2 = 1, T = T_0 = 0$, and $\rho = 0.6$. As the lower layer thickness increases at each fixed k, the parameter a_2 tends to some limiting value that characterises the ratio of the amplitude of the interface deviation to the amplitude of the free surface deviation for the wave with frequency ω_2. The free surface deviations with frequency ω_2 are larger than the interface deviations, which is typical for surface waves.

The parameter a_1 for the waves with frequency ω_1 strongly depends on the density ratio. When the density ratio tends to unity, the parameter a_1 decreases (Fig. 6.11). Such a behaviour is typical for the internal waves that originate owing to the fluid density difference. Note that for the waves with frequency ω_2 the parameter a_2 does not significantly depend on density ratio, and its dependence on the lower layer thickness is close to that shown in Fig. 6.10b.

It can easily be proved that the wave frequencies satisfy the inequality $\omega_1 < \omega_2$, i.e. the internal wave with frequency ω_1 is slower than the surface wave with frequency ω_2. Therefore, their superposition is not stationary in time and there is the energy transfer from the surface wave to the internal wave and vice versa.

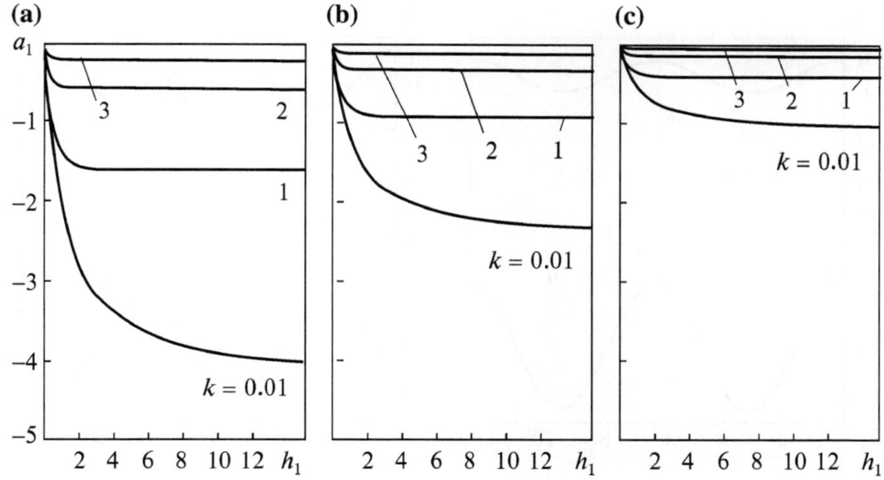

Fig. 6.11 Parameter a_1 as a function of the lower layer thickness h_1 for the density ratios **a** $\rho = 0.7$, **b** $\rho = 0.8$, **c** $\rho = 0.9$ and for the wave numbers $k = 0.01$, 1, 2, 3 in the case when $h_2 = 1$ and $T = T_0 = 0$

In the first approximation, the interface deviation and the free surface deviation can represented as the sum of two harmonics:

$$\eta = A\,\cos(kx - \omega_1 t) + Ba_2 \cos(kx - \omega_2 t),$$
$$\eta_0 = Aa_1 \cos(kx - \omega_1 t) + B\,\cos(kx - \omega_2 t),$$

where A and B are the arbitrary constants that describe the amplitudes of the harmonics with frequencies ω_1 and ω_2. Figure 6.12 shows the free surface deviation $\eta_0(x,t)$ and the interface deviation $\eta(x,t)$ in the case when $A = 0.2$, $B = 0.09$, $k = 1$, $\rho = 0.7$, $h_1 = 3$ and $h_2 = 1$ for the moments $t = 0$ (a) and $t = 6$ (b). At the moment $t = 0$, the free surface amplitude is smaller than $B = 0.09$, whereas the interface amplitude is larger than $B = 0.09$. The opposite situation is observed at the moment $t = 6$.

Note that in introducing dimensionless variables (6.62) the free surface deviation and the deviation at the interface of two fluid layers were normalised by the same scaling factor a. However, the amplitudes of internal and surface waves can be significantly different under realistic conditions. Therefore, there is good reason to introduce two dimensionless factors separately for the internal and surface waves. Moreover, by selecting the characteristic length equal to the upper layer thickness we limited the domain of applicability of our model to the wavelengths approximately equal to the upper layer thickness. Nevertheless, the qualitative relationship between the internal and surface waves should remain the same as was presented here.

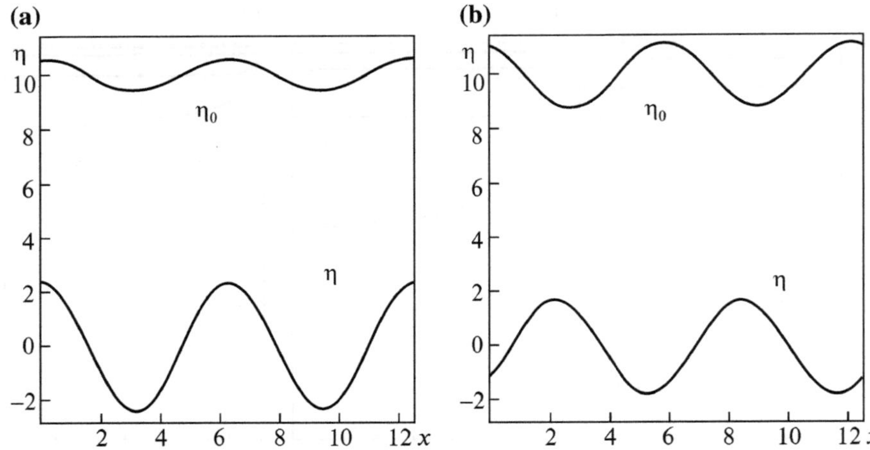

Fig. 6.12 The first approximations of the free surface deviation $\eta_0(x, t)$ and the interface deviation $\eta(x, t)$ for the moments **a** $t = 0$ and **b** $t = 6$

References

1. Ablowitz, M.J., Segur, H.: Long internal waves in fluids of great depth. Stud. Appl. Maths. **62**, 249–262 (1980)
2. Aida, I.: Numerical simulation of tsunami generated by landslide of Mt. Mayuyama in 1792. Earthquake **28**, 449–460 (1970)
3. Avramenko, O.V., Gurtovyi, Yu.V., Selezov, I.T.: Kharakternye svoistva rasprostraneniya volnovykh paketov v dvukhsloinoi zhidkosti (Typical features of wave train propagation in a two-layer fluid). Prikladnaya gidromekhanika **11**(4), 3–8 (2009)
4. Avramenko, O.V., Selezov, I.T.: Nonlinear wave propagation in a fluid layer based on a semi-infinite fluid. Dopovidi NAN Ukrainy **10**, 61–66 (1997)
5. Bakhanov, V.V., Kropfli, R.A., Ostrovsky, L.A.: On the effect of strong internal waves on surface waves. In: Proceedings of the Geoscience and Remote Sensing Symposium IEEE 1999, vol. 1(1), pp. 170–172 (1999)
6. Balas, C.E.: Risk assessment for Tuzla naval base breakwater. China Ocean Engineering **17** (3), 427–438 (2003)
7. Benjamin, T.B.: Internal waves of finite amplitude and permanent form. J. Fluid Mech. **25**, 241–270 (1966)
8. Benjamin, T.B.: Internal waves of permanent form of great depth. J. Fluid Mech. **29**, 559–592 (1967)
9. Benney, D.J.: Long nonlinear waves in fluid flows. Stud. Appl. Math. **45**, 52–63 (1966)
10. Bhatnagar, P.L.: Nonlinear Waves in One-Dimensional Dispersive Systems. Clarendon Press, Oxford (1979)
11. Braddock, R.D., Van Den Driessche, P., Peady, G.W.: Tsunami generation. J. Fluid Mech. **59**(4), 817–828 (1973)
12. Brevdo, L.: Three-dimensional logarithmic resonances in a homogeneous elastic wave guide. Eur. J. Mech. A. Solids **19**, 121–137 (2000)
13. Buick, J.M., Martin, A.J.: Comparison of a lattice Boltzmann simulation of steep internal waves and laboratory measurements using particle image velocimetry. Eur. J. Mech. B/Fluids **22**(1), 27–38 (2003)

14. Čadek, O., Yuen, D.A., Čížková, H.: Mantle viscosity inferrred from geoid and seismic tomography by genetic algorithms: results for layered mantle flow. University of Minnesota Supercomputing Inst. Research. Report UMSI 97/118 (1997)
15. Cagniard, L.: Réflexion et réfraction des ondes séismiques progressives. Gauthier-Villard, Paris (1939), English translation: Flinn, E.A., Dix, C.H. (eds.) Reflection and Refraction of Progressive Seismic Waves. McGraw-Hill, New York (1962)
16. Camassa, R., Choi, W., Michallet, H., Rusas, P.-O., Sveen, J.: On the realm of validity of strongly nonlinear asymptotic approximations for internal waves. J. Fluid Mech. **549**, 1–23 (2006)
17. Carr, M., Davies, P.A.: The motion of an internal solitary wave of depression over a fixed bottom boundary in a shallow, two-layer fluid. Phys. Fluids **18**, 016601 (2006)
18. Cherkesov, L.V.: Gidrodinamika voln (Hydrodynamics of Waves). Naukova Dumka, Kiev (1980)
19. Choi, J.W., Sun, S.M., Shen, M.C.: Internal capillary-gravity waves of a two-layer fluid with free surface over an obstruction—Forced extended KdV equation. Phys. Fluids **8**(2), 397–404 (1996)
20. Choi, W., Camassa, R.: Fully nonlinear internal waves in a two-fluid system. J. Fluid Mech. **396**, 1–36 (1999)
21. Davis, R.E., Acrivos, A.: Solitary internal waves in deep water. J. Fluid Mech. **29**, 593–607 (1967)
22. Debsarma, S., Das, K.P.: Fourth-order nonlinear evolution equations for a capillary-gravity wave packet in the presence of another wave packet in deep water. Phys. Fluids **19**, 097101 (2007)
23. Dias, F., Il'ichev, A.: Interfacial waves with free-surface boundary conditions: an approach via model equation. Physica D **150**, 278–300 (2001)
24. Ditkin, V.A., Prudnikov, A.P.: Spravochnik po operatsionnomu ischisleniyu (A Handbook on Operational Calculus). Vysshaya Shkola, Moscow (1965)
25. Donato, A.N., Peregrine, D.H., Stocker, J.R.: The focusing of surface waves by internal waves. J. Fluid Mech. **384**, 27–58 (1999)
26. Dotsenko, S.F.: Generatsiya poverkhnostnykh voln pri finitnykh deformatsiyakh dna basseina (Generation of surface waves at finite deformations of the basin bottom). Izvestiya RAN. Mekhanika zhidkosti i gaza **2**, 151–156 (1996)
27. Dzwinel, W., Yuen, D., Kaneko, Y., Boryczko, K., Ben-Zion, Y.: Multi-resolution clustering analysis and 3-D visualization of multitudinous synthetic earthquakes. Vis. Geosci. **8**, 12–25 (2003)
28. Elsasser, W.M.: Convection and stress propagation in the upper mantle. In: Runcorn, W.K. (ed.) The Application of Modern Physics to the Earth and Planetary Interiors, pp. 223–246. Interscience, New York (1969)
29. Ezerskii, A.B., Papko, V.V.: Laboratornoe issledovanie krupnomasshtabnykh techenii, indutsirovannykh paketom poverkhnostnykh voln (A laboratory investigation of large-scale flows induced by a surface wave train). Izvestiya AN SSSR **22**(9), 979–986 (1986)
30. Feng, W.: Resonant generation of internal waves on the soft bed by a surface water wave. Phys. Fluids **7**(8), 1915–1922 (1995)
31. Geist, E.L., Titov, V.V., Synolakis, C.E.: Tsunami: wave of change. Sci. Am. **294**, 56–63 (2006)
32. Goto, C., Tanimoto, K.: A study on the effect of tsunami breakwater by numerical simulation. In: Raufaste, N.J. (ed.) Proceedings of the 23rd Joint Meeting of the U.S.-Japan Cooperative Program in Natural Resources. Panel on Wind and Seismic Effects, vol. 820, pp. 345–349. National Institute of Standards and Technology, Washington (1991)
33. Grigoriev, A.I., Fedorov, M.S., Shiryaeva, S.O.: Volnovoe dvizhenie v pole sily tyazhesti na svobodnoi poverkhnosti i na granitse stratifikatsii sloisto-neodnorodnoi zhidkosti. Nelineinyi analiz (Wave motion in the gravity force field on a free surface and at the stratification interface of a layered-inhomogeneous fluid. Nonlinear analysis). Izvestiya RAN. Mekhanika zhidkosti i gaza **5**, 130–140 (2010)

34. Grimshaw, R.H.J.: The modulation of an internal gravity-wave packet, and the resonance with the mean motion. Stud. Appl. Math. **56**, 241–266 (1977)
35. Grimshaw, R., Pelinovsky, E.: Vzaimodeistvie uedinennykh poverkhnostnykh i vnutrennikh voln s begushchimi vozmushcheniyami (Interactions of surface and internal solitary waves with progressive disturbances). Doklady RAN **344**(3), 394–396 (1995)
36. Grimshaw, R., Pelinovsky, E., Poloukhina, O.: Higher-order Korteweg–de Vries models for internal solitary wave in a stratified shear flow with a free surface. Nonlinear Process. Geophys. **9**, 221–235 (2002)
37. Hammack, J.L., Segur, H.: Modelling criteria for long water waves. J. Fluid Mech. **84**(2), 359–373 (1978)
38. Hamzah, M.A.: Solitary wave pressure on a barrier. In: Proceedings of the 10th International Offshore and Polar Engineering Conference, USA, Seattle, 28 May–2 June, 2000, pp. 519–523
39. Hashizume, Y.: Interaction between short surface waves and long internal waves. J. Phys. Soc. Japan **48**(2), 631–638 (1980)
40. Hiraishi, T.: Characteristics of wave overtopping in a harbor induced by Typhoon 9918. In: Proceedings of the 11th International Offshore and Polar Engineering Conference, Stavanger, Norway, 17–22 June 2001, pp. 553–558
41. Holloway, P., Pelinovsky, E., Talipova, T.: Internal tide transformation and oceanic solitary waves. In: Grimshaw, R. (ed.) Environmental Stratified Flows, chap. 2. Kluwer Academic Publishers (2000)
42. Jamali, M., Seymour, B., Lawrence, G.A.: Asymptotic analysis of a surface-interfacial wave interaction. Phys. Fluids **15**(1), 47–55 (2003)
43. Kajura, K.: The leading wave of a tsunami. Bull. Earthquake Res. Inst. **42**, 535–571 (1963)
44. Karambas, T.V., Tozer, N.P.: Breaking waves in the surf and swash zone. J. Coast. Res. **19** (3), 514–528 (2003)
45. Kubota, T., Ko, D.R.S., Dobbs, L.D.: Propagation of weakly nonlinear internal waves in a stratified fluid of finite depth. J. Hydronaut. **12**(4), 157–165 (1978)
46. Ladbury, R.: Energy budget of deep-focus earthquakes suggests they may be slip-sliding away. Phys. Today **51**(4), 19–21 (1998)
47. Liu, C.-M.: Second-order random internal and surface waves in a two-fluid system. Geoph. Res. Lett. **33**, L06610 (2006)
48. Lu, D.Q., Chwang, A.T.: Interfacial waves due to a singularity in a system of two semi-infinite fluids. Phys. Fluids **17**, 102107 (2005)
49. Makarenko, N.I.: Mal'tseva, Zh.L.: Asymptotic models of internal stationary waves. J. Appl. Mech. Tech. Phys. **49**(4), 646–654 (2008)
50. Matsuno, Y.A.: Unified theory of nonlinear wave propagation in two-layer fluid system. J. Phys. Soc. Japan **62**(6), 1902–1916 (1993)
51. Murty, T.S.: Seismic Sea Waves: Tsunamis. Department of Fisheries and the Environment, Fisheries and Marine Service, Ottawa (1977)
52. Nayfeh, A.H.: Second-harmonic resonance in the interaction of a stream with capillary-gravity waves. J. Fluid Mech. **59**, 803–816 (1973)
53. Nayfeh, A.H.: Nonlinear propagation of wave-packets on fluid interface. Trans. ASME. Ser. E. **43**(4), 584–588 (1976)
54. Nikolayevskii, V.N., Ramazanov, T.K.: Theory of fast tectonic waves. J. Appl. Math. Mech. **49**(3), 356–362 (1985)
55. Ono, H.: Algebraic solitary waves in stratified fluids. J. Phys. Soc. Japan **39**, 1082–1091 (1975)
56. Papanicolaou, G.: Mathematical problems in geophysical wave propagation. In: Proceedings of the International Congress of Mathematicians, Berlin, 18–27 August 1998. Documanta Mathematica, vol. ICM 98 I, pp. 241–265
57. Papoulis, A.: A new method of inversion of the Laplace transform. Quart. Appl. Math. **14**, 405–414 (1957)

58. Pelinovskii, E.N.: Gidrodinamika voln tsunami (Hydrodynamics of Tsunami Waves). Institut Prikladnoi Fiziki, Nizhnii Novgorod (1996)
59. Potetyunko, E.N.: Volnovye dvizheniya zhidkosti so svobodnymi granitsami (Wave Motion of a Fluid with Free Boundaries). ROP SNIO, Rostov-na-Donu (1993)
60. Schneider, G., Wayne, C.E.: On the validity of 2D-surface water wave models. GAMM Mitt. Ges. Angew. Math. Mech. **25**(1–2), 127–151 (2002)
61. Segur, H.: The Korteweg-de Vries equation and water waves. Solutions of the equation. Part 1. J. Fluid Mech. **59**(4), 721–736 (1973)
62. Segur, H., Hammack, J.L.: Soliton models of long internal waves. J. Fluid Mech. **118**, 285–304 (1982)
63. Selezov, I.T.: Modelirovanie volnovykh i difraktsionnykh protsessov v sploshnykh sredakh (Modeling the Wave and Diffraction Processes in Continuous Media). Naukova Dumka, Kiev (1989)
64. Selezov, I.T.: Interaction of water waves with engineering constructions and topography in coastal area. In: Proceedings of the 5th International Conference on Coastal and Port Engineering in Developing Countries COPEDEC 5, Cape Town, South Africa, 19–23 April 1999, vol. 1, pp. 1–12
65. Selezov, I.T.: Some hyperbolic models for wave propagation. In: Fey, M., Jeltsch, R. (eds.) International Series of Numerical Mathematics. Hyperbolic Problems: Theory, Numerics, Applications, vol. 130, pp. 833–842. Birkhauser Verlag, Basel (1999)
66. Selezov, I.T.: Issledovanie neustanovivshykhsya volnovykh dvizhenii v gidrouprugikh sistemakh obolochka-zhidkost (Investigating the transient wave motions in hydroelastic shell-fluid systems). In: Prikladnye problemy mekhaniki tonkostennykh konstruktsii (Applied Problems in the Mechanics of Thin-Wall Constructions), pp. 286–305. Izdatelstvo Moskovskogo Universiteta, Moscow (2000)
67. Selezov, I.T.: Wave processes in fluids and elastic media. Int. J. Fluid Mech. Res. **30**(2), 219–249 (2003)
68. Selezov, I.T.: Tsunami wave excitations by a local floor disturbance. In: Submarine Landslides and Tsunamis. NATO Sci. Ser. IV. Earth and Environmental Sci., vol. 21, pp. 139–150. Kluwer Academic Publishers, Netherlands (2003)
69. Selezov, I.T.: Some degenerate and generalized wave models in elasto- and hydrodynamics. J. Appl. Math. Mech. **67**(6), 871–877 (2003)
70. Selezov, I.T.: Modeling of tsunami wave generation and propagation. Int. J. Fluid Mech. Res. **33**(1), 44–54 (2006)
71. Selezov, I.T., Avramenko, O.V.: Evolyutsiya nelineinykh volnovykh paketov s uchetom poverkhnostnogo natyazheniya na poverkhnosti kontakta (Evolution of nonlinear wave trains with allowance for the surface tension at the contact surface). Matematychni metody i fizyko-mekhanichni polia **44**(2), 113–122 (2000)
72. Selezov, I.T., Avramenko, O.V.: Evolyutsionnoe uravnenie tretego poryadka dlya nelineinykh volnovykh paketov pri okolokriticheskikh volnovykh chislakh (A third-order evolution equation for nonlinear wave trains at near-critical wave numbers). Dinamicheskie sistemy **17**, 58–67 (2001)
73. Selezov, I.T., Avramenko, O.V., Gurtovyi, Yu.V.: Osobennosti rasprostraneniya volnovykh paketov v dvukhsloinoi zhidkosti konechnoi glubiny (Peculiarities of wave train propagation in a two-layer fluid of finite depth). Prikladnaya gidromekhanika **7**(1), 80–89 (2005)
74. Selezov, I.T., Avramenko, O.V., Gurtovyi, Yu.V.: Nelineinaya ustoichivost rasprostraneniya volnovykh paketov v dvukhsloinoi zhidkosti (Nonlinear stability of wave train propagation in a two-layer fluid). Prikladnaya gidromekhanika **8**(4), 60–65 (2006)
75. Selezov, I.T., Avramenko, O.V., Gurtovyi, Yu.V.: Rasprostranenie nelineinykh volnovykh paketov pri okolokriticheskikh volnovykh chislakh v dvukhsloinoi zhidkosti konechnoi glubiny (Propagation of nonlinear wave trains at near-critical wave numbers in a two-layer fluid of finite depth). Matematychni metody i fizyko-mekhanichni polia **50**(1), 91–97 (2007)

76. Selezov, I.T., Avramenko, O.V., Gurtovyi, Yu.V., Naradovyi, V.V.: Nonlinear interaction of internal and surface gravity waves in a two-layer fluid with a free surface. J. Math. Sci. **168** (4), 590–602 (2010)
77. Selezov, I.T., Avramenko, O.V., Naradovyi, V.V.: Osobennosti rasprostraneniya slabonelineinykh voln v dvukhsloinoi zhidkosti so svobodnoi poverkhnostyu (Peculiarities of weakly nonlinear wave propagation in a two-layer fluid with a free surface). Dinamicheskie sistemy **1**(1), 53–68 (2011)
78. Selezov, I., Avramenko, O., Nayfeh, A., Huq, P., Zeegers, N.: Propagation of water wave-packets at the interface of layer and half-space fluid. In: Nayfeh, A., Rozhdestvensky, K. (eds.) Proceedings of the 2nd International Conference on Asymptotics in Mechanics, St.-Petersburg, 13–16 October 1996, pp. 245–252. St.-Petersburg State Marine Technical University, St.-Petersburg (1997)
79. Selezov, I., Huq, P.: Asymptotic-heuristic analysis of nonlinear water wave propagation in two- and three-layer fluids. In: Nayfeh, A., Rozhdestvensky, K. (eds.) Proceedings of the 2nd International Conference on Asymptotics in Mechanics, St.-Petersburg, 13–16 October 1996, pp. 237–244. St.-Petersburg State Marine Technical University, St.-Petersburg (1997)
80. Selezov, I.T., Korsunsky, S.V.: Transformation of plane waves over moving or elastic bottom inhomogeneity. Rozprawy Hydrotechniczne **54**, 49–54 (1991)
81. Selezov, I.T., Kuznetsov, V.V., Chernikov, D.O.: Generation of surface gravity waves by bottom time-repetitive pulses. J. Math. Sci. **171**(5), 596–602 (2010)
82. Selezov, I.T., Mironchuk, M.V., Huq, P.: Evolution equation for waves forced by a slender obstacle in a two-layer fluid. Dopovidi NAN Ukrainy **4**, 77–82 (1999)
83. Selezov, I.T., Ostroverkh, B.N.: Modelling of seismic underwater centers of generation and transformation of tsunami waves in seismic active regions. Marine Geophys. J. **1**, 66–77 (1996)
84. Selezov, I.T., Sidorchuk, V.N., Yakovlev, V.V.: Transformatsiya voln v pribrezhnoi zone shelfa (Wave Transformation in the Coastal Shelf Zone). Naukova Dumka, Kiev (1982)
85. Selezov, I., Spirtus, V.: Propagation of disturbances along axial lines of tectonic flows under conditions of continuum destruction. In: Maruszewski, B.T., Muschik, W., Radowicz, A. (eds.) Proceedings of the International Symposium on Trends in Continuum Physics, Poznan, 17–20 August 1998, pp. 322–333. World Scientific, Singapore (1998)
86. Selezov, I.T., Volynskii, R.I., Suzdaltsev, A.I.: Chislennoe modelirovanie dinamiki tverdoi chastitsy v poverkhnostnykh gravitatsionnykh volnakh (Numerical modeling of the dynamics of a solid particle in surface gravity waves). Gidromekhanika **67**, 67–71 (1993)
87. Selezov, I.T., Zheleznyak, M.I., Tkachenko, V.A., Yakovlev, V.V.: On the numerical modeling of tsunami wave generation and propagation. Mar. Geodesy **6**(2), 149–165 (1983)
88. Serebryanyi, A.N., Furduev, A.V., Aredov, A.A., Okhrimenko, N.N.: Shum vnutrennei volny bolshoi amplitudy v okeane (Noise from a large-amplitude internal wave in the ocean). Doklady RAN **402**(4), 543–547 (2005)
89. Shingareva, I., Celaya, C.L.: On frequency-amplitude dependences for surface and internal standing waves. J. Comput. Appl. Math. **200**(2), 459–470 (2007)
90. Soloviev, S.L.: Tsunami (Tsunamis). Zemlya i Vselennaya **3**, 12–16 (1980)
91. Sutherland, B.S., Nault, J.T.: Intrusive gravity currents propagating along thin and thick interfaces. J. Fluid Mech. **586**, 109–118 (2007)
92. Tanimoto, K.: On the hydraulic aspects of tsunami breakwaters in Japan. In: Iida, K., Iwasaki, T. (eds.) Tsunamis: Their Science and Engineering, pp. 423–435. Terra Scientific Publishing Co., Tokyo (1983)
93. Vincze, M., Kozma, P., Gyüre, B., Jánosi, I.M., Szabó, K.G., Tél, T.: Amplified internal pulsations on a stratified exchange flow excited by interaction between a thin sill and external seiche. Phys. Fluids **19**, 108108 (2007)
94. Volynski, R., Azmon, E., Selezov, I., Suzdaltsev, A.: Computer simulation of small particles transport in waves. In: Proceedings of the 26th Israel Conference on Mechanical Engineering, Technion City, Haifa, 21–22 May 1996, pp. 234–236

95. Wang, Y.-Sh., Yu, G.-L.: Re-polarization of elastic waves at a frictional contact interface—II. Incidence of a P or SV wave. Int. J. Solids and Struct. **36**, 4563–4586 (1999)
96. Watson, K.M.: The coupling of surface and internal gravity waves revised. J. Phys. Oceanogr. **20**, 1233–1248 (1990)
97. Weyl, P.K.: Oceanography. An Introduction to the Marine Environment. John Wiley & Sons, New York (1970)
98. Whitham, G.B.: Linear and Nonlinear Waves. John Wiley & Sons, New York (1999)
99. Yuen, H.C., Lake, B.M.: Nonlinear dynamics of deep-water waves. Adv. Appl. Mech. **22**, 33–45 (1982)
100. Zhou, C.P., Lee, J.H.W., Cheung, Y.K.: Instabilities and bifurcations of interfacial water waves. Phys. Fluids A **4**(7), 1428–1438 (1992)

Conclusion

In this book, we have reviewed the current state of the theory of wave propagation and diffraction and have described main results of particular studies in this field. A special focus was made on the mathematical methods that are frequently used in studying the wave propagation and diffraction. These methods include the method of separation of variables, method of power series, method of spline functions, method of an auxiliary boundary, Laplace transform and its inversion, method of multiple scales and method of Fourier series.

The theory of wave propagation on the fluid surface has direct engineering applications in the field of wave energy take-off and conversion because ocean waves represent a viable source of renewable energy. Wave energy can be converted either to mechanical work or electric power for their use on site or in the grid. The nonlinear theory of wave propagation is of crucial importance in understanding the mechanisms of the formation of giant (freak) waves on the sea/ocean surface. Such waves pose significant hazard for the ships and off-shore marine structures. Studying the evolution of narrow-band wave trains and solitons in different media have direct engineering applications in the problems dealing with information transfer on large distances.

Studying the wave refraction on water over real topography is of great practical interest. The ray theory, which is based on the concepts of geometrical optics, is used as a tool for the quick and efficient computation of wave propagation over the bottom with arbitrary topography based on the data that can be retrieved from the water depth measurements by an echo sounder. The influence of nonlinear effects is also significant when the rays characterising the wave energy transfer deviate from the normal incidence on the shore. Based on the methods presented in this book, it is possible to determine the convergence (focusing) and divergence (defocusing) zones of wave energy in real water basins, which is essential for choosing the location of harbours, quays and various apparatuses that take off the wave energy.

Wave diffraction is a process that is localised in space and results in problems that are much more difficult to solve from the mathematical point of view. The development of analytical and numerical methods for investigating the problems of nonstationary wave diffraction is one of the topical demands in this field. It especially refers to the case of initial-boundary-value problems. Analytical solutions can

I.T. Selezov et al., *Wave Propagation and Diffraction*,
Foundations of Engineering Mechanics, DOI 10.1007/978-981-10-4923-1

be used as benchmarks for testing the corresponding computational programs and are useful in the analysis of experimental data. The wave interaction with obstacles of various shapes strongly depends on the shape of the scatterers and their internal properties. The results of calculating the wave field diffracted by the objects of various shapes can be used in the acoustic sounding of marine objects for data analysis. Methods for calculating the diffraction of electromagnetic waves can be used to model various configurations of mobile communication antennas in order to improve the mobile communication.

Tsunami waves represent a real challenge to mankind. They originate mainly due to underwater earthquakes. In this regard, studying the mechanisms of tsunami wave generation is of great importance. They highly depend on the nature of bottom disturbances—their spatial localisation and temporal pattern. Of particular interest is the case of repeated shocks observed in real situations during the excitation of tsunami waves. Such shocks significantly affect the generation of tsunami waves. The triggering mechanism of underwater (and ordinary) earthquakes corresponds to a singular degeneracy in the vicinity of Morse points that predicts jumps in structural parameters. In this regard, a particular attention should be given to the indeterminacy of tsunami wave excitation and, as a consequence, to the indeterminacy in the formulations of the corresponding initial-boundary-value problems. Biological objects such as animals and plants have the ability to sense the upcoming triggering event. This fact can apparently be associated with electromagnetic radiation generated by very complex and diverse processes in the mantle that are accompanied by the propagation and diffraction of mantle disturbances. For example, jellyfish move into deep water long before the triggering of an underwater earthquake. The evolution equations derived for the case of an arbitrary disturbance of an inhomogeneous bottom demonstrate that the bottom elasticity affects the propagation speed of tsunami waves, which can decrease under certain conditions.

Studying the wave propagation in a two-layer liquid is of great interest due to the fact that in the oceans and seas there is an interface (a thin layer called a pycnocline) separating the upper and lower layers with different properties (such as density, temperature and salinity) at some depth. A small difference in densities makes this system very sensitive to disturbances (triggering mechanism) leading to the generation and propagation of internal waves. As a result, the mixing of the aqueous medium facilitates the heat exchange with the atmosphere that significantly affects the climate, flora and fauna.

We hope that this book will stimulate further development of the theory of wave propagation and diffraction that would lead to new mathematical, physical and engineering applications.

Index

I.T. Selezov et al., *Wave Propagation and Diffraction*,
Foundations of Engineering Mechanics, DOI 10.1007/978-981-10-4923-1

MIX
Papier aus verantwortungsvollen Quellen
Paper from responsible sources
FSC® C105338

If you have any concerns about our products,
you can contact us on
ProductSafety@springernature.com

In case Publisher is established outside the EU,
the EU authorized representative is:
Springer Nature Customer Service Center GmbH
Europaplatz 3, 69115 Heidelberg, Germany

Printed by Libri Plureos GmbH
in Hamburg, Germany